Milestones in Systematics

The Systematics Association Special Volume Series

Series Editor

Alan Warren

Department of Zoology, The Natural History Museum,
Cromwell Road, London SW7 5BD, UK.

The Systematics Association promotes all aspects of systematic biology by organizing conferences and workshops on key themes in systematics, publishing books and awarding modest grants in support of systematics research. Membership of the Association is open to internationally based professionals and amateurs with an interest in any branch of biology including palaeobiology. Members are entitled to attend conferences at discounted rates, to apply for grants and to receive the newsletters and mailed information; they also receive a generous discount on the purchase of all volumes produced by the Association.

The first of the Systematics Association's publications *The New Systematics* (1940) was a classic work edited by its then-president Sir Julian Huxley, that set out the problems facing general biologists in deciding which kinds of data would most effectively progress systematics. Since then, more than 70 volumes have been published, often in rapidly expanding areas of science where a modern synthesis is required.

The *modus operandi* of the Association is to encourage leading researchers to organize symposia that result in a multi-authored volume. In 1997 the Association organized the first of its international Biennial Conferences. This and subsequent Biennial Conferences, which are designed to provide for systematists of all kinds, included themed symposia that resulted in further publications. The Association also publishes volumes that are not specifically linked to meetings and encourages new publications in a broad range of systematics topics.

Anyone wishing to learn more about the Systematics Association and its publications should refer to our website at www.systass.org.

Forthcoming titles in the series:

Organelles, Genomes and Eukaryote Phylogeny:
An Evolutionary Synthesis in the Age of Genomics

Edited by R.P. Hirt and D. S. Horner

Other Systematics Association publications are listed after the index for this volume.

The Systematics Association Special Volume Series 67

Milestones in Systematics

Edited by
David M. Williams

Department of Botany
The Natural History Museum
London

Peter L. Forey

Department of Palaeontology
The Natural History Museum
London

CRC PRESS

Boca Raton   London   New York   Washington, D.C.

Cover Illustration

The illustration depicted on the cover is of the Chironomid *Zelandochlus latipalpis* Brundin, the "ice-worm," taken from Lars Brundin's 1966 monograph (*Kungliga Svenska Vetenkskapsakademiens Handlingar, fjärde serien,* 4 (11):Fig. 35, reproduced with permission). Brundin's monograph was the impetus for much of the cladistic revolution, particularly with respect to progress in biogeography and as a major inspiration for the "reform" of palaeontology. Brundin's influence can be found in Chapters 6, 7 and 10. His influence on the development of systematics and biogeography has yet to be fully explored.

Library of Congress Cataloging-in-Publication Data

Milestones in systematics / edited by David M. Williams, Peter L. Foley
    p. cm. (Systematics Association special volume; no. 67)
Essays from a symposium held within the 3rd Systematics Association Biennial Meeting, September 2001.
Includes bibliographical references and index. (p.).
ISBN 0-415-28032-X
1. Biology--Classification--History--Congresses. I. Williams, David M. (David Malcolm), 1940- II. Forey, Peter L. III. Series.
QH83.M64 2004
578'.01'2—dc22   2003065587

Visit the CRC Press Web site at www.crcpress.com

© 2004 by Systematics Association

No claim to original U.S. Government works
International Standard Book Number 0-415-28032-X
Library of Congress Card Number 2003065587
Printed in the United States of America 1 2 3 4 5 6 7 8 9 0
Printed on acid-free paper

Introduction

Peter Forey and David Williams

This collection of essays arose out of a symposium under the same title of this book, held within the 3rd Systematics Association Biennial Meeting (September 2001). As organizers we tried to invite contributors who had first-hand experience of the changes in systematic practice that were taking place during the last third of the twentieth century. This period was pivotal for the position in which we now find ourselves. Systematic methodologies were being scrutinized mainly through the meaning of relationship and how that meaning was to be expressed in classification — the vehicle reflecting our understanding of the order in the natural world. The concept of relationship is so basic to biology that there has been constant upheaval and realignments of ideas, so to single out authors active during the past 30 years may seem eclectic. However, these authors do provide a link with the more distant past and they are a part of that history, ostensibly better equipped to document the contemporary background than authors one hundred years from now. At the same time, our authors also approached their subject with their own reading of history so we cannot claim objectivity.

The first two chapters come nearest to this endeavor since they are written by historians of science, skilled in the art of unprejudiced commentary. Mary (Polly) Winsor (Chapter 1) has some rather penetrating statements on scientists writing history, claiming that some do so to influence the future. Her case is compelling even though it may sound uncomfortable to some and is exemplified by the writings of Ernst Mayr and Peter Sneath who, she claims, set up Darwin and Adanson, respectively, as forefathers of their own views.

In general, systematists have not been good at organizing themselves to speak with one voice, which is seen today as counterproductive to securing the financial underpinning of the discipline. This is true. There have always been deep divisions in philosophical stances, objectives and methods in trying to make sense of the order of life. Richard Blackwelder's account of the formation of the Society of Systematic Zoology (SSZ) in 1947, ostensibly in opposition to the Society for the Study of Evolution, has always been viewed as just another instance of an ideological tiff. Joe Cain's fascinating retelling of events (Chapter 2) leading up to the formation of the SSZ draws on previously ignored documentation and suggests a more complex history, steeped in postwar reconstruction of scientific dialogue. He suggests that Blackwelder's account is distorted by his dislike of the evolutionary systematics championed by Ernst Mayr.

In the early days of spirited debate between cladists and the then traditional taxonomists, Walter Bock was one of the first to argue the case for evolutionary taxonomy from a philosophical point of view. Part of those early arguments concerned Karl Popper's distinction between science (theories established that could be falsified through some test) and nonscience (theories that could not generate tests leading to potential falsification). Cladists had claimed that evolutionary taxonomy fell into the latter category, although none claimed that it was unscientific in the more general sense of that word. In Chapter 3 Bock returns to this subject in distinguishing between nomological deductive explanations (those arrived at through Popperian science) and historical narrative explanations (instances of singular events such as those documented in phylogeny reconstruction). His thesis is that historical narrative explanations, once proposed, are corroborated by more evidence rather than falsified. Nevertheless, the formulation of those historical narratives must of necessity involve nomological deductive explanations such as several facets of evolutionary theory.

The influence of Popper's ideas on the demarcation of science from nonscience continued to wax and wane throughout the past 25 years, at one time dropping out of sight altogether only to be recently resurrected, this time in defense of statistical methods of phylogeny reconstruction such as maximum likelihood. Olivier Rieppel, a previous champion of the more usual hypothetico-deductive understanding of the application of Popper's ideas, presents a general attack on falsification as applied to systematics, damning its use in both camps, hypothetico-deductive and statistical (Chapter 4). Rieppel's arguments are complex but repay a careful read. If, as is suggested, congruence might be a questionable enterprise, then what, if anything, might there be left for the systematist with which to evaluate their data? It is also evident that statistical methods are not applicable to the falsification arguments either. Rieppel presented suggestions for the future elsewhere, but the issues he raises here are of general concern, not least that it might be surmised that Willi Hennig, the father of cladistic analysis, may indeed have had some sympathy for Rieppel's general conclusions.

The method of phylogenetic reconstruction proposed by Willi Hennig and known as phylogenetic systematics is rather different from the cladistic methods in use today. Hennig formulated his ideas before the era of computer-generated trees and the strict application of parsimony algorithms. He suggested that characters be polarized into primitive (plesiomorphic) and derived (apomorphic) states before analysis; he did not believe in character weighting in the sense that it is applied these days and he was probably unaware of such things as unrooted networks and the possibilities of finding the root by some outgroup criterion. Johann-Wolfgang Wägele (Chapter 5), in describing the transmutation of Hennig's method, suggests that we might be better served by developing some of Hennig's original ideas to improve our methods of phylogeny reconstruction. Most importantly he suggests that considerably more attention be paid to the identification, description and codification of characters before entering them into the computer. He calls this crucial stage character analysis (a term more usually reserved for the computational stage in most cladistic analyses). Wägele is a morphologist who appreciates the richness of morphological variability and proposes that characters can be polarized, assessed and weighed by their quality and discusses ways in which this might be done.

Gary Nelson, one of our contributors who played a dominant role in the cladistic revolution, addresses the concept of ancestry in systematics (Chapter 6). He recounts that the difficulties posed by trying to recognize ancestors were started by paleontologists at the turn of the twentieth century and are continued today in the oft-quoted contention that dinosaurs are ancestral birds — a view, he argues, that is entirely mistaken. He has more to say about optimization, as used by modern computerized parsimony analysis, finding that it falls far short of what systematists should be doing — enquiring about relationships. Alarmingly, he identifies that current computer programs fail to find relationships (nodes on cladograms) where they should be found and finds them where they should not. If this happens, then the addition of more data might simply lead to a path of least distortion rather than a path to enlightenment.

Peter Forey (Chapter 7) covers some of the same ground as Nelson, tracing the paleon-tologists' quest for ancestors at the end of the nineteenth century and its unhelpful influence on systematics, with the establishment of grade groups and the preoccupation with time. Two strands of evolutionary theory (adaptation and genealogy) caused difficulties for the paleontological contribution to phylogeny reconstruction, which became uncritical of the evidence and generous with evolutionary scenarios. This approach came under attack by cladists who questioned the use of time in phylogeny reconstruction as well as the ability to recognize ancestors, two uniquely paleontological contributions. Recently, both time and ancestors have resurfaced in the latest discussions of phylogeny reconstruction.

A.W.F. Edwards recounts the early days of computer algorithms used to reconstruct phylogenetic relationships (Chapter 8). He was in the 1960s circle of those debating the use of parsimony analysis and the theoretical justification for its implementation. He argued that the most plausible estimate of an evolutionary tree is that which invokes the minimum amount of evolution. Naturally this approach was not accepted by all since "plausible" and "amount of evolution" are constrained by particular models of evolution. The algorithms developed by Edwards and contemporaneous workers needed to address these uncertainties. Edwards' chapter is interesting because similar discussions of model parameters surround the use of current algorithms used to reconstruct phylogeny. As an aside it is interesting to note that in these early days (1964) computing power restricted analyses to just four taxa!

David Williams (Chapter 9) traces the history of the concept of homology, the most fundamental concept in comparative biology, at the heart of the meaning of relationship. No wonder that, with the exception of the species, it has been the most discussed word in biology. He reaches back to the middle of the nineteenth century when Richard Owen was one of the first to clearly articulate the meaning of homology and to distinguish it from other kinds of relationships (e.g., analogy). After Darwin, Ray Lankester translated the idea of homology to historical continuity (evolution) while George Mivart resurrected the embry-ological spin so that at the turn of the nineteenth century ideas of homology multiplied and became entangled with ideas of pattern and process. This led to a plethora of terms that were used in subtly different ways by different authors. Williams identifies a second milestone in the 1970s. Here, the distinction between transformational (process) homology and taxic (pattern) homology was made clear and led to the third milestone, identified by Williams as the concept of homology as a relation between groups. As he says, this last is yet to have its full impact.

Homology has been applied not just to morphological features but to molecular data, behavioral data and many other kinds of information that can be applied to comparative biology. It remains a little surprising, then, that the concept of biogeographic homology has been little explored, in spite of the resurrection of biogeography as an analytical science in the late 1970s and 1980s. Chris Humphries (Chapter 10) continues his exploration of Leon Croizat's dictum that "Earth and Life evolved together," captured in the revised threefold parallelism, "space, time, form." Humphries examines much new historical material, again showing that the notion of a stable Earth was challenged on many occasions, resisted almost wholly because of prevailing trends, rather than careful assessment of the available data. Darwin's natural selection, as Humphries points out, was essentially an ecological theory concerning the origin of life and had little to do with historical development of the Earth those same organisms inhabited. Humphries reviews the methods for analytical biogeogra-phy, especially those developed in the past 30 years. Surprisingly, almost all researchers neglect the concept of biogeographical homology save those who deal directly with taxon cladograms. Some noted 25 years ago that biogeography was in its infancy. Humphries' chapter suggests that rather than reaching maturity, childhood is still some way off.

Peter Holland (Chapter 11) documents the revival of the link between studies of evolution and embryology, colloquially known as evo-devo. These two endeavors had drifted apart during the first half of the twentieth century, each searching for answers to different questions: evolutionary biology became concerned with gene frequencies, selection coefficients, etc., while embryology became less of a comparative discipline and concentrated on experimentation in a few tame lab animals. The renewed link has been made through the intermediary of molecular biology and phylogenetics, new techniques in tracing cell lineages as well as the manifestation and understanding of gene action. These techniques have allowed broad comparisons to be made between many species. Systematics is central to evo-devo because it provides the patterns of morphological variation to be explained, articulates the questions to be asked as well as directing the selection of species to be investigated. In turn, patterns of development that were key to systematics in the nineteenth century are becoming known in ever finer detail, such that our conjectures of homology may be more tightly constrained.

One area in which we feel some regret is the absence of a chapter dealing with the rise and development of molecular systematics, which was almost exponential in the last quarter of the twentieth century. While the subject was addressed during the symposium, perhaps it is too soon to assess a history that has not yet had time to get old. Nevertheless, its history is intermingled with the development of the molecular clock, events that have been captured in a recent essay by Morgan (1998). Hillis et al. (1996) provide some historical details pertinent to the practical techniques that have made molecular systematics routine enquiry, and Page and Holmes (1998) cover many of the analytical techniques used in inferring phylogeny from various kinds of molecular data.

Finally, we express our thanks to all of our colleagues who refereed the contributions; we hold them in no way responsible for the final outcome.

References

Hillis, D.M., Moritz, C., and Mable, B.K., Eds., *Molecular Systematics,* 2nd ed., Sinauer Associates, Sunderland, MA, 1996.

Morgan, G.M., Emile Zuckerkandl, Linus Pauling, and the molecular evolutionary clock 1959-1965, *J. Hist. Biol.,* 31, 155–178, 1998.

Page, R.D.M. and Holmes, E.C., *Molecular Evolution,* Blackwell Science, Oxford, 1998.

Editors

David Williams is a researcher of diatoms — freshwater, marine, and fossil, at The Natural History Museum, London. His primary interests are in the systematics, evolution, and biogeography of freshwater diatoms. Diatoms, a group of photosynthetic eukaryotes, have often been recognized as a ubiquitous group of organisms, without any discernable geographic differentiation. Williams has endeavored to penetrate this myth, and has contributed to the beginning of serious diatom biogeography. His other interests are the history, theory and practice of systematics, classification, and biogeography, studies that help to expose other myths — some harmless, others not. He has co-authored two books on cladistic methods in systematics and co-edited a book on phylogenetic methods.

Peter Forey is a researcher in fossil fishes at the Natural History Museum, London, where he undertakes research into the anatomy and relationships of fishes, in particular coelacanths and primitive teleost fishes. While most of his research is specimen based, there is inevitably a theoretical component concerned with how relationships are discovered and how the results are expressed in diagrams and classifications. Within the field of paleontology there is division between those who advocate that the "present is the key to the past," and those who believe that the "past is the key to the present." Forey sides with the former, and explores ways in which the fossil record is best able to supplement our explanations of present diversity. He has contributed to and edited several volumes of essays concerned with such diverse subjects as the theory and practice of cladistics, the relationship between systematics and conservation, and the kinds of observations that can usefully reveal the paths of evolution.

Contributors

Walter Bock, Department of Biological Sciences, Columbia University, New York, NY

Joseph Cain, Department of Science and Technology Studies, University College London, London, UK

A.W.F. Edwards, Gonville and Caius College, Cambridge University, Cambridge, UK

Peter Forey, Department of Palaeontology, The Natural History Museum, London, UK

Peter Holland, Department of Zoology, Oxford University, Oxford, UK

Chris Humphries, Department of Botany, The Natural History Museum, London, UK

Gareth Nelson, School of Botany, University of Melbourne, Victoria, Australia

Olivier Rieppel, Department of Geology, The Field Museum, Chicago, IL

Johann-Wolfgang Wägele, Fakultät Biologie, Ruhr-Universität Bochum, Bochum, Germany

Mary Winsor, Institute for the History and Philosophy of Science and Technology, Victoria College, University of Toronto, Toronto, ON

David Williams, Department of Botany, The Natural History Museum, London, UK

Table of Contents

Chapter 1

Setting Up Milestones: Sneath on Adanson and Mayr on Darwin

Mary P. Winsor

Abstract

History is written by people, and whether those people are historians, scientists, or philosophers makes a difference in what they want from the past. Usually, scientists hope to foretell or even influence the future, a motive uncongenial to historians. Two instances of biologists setting up historical figures as milestones, heroic forerunners of their own views, exemplify important issues about the writing of history. In 1957 Peter Sneath, a founder of numerical taxonomy, identified Michel Adanson as his precursor, proposing the term *Adansonian* for principles Sneath advocated. Sneath did not realize that statements about Adanson by previous scientists, including Francis Bather and Georges Cuvier, misrepresented his methods. Sneath's historical interpretation was immediately challenged by botanists who had paid close attention to Adanson's writings. With co-author Robert Sokal, Sneath appealed to historians of science to adjudicate the question, but none responded. In 1957 Ernst Mayr announced that Charles Darwin had replaced typological thinking by population thinking. Mayr's claim about the dominance of typology acquired new lustre when melded with Karl Popper's coinage, "essentialism." Mayr's historical interpretation reflected 20th-century concerns but was supported by scant historical evidence. The example of Adanson undermines Mayr's claim, but here too historians failed to supply an effective evaluation. Both stories warn that scientists must take responsibility for their own history.

> Several recent occurrences indicate a revival of interest in systematic biology and the broad problems of classification … . [Their effect is] to disturb our confidence in the concepts with which we have worked so long and to make us wonder whether any System at all can be based on the Theory of Descent. What, we fearfully enquire, is to be the future of Biological Classification? Is the Linnean nomenclature breaking down? (Bather, 1927:lxii–lxiii)

Although sounding so modern, these words date back to 1927, when a distinguished British Museum palaeontologist, Francis Bather, addressed his colleagues on the subject of the past, present, and future of systematics. Many systematists still believe, as he did, that investigating history illuminates fundamental issues about systematics, issues that are central to choices being made about future taxonomic practice. "To foretell the future," Bather declared, "it is necessary to understand the past" (p. lxiii).

Scientists' interest in the past is usually, perhaps always, intimately tied up with their interest in the future. Science in its innermost nature is heavily oriented toward the future.

1

Unanswered questions cry out for action, debates demand resolution on the basis of new evidence rather than mere rhetoric, and more research is needed. Arthur Cain was very clear, in his several historical articles of the late 1950s and early 1960s, that his motive for doing history was to reorient present thinking in hopes of a better future (Winsor, 2001). When he said in 1958, "I think that we are about to see a considerable revision of the whole basis of taxonomic theory" (Cain, 1959b:241), he was pointing to the infant field of numerical taxonomy arising from his own work with Harrison and the work of Sokal, Michener, and Sneath.

At that time Cain was pleased to discover, as he then thought, that Linnaeus's understanding of nature was hopelessly bound up with inappropriate Aristotelian logic. Cain's superficial smattering of philosophy dazzled biologists and historians alike, and no philosopher troubled to chastise him. Forty years later Cain in his retirement began to make more careful readings of Linnaeus, but by then he had quite given up on the idea that investigations into the past were useful guides to the future (Winsor, 2001).

Peter Sneath's excursion into history was likewise part of his effort to shape the future of taxonomic theory and practice. Searching for a rationally based systematics, he recognized a kindred spirit when he read John Gilmour's essay "Taxonomy and Philosophy" in the Systematics Association volume *The New Systematics* (Gilmour, 1940; Winsor, 2000). The first item in Gilmour's list of references was Bather's address, which Sneath looked up and found very useful. There (Bather, 1927: lxx) Sneath read about an eighteenth-century botanist, Michel Adanson, who had, in Bather's words, "set to work to tabulate all possible characters, basing a classification on each. Then, setting his 65 classifications side by side, he found certain groupings to occur most frequently, and those he took as his families." This sounded like some kind of mechanical tabulation, so it immediately resonated with Sneath, who at that time was tabulating similarities between strains of bacteria (Vernon, 1988, 2001). How could Sneath have guessed that his guide, Bather, had himself been misled about what Adanson's method was? Adanson's own two-volume *Familles des Plantes*, though hard to interpret, can give the impression of fitting Bather's description, for it contains tables of his 65 artificial classifications as well as his 58 natural families. In fact, however, Sneath was about to step on an old landmine, laid during a long-forgotten war.

By the time of Adanson's death in 1806, Georges Cuvier had established himself at the center of French biology, and, as Secretary of the Institute, it was up to him to write the naturalist's obituary (Outram, 1978:162). An important part of Cuvier's claim to fame was his new and supposedly rational approach to systematics, an approach distinctly at odds with the raw empiricism Adanson had espoused. Cuvier's bias is revealed in his comment that Adanson's method produced an estimate of affinity "independent of the rational, physiological knowledge of the influence of organs ..." (Cuvier, 1807:282), in other words, knowledge of the kind required by Cuvier's principle of the subordination of characters. Adanson had clearly argued in favor of a natural classification, but to him that simply meant one that takes into account all features of organisms, in contrast to an artificial method, which uses only a few features.

Adanson said that he had constructed several artificial classifications in his early youth, but later decided that none would ever mirror nature because the features important for some groups were not the important ones for other groups. He nevertheless continued to make one-character systems, while also seeking nature's own groups. The result of his search was 58 natural families, given in detail in Vol. 2 of *Familles des Plantes*, published in 1763. In Vol. 1 (not actually published until the following year [Stafleu, 1963:238—239]), Adanson sketched his artificial systems, followed by a table that measured the "degree of goodness" of each, in comparison with the 58 natural families. This exercise was designed to show how far each of the 65 fell short, and it also suggested which characters were less bad than others.

Adanson did explain how he had arrived at his 58 natural families:

> First I made a complete description of each plant, putting into the description each of its parts in separate articles, in all its details; and to the extent that some new species appeared that had a relationship to ones I had described, I described them on the side, not mentioning the resemblances but only noting their differences. It was by the totality of these compared descriptions that I perceived that plants naturally arranged themselves under classes or families which were neither systematic nor arbitrary, not being based on one or a few parts that can only change within certain limits, but on all parts ... . (Adanson, 1763–1764:clviij)[1]

Probably Cuvier, who did all his work with notorious speed, was really unconscious of how inadequate was his comprehension of Adanson. The atmosphere Adanson had breathed as a student of Bernard de Jussieu in the 1740s, when Linnaeus's artificial classes and orders were the subject of passionate debate among botanists, was a thing of the past when the young zoologist Cuvier was a student of C. F. Kielmeyer in the 1780s. After another quarter of a century, the mature Cuvier began by reporting what Adanson had done.

> Considering each organ by itself, he made a system of division based on its different modifications. He arranged in this system all the organisms known. Repeating the process for each of a great many organs, he constructed a number of such systems, all of them artificial and each based on a single organ chosen arbitrarily.

Then Cuvier went on to offer a speculative account of what a reasonable man could possibly do with such a large number of artificial systems, an account any trusting reader would wrongly imagine to be based on something Adanson had said.

> It is evident that entities which are classed together in every one of these systems are exceedingly close to one another, since they resemble each other in all their organs. The relationship is less when the entities are placed in different classes by some of the systems. And finally, the most distant are those entities which are not grouped into the same class by any of these systems. This method thus gives a precise estimate of the degree of affinity between the organisms ... . (Cuvier, 1807:282)[2]

---

[1] "Je faisois d'abord une description entière de chaque Plante, en metant dans autant d'articles séparés, chacune de ses parties, dans tous ses détails; & à mesure qu'il se présentoit de nouveles Espèces qui avoient du raport à celles déja décrites, je les décrivois à côté, en suprimant toutes les ressemblances, & en notant seulement leurs diférences. Ce fut par l'ensemble de ces descriptions comparées que je m'aperçus que les Plantes se ranjoient naturelement d'elles-mêmes sous des Classes ou Familles, qui ne pouvoient être systématiquès ni arbitrères, n'étant pas fondées sur 1 ou quelques parties qui dussent chanjer à de certaines limites, mais sur toutes les parties ... ." Adanson believed that French spelling should be made more phonetic, which did nothing to make his text more attractive to his contemporaries.

[2] "Considérant chaque organe isolément, il forma de ses différentes modifications un système de division dans lequel il rangea tous les êtres connus. Répétant la même opération par rapport à beaucoup d'organes, il construisit ainsi un nombre de systèmes, tous artificiels, et fondés chancun sur un seul organe arbitrairement choisi. Il est évident que les êtres qu'aucun de ces systèmes ne séparerait, seraient infiniment voisins, puisqu'ils se ressembleraient par tous leurs organes; la parenté serait un peu moindre dans ceux que quelques systèmes ne rassembleraient pas dans les mêmes classes; enfin, les plus éloigns de tous seraient ceux qui ne se rapprocheraient dans aucun système. Cette méthode donnerait donc une estimation précise du degré d'affinité des êtres ... ." The translation is Sneath's (1965:482).

Yet Adanson said no such thing. He never explained exactly what use he made of his 65 artificial systems while seeking his natural families. The closest he came were these two statements:

> [After finding that none of my artificial systems worked] I only used them in the search for the natural method, in which they all helped me greatly.[3]

> [Among the advantages of my set of artificial systems is:] Taken together, they give all existing or observed relations among all the parts of plants, relations from which our 58 families are formed. (Adanson, 1763–1764:cciij)[4]

Adanson never pointed to any entities "classed together in every one" of his 65 systems. His testimony that he had perceived how plants naturally arranged themselves, after he had written as full a description as he could, suggests that his artificial systems were useful to him as a data bank, a kind of index in which he could look up whether a given character was found in a particular plant. We must assume that by the time he was writing his book, every species he knew had been entered in its proper place within each of his 65 systems, but he did not put into print this complete catalog because, he said, with 1615 genera, it would have taken too much space. What he gave instead was, in each artificial system, the names of his natural families, distributed among the several sections. This was enough for him to then compute the extent to which his natural families are broken up or left intact by each system, but obviously he cannot have portrayed his systems in this way until he had decided upon his natural families (Nelson, 1979:16–17).

Cuvier's student Augustin-Pyramus de Candolle repeated Cuvier's interpretation in his 1813 *Théorie Élémentaire de la Botanique*, adding the information, not mentioned by Cuvier, that the number of Adanson's artificial systems was 65. de Candolle had the wit to recognize how unlikely it was that anyone could really have done what Cuvier claimed, so he added that Adanson had probably formed his families "as much by means of feeling his way as by his own method" (de Candolle 1813: 71).[5] de Candolle's reservation as to its practicality did not call into question Cuvier's version of the method itself, however. As the years rolled by, Cuvier's name acquired the luster of immortal fame, while eighteenth-century theoretical disputes dropped further beneath the horizon. For many later writers, including Bather, the lucid historical section of de Candolle's widely reprinted textbook was the source for their mention of Adanson.

To Sneath it looked obvious that a person whose method was to make 65 different classifications, each based on a different character, and then to combine them, must be giving equal weight to each character. Indeed de Candolle had made the same assumption. In 1957 Sneath proposed that a classification "based on giving every feature equal weight ... may conveniently be called 'Adansonian'" (Sneath, 1957:196, 1958). This interpretation was fully endorsed by Arthur Cain (1959a), who was misled by the same sources (Bather and de Candolle) and for the same reasons (he too thought taxonomy needed to become more objective). The founding textbook of numerical taxonomy that Sneath co-authored with Robert Sokal relayed this mistaken view of Adanson and expanded the scope of the term *Adansonian* (Sokal and Sneath, 1963:50).

When Sneath was invited to speak at a well-funded conference at the Hunt Botanical Library in Pittsburgh, he was so far from recognizing the ticking bomb set by Cuvier that

---

[3] "... je ne les emploiai que pour la recherche de la Métode naturele, à laquele leur ensemble m'aida beaucoup."
[4] "Leur ensemble done tous les raports existans ou observés entre toutes les parties des plantes, raports d'où se sont formées nos 58 Familles."
[5] "... qu'il a peut-être formées, autant par voie de tâtonnement, que par sa propre méthode ..."

he said (in August 1963) "What were the main features of Adanson's method? One cannot do better than quote from Cuvier's *Éloge* ..." (Sneath, 1965:482). Yet in the meanwhile Sneath had been studying Adanson's own works, the earlier work on the molluscs of Senegal as well as the botanical volumes, so how could he still accept Cuvier's version as sound? Part of the reason is doubtless Adanson's own lack of clarity; Sneath was a microbiologist, and Adanson's examples are hard even for a malacologist or botanist to penetrate, because so many taxon names have changed. Perhaps also, Sneath imagined that Cuvier, a contemporary of Adanson's though 42 years his junior, was more likely to understand Adanson's intent than a twentieth-century reader could. It must also be admitted that Sneath's interest in having a colourful precursor for his new scientific enterprise may have dulled his critical faculties. Sneath had a much sharper conception than de Candolle had had of the practical impossibility of the method Cuvier described, yet still it did not occur to him, any more than it had to de Candolle, that the problem might lie with Cuvier, not with Adanson. After quoting Cuvier, Sneath said,

> [Adanson] then apparently counted the number of times that a pair of entities fell together in the subdivision. In effect this procedure counts the number of disagreements in the characters used to make the divisions, and if carried out systematically would have yielded a table of the comparisons between each organism and every other, which would have been, in effect a similarity matrix. Whether Adanson ever proceeded in this systematic way is very doubtful: the number of pairwise comparisons between the 1615 genera in the *Familles des plantes* total over a million. It is more likely that he counted the disagreement for some of the comparisons only, but did enough to obtain a fair idea of the salient relations between the organisms. (Sneath, 1965:483)

The number "over a million" shows that Sneath had really not given close attention to the statement he quoted from Cuvier, much less to Adanson's statements. Sneath's figure comes from making one act of comparison between each of the 1,615 genera and all the others, which is 1,615 squared, minus the self-comparisons and minus the duplicates (1,615 x 1,614)/2 = 1,303,305. However, the method described by Cuvier would have required Adanson to also ask, for each pair being compared, whether or not it is found united in each of his 65 artificial systems. This extra step, without even wondering how the answer to that question could generate the natural families, forces the 1,303,305 to be multiplied by 65, so even being conservative, Sneath should have said "over 84 million" (for 84,714,825). Actually, neither number would be so terrible if one had set up the data such that comparisons could be made at a glance. One comparison every 5 seconds, working an 8-hour day, resting on Sundays, would allow one to cruise through the task set by Sneath in less than 9 months and finish the task described by Cuvier in 47 years, not fun but feasible.

The explosion Sneath met with in Pittsburgh was an explosion of historical information. Jean-Paul Nicholas, financed out of the deep pockets of Roy Arthur Hunt to make a full-time study of the manuscripts Hunt had purchased from Adanson's heirs, traced his life in minute detail (Nicholas, 1963). Frans Stafleu, the Dutch botanist and historian of botany, had gone through Adanson's *Familles des Plantes* with a fine-toothed comb and went to great lengths in the published version of his Pittsburgh lecture to contradict all of Sneath's historical claims (Stafleu, 1963:195, 201). The proceedings of the Hunt conference were promptly published, soon followed by a reprinting of Adanson's *Familles des Plantes* with a preface in which Stafleu insisted that any taxonomy using the unbiased inductive method ... has always been Adansonian. To use the term Adansonian for numerical taxonomy alone is not conducive to a better understanding of the latter subject. Adanson was anything but a computer avant la-lettre, and a study of his systems and basic statements reveals that he did not rest content with equal weighting. (Stafleu, 1966:x)

Stimulated by the publication of the proceedings of the Pittsburgh conference, several other botanists, unsympathetic to numerical taxonomy, weighed in with complaints against Sneath and Sokal's "wish to adorn their movement with a historical figure: poor Adanson" (Jacobs, 1966:55). B. L. Burtt of the Edinburgh Botanical Garden seized on Sneath's admission that to really do the matrix implied by Cuvier would require over a million comparisons (Burtt, 1966). Michel Guédès, botanist at the Muséum National d'Histoire Naturelle, Paris compiled a thorough review and published it in a historical journal. It was he who first identified Cuvier as the source of the mistake (Guédès, 1967).

Sokal and Sneath now faced a dilemma. The botanists cited above had put more time into reading Adanson than this zoologist and this microbiologist were willing to devote to the issue, but to simply drop Adanson from their account would be too much like admitting their opponents' case, which was not only bad strategy but not what they felt. Unsure of just how strong a case might be made for or against calling their approach Adansonian, they chose to keep him in the story but with less emphasis. In the 1973 revision of their textbook, rather than repeating their 1963 statement that "numerical taxonomy is based on the ideas first put forward by Adanson" (p. 50), they made the weaker claim that the principles of numerical taxonomy "embody concepts that can be traced to Michel Adanson" (Sneath and Sokal, 1973:5). If you press hard for the meaning of either statement, you find yourself asking things such as, for a later idea to be "based on" an earlier one, does the later author have to be familiar with the earlier author, or is it enough if the later author has come upon a similar idea independently? Are ideas like living things, which must maintain an unbroken chain of reproduction to maintain their identity, or do they enjoy some kind of ideal existence, like mathematical relations, so that different people may call to mind the same idea?

Sokal and Sneath also toned down their 1963 statement that an idea, like the equal weighting of characters, "may be called Adansonian," wording that seems to give permission and perhaps encourages that usage; however, in 1973 they just say that such ideas are "frequently called neo-Adansonian," which seems to report usage without judging it. The difference between the euphonious *Adansonian* and the longer *neo-Adansonian* is fraught with — what? politics? rhetoric? optics?

In their second edition, Sokal and Sneath retreat to pretended neutrality on the question of whether Sneath "is wrong in proclaiming Adanson to be the father of numerical taxonomy" (Sneath and Sokal, 1973:23). As if to wash their hands of the whole question, they declared,

> We prefer to let historians of science pursue this argument. For although it was — and remains — important to trace the roots of the historical origins of an idea in science, the development of numerical taxonomy has so far outpaced the early primitive ideas on this subject that to have to rely on Adanson's views for a validation of modern numerical phenetics seems as irrelevant as to rely on Mendel's writings for a validation of the findings of the molecular geneticists. (Sneath and Sokal, 1973:23–24)

Their defensiveness is obvious, for surely no one who used the term *Adansonian* pretended that the views of that dead botanist certified the validity of numerical taxonomy.

These remarkable statements seem to me to point to some significant issues concerning the status of history in relation to science. Sokal and Sneath said:

1. History is important.
2. Old views cannot validate current science.
3. Historians of science rather than scientists should sort history out.

I believe the apparent inconsistency between issue 1 and 2 can be explained by a careful consideration of what history is and how the present relates to the past. I believe that both statements are deeply true and I intend to address this topic explicitly elsewhere. Here, I shall address only issue 3. Sokal and Sneath seem to assume the existence of an unbiased higher court populated by "historians of science." They imply that the critics of their view of Adanson, who were biologists by profession rather than historians, had an axe to grind and were finding fault with Sneath's references to Adanson as a way of questioning the soundness of numerical taxonomy itself. That concern about objectivity is valid, though of course it also points to the possibility that their own interest in Adanson was likewise liable to be biased.

The point that interests me, as a professional historian of science, is the fact that I and my fellow historians of science responded to this direct appeal for our judgement with stony silence. (The only non-biologist who offered a judgement, as far as I know, was Réjane Bernier [1984], but that was almost 20 years later, and her professional home is philosophy rather than history.) The guild to which we historians belong has membership rules that strongly discourage, indeed practically forbid us, from accepting such an invitation. We are taught from our first days of graduate school that when we do history, we must view past events in their own context. We could take up questions such as What did Adanson actually say? What did he do? What was he probably thinking? Who read his work and what did they think of it? We are carefully taught, though, to avoid pronouncing on questions of the form, "Of what modern branch of biology was Adanson the forefather?" These professional rules are exactly parallel to what graduate students in research science learn, that a biologist can report on what kinds of organisms live in a particular place, and how they interact with one another, and even what their ancestors may have looked like, but the pressing question of whether the property owner has the moral or legal right to pave the place over is beyond the realm of pure or basic science. The professional training of most historians of science pledges us to do pure history. When scientists have serious disagreements, acting as referee is the last thing the historian wants to do. This is not to say that the scientists' appeal to our expertise is unreasonable, nor that our tendency to stay aloof is a virtue, only to say that the structure of academic disciplines has had this effect.

You may be inclined to doubt the accuracy of my portrayal, because you are thinking of several people who write the history of biology who have no hesitation at taking part in current debates, among whom are Michael Ghiselin, David Hull, and Marc Ereshefsky (Ghiselin, 1997). These, however, are people who do history without being members of our guild. Ghiselin is a biologist; Hull and Ereshefsky are philosophers.[6] The distinction between a philosopher and a historian is just as real as that between a mouse and a finch. The fact that we both talk about the past (which the philosophers do only some of the time, historians

---

[6] I am oversimplifying a bit. David Hull, when a student in a program in history and philosophy of science, did do some graduate work in the history of science and sees himself as interdisciplinary, saying that "historians think I do philosophy, scientists think I do philosophy, while philosophers think I do science" (personal communication, October 2001). He also testifies, however, that even in programs like Indiana's, students either identified themselves as historians or philosophers, not as hybrids (Hull, 2002).

full time) should not lead you into mistaking one kind for the other any more than the fact that both finch and mouse eat seeds. The guild to which philosophers belong does not forbid them from becoming involved in scientists' quarrels, although their professional rewards depend upon them returning from that excursion with some spoils that their fellow philosophers will think worthwhile. Perhaps you consider my ecological analogy far-fetched, because when animals use a resource they consume it, whereas when people look at historical material, the material remains available (indeed may become even more available through quotation and reprinting). Yet the professional demand that a publication must say something new means that if a historical event is well described, it becomes unavailable to another historian looking for an original topic. There may even be a process analogous to character displacement, for when historians of science emerged as a distinct new species, in the late 1950s, they defined their questions so as to be distinct from the questions of priority, precursors, and guidance for the future that scientists had always addressed through history.

A more sophisticated analysis of such guild distinctions than my ecological analogy is Elihu Gerson's application of the sociological concept of social worlds (Gerson, 1983). Scholarly research, even though one experiences it as the very lonely reading of an old book no one else has looked at, is truly a social activity, because one carries into the library one's expectations of the reactions of one's peers, to whom any significant findings must be communicated. Unfortunately the consequence is often that workers in each discipline respond only to the work of their peers working within the same discipline instead of giving appropriate attention to all relevant publications. For example, writings on Adanson by biologists, even such scholarly ones as Stafleu, Guédès, and Nelson, have been ignored by some later philosophers (Tort, 1989).

Biologist Ernst Mayr has never been content to leave anything he cares about to the historians to settle. Indeed he has been eloquent in support of biologists' right to do their own historical research and to make their own claims about the past. When professional historians, in reviewing his book *The Growth of Biological Thought*, criticized Mayr's Whiggishness; that is, his willingness to focus only on those past developments that led up to the modern state of things, he vigorously defended his style of history (Mayr, 1990). Yet that monumental book, and his hundreds of historical statements in his other books and articles and lectures, have not been enough to make him into a historian in the judgement of the inner elite of my discipline, any more than Hull's presidency of the Society for Systematic Zoology made him into a zoologist. The subject matter is one thing, but the guild is quite another.

At first, Mayr added history to his exposition of biology even more cautiously than Sneath had done, although his excursions into the past were just as clearly linked to the vision Mayr cherished for the future of systematics as Sneath's would later be to his own vision. Mayr barely mentioned pre-twentieth-century authors in his landmark 1942 book *Systematics and the Origin of Species*. Step by small step, however, in the 1950s he began to develop a historical narrative to add authority to his biological arguments about species (Junker, 1996). There was a very short historical section in the textbook he co-authored with Gorton Linsley and Robert Usinger in 1953 (Mayr et al., 1953). The next year he was invited to a celebration of Karl Jordan's 94th birthday, and for that occasion he reviewed Jordan's publications of the late nineteenth century (Mayr, 1955). In 1955, Mayr organized a symposium on the species question for the American Association for the Advancement of Science (AAAS), and he opened the day with a historical review (Mayr, 1957) that was based on previous surveys of the species question by German biologists.

It was quite clear that Mayr hoped that history would yield something to assist biologists in their current struggles to understand species. At first the historical material seemed to offer no outstanding forefathers. Karl Jordan was the closest thing to a noble precursor of Mayr's view of species, but Jordan would never be well known outside of entomology. Mayr had to contend with the problem that the most attractive and apparently central hero in the history of the evolutionary view of species, Charles Darwin, seemed to belong to the wrong camp. The view of the biological species Mayr was pushing featured its status as a real natural entity, something more substantial than a man-made category, more than a set of individuals. It was a population, "held together by a supraindividualistic bond" (Mayr, 1957:8); the geneticists who thought of taxonomists as mere stamp collectors had missed this central fact of evolutionary biology. Species are not categories, they are real entities, the subjects of the verb *evolve*. Linnaeus, in spite of being a creationist, was a positive landmark from Mayr's viewpoint, because he had believed in the objective existence of species, whereas Darwin, in spite of his virtues as an evolutionist, had said things about species that made them dangerously subjective. Mayr said at the 1955 AAAS symposium,

> In Darwin, as the idea of evolution became firmly fixed in his mind, so grew his conviction that this should make it impossible to delimit species. He finally regarded species as something purely arbitrary and subjective. "I look at the term species as one arbitrarily given for the sake of convenience to a set of individuals closely resembling each other, and that it does not essentially differ from the term variety which is given to less distinct and more fluctuating forms ... .The amount of difference is one very important criterion in settling whether two forms should be ranked as species or variety." And finally he came to the conclusion that "In determining whether a form should be ranked as a species or a variety, the opinion of naturalists having sound judgment and wide experience seems the only guide to follow" (Darwin, 1859). (Mayr, 1957:4)

This subjective view of species, which a philosopher would call nominalist, was one of the dragons Mayr was determined to slay. It was an attitude still very common among many biologists, who regarded taxonomists as librarians of a kind, who erected classifications of their own invention for the sake of convenient retrieval of information. Mayr felt it was necessary to contradict this nominalist view if systematists were to be promoted from their lowly position of servants up to the status of real scientists.

There was at the same time a second dragon standing in the way of this promotion of systematics. Herbarium and museum taxonomists, although many or most believed the taxa they named had some kind of substantial reality, were vague or worse when it came to articulating that reality. At the species level, where they should have cared about the reproductive dynamics that Darwin had discussed and the breeding networks geneticists such as Theodosius Dobzhansky described as a gene pool, herbarium and museum taxonomists limited their attention to characters preserved in their dead specimens, that is, morphological characters, explicitly declaring out of bounds features of living organisms such as breeding habits, that is, biological characters. They had an elaborate system of voucher specimens they called "types," which the more enlightened taxonomists often had to remind their dimmer associates did not need to be typical of the species for which they were the name bearers. With respect to the higher categories, some taxonomists were frankly nominalist, declaring all groups above species arbitrary, but among the antinominalists there were some, most of whom were anti-Darwinians, who advocated idealistic theories about abstract archetypes, heirs of the nineteenth-century science of morphology. Mayr lumped together all these unconnected, misguided beliefs to make one coherent enemy, which he named "typology" or "typological thinking."

In his 1955 AAAS paper, Mayr's historical section identified the real species of local naturalists as the Linnaean stream, which had flowed through time distinct from the nominalist stream flowing from Darwin. Then, leaving history, Mayr proposed a classification of current species concepts, in which the typological species concept was introduced as a possible mistake to be avoided. Although in his discussion the name of an ancient Greek is prominent, Mayr was alluding to a timeless logical position rather than proposing a concrete historical development.

> *The Typological Species Concept.* This is the simplest and most widely held species concept. Here it merely means "kind of." ... This simple concept of everyday life was incorporated in a more sophisticated manner in the philosophy of Plato. Here, however, the word *eidos* (*species*, in its Latin translation) acquired a double meaning that survives in the two modern words "species" and "idea" both of which are derived from it. According to Plato's thinking objects are merely manifestations, "shadows," of the *eidos*. By transfer, the individuals of a species, being merely shadows of the same type, do not stand in any special relation to each other, as far as a typologist is concerned. Naturalists of the "idealistic" school endeavor to penetrate through all the modifications and variations of a species in order to find the "typical" or "essential" attributes. Typological thinking finds it easy to reconcile the observed variability of the individuals of a species with the dogma of the constancy of species because the variability does not affect the essence of the *eidos*, which is absolute and constant. Since the *eidos* is an abstraction derived from individual sense impressions, and a product of the human mind, according to this school, its members feel justified in regarding a species "a figment of the imagination," an idea. Variation, under this concept, is merely an imperfect manifestation of the idea implicit in each species. If the degree of variation is too great to be ascribed to the imperfections of our sense organs, more than one *eidos* must be involved. Thus species status is determined by degrees of morphological difference. The two aspects of the typological species concept, subjectivity and definition by degree of difference, therefore depend on each other and are logical correlates.
>
> The application of the typological species concept to practical taxonomy results in the morphological defined species ... . (Mayr, 1957:11–12)

Mayr's portrayal of Plato was somewhat inaccurate. The Platonic *eidos* was not a product of human thinking; rather, human intuition allows us to recognize the really existing *eidos*. Species, far from being figments of the imagination, are more real than individuals. The imperfections of our sense organs have nothing to do with variability for a Platonist. Contrary to Mayr's polemic, no biologist would contend that the individuals of a species "do not stand in any special relation to each other," for they have always been understood to be related as blood relatives. The bonds of reproduction that link all members of a biological species together has been recognized from time immemorial. Plato's student Aristotle discussed it at length, and from the first days of the revival of natural history in the Renaissance, the idea was commonplace that what the naturalist was doing when characterizing a species by its morphology, that is, its outward form, was seeking marks by which blood relatives could be recognized. Mayr was creating an ugly category from which all museum workers would want to distance themselves.

Immediately after sketching the typological species concept, Mayr admitted that few people believed in it: "Most systematists found this typological-morphological concept inadequate and have rejected it," (1957:12). The whole exercise would be odd indeed if this were a disinterested historical narrative. When we recall its context, introducing a symposium he had organized, a symposium that was part of his two-pronged efforts to get systematists to adopt the biological species concept and nonsystematists to accord them more respect, it makes perfect sense that Mayr would conjure up the typological dragon, branded with the name of a long-dead philosopher.

Mayr's dragon of typology has been breathing flames so frighteningly for close to 50 years now, that it may be surprising to recognize it as a specter of his own creation. His 1942 *Systematics and the Origin of Species* had criticized the morphological concept of species without anywhere alluding to typological thinking. That is, he criticized the practice of taxonomists who limited their attention to physical features without identifying them as followers of the wrong philosophy. Likewise, the authors contributing to the 1940 collection *The New Systematics* (Huxley, 1940) seemed unaware of this dangerous mode of thought, and so was G. C. Robson, whose 1928 book *The Species Problem* focused on naturalists and ignored philosophers and other nonscientists (Robson, 1928). Bather had been oblivious to the problem in 1927, when he said, "... I propose to pass by the old logical methods of classification, because they have no practical bearing on the arrangement of any natural objects, least of all those endowed with life" (lxiv). Darwin, while writing his most famous book, put concentrated effort into surveying the views of fellow naturalists about the nature of species and the nature of classification, and though he found that many of them harbored vague and inconsistent concepts, including "God's plan," the Platonic *eidos* was not an idea Darwin felt he needed to slay.

Mayr had been interested in Greek philosophy in his youth[7] and had heard that Plato was the enemy of empirical science. He learned from his teacher Erwin Stresemann that idealism had diverted ornithology from its healthy progress. In the 1930s and 1940s he was concentrating fully on the demanding job of being a museum taxonomist. Arthur O. Lovejoy's 1936 book *The Great Chain of Being* may not have come to Mayr's attention until its 1960 reprinting (Lovejoy, 1960). This book, which he greatly admired (Mayr, 1976:254), encouraged Mayr to view the history of thought as a long story of persistent themes, and one of the themes identified by Lovejoy was Plato's rationalism, the conviction that the world was put together in such a way that reason could penetrate it. Lovejoy argued that people, and generations of people, can be following assumptions or habits of thought of which they are unconscious, and such ideas can be general and vague.

Viewing the history of concepts in terms of grand themes allowed Mayr to give his story the hero it had lacked. As a promoter of systematics, the situation he found himself in was a competitive one, as he explained in the opening sentences of *Systematics and the Origin of Species*.

> The rise of genetics during the first thirty years of this century had a rather unfortunate effect on the prestige of systematics. The spectacular success of experimental work in unraveling the principles of inheritance and the obvious applicability of these results in explaining evolution have tended to push systematics into the background. There was a tendency among laboratory workers to think rather contemptuously of the museum man, who spent his time counting hairs or drawing bristles, and whose final aim seemed to be merely the correct naming of his specimens. (Mayr, 1942:3)

As he later put it, "a peculiar myth" had arisen, to the effect that "mathematical population genetics is the source of population thinking" (Mayr, 1976:307). His reading of the early papers of Karl Jordan had made clear that this distinguished old taxonomist had recognized the importance of the wide range of variability in natural populations in 1905. Mayr also knew that many taxonomists, "counting hairs or drawing bristles," still regarded

[7] E. Mayr letter to Karl Popper, 13 February 1978 (Harvard University Archives, Mayr Papers, Box 26, file 1297).

such variability as an inconvenience to be suppressed rather than an opportunity to investigate evolutionary processes, and he had to admit that Jordan had been far in advance of his peers. It was strategically essential that Mayr give taxonomists rather than geneticists priority for the central evolutionary concept of the variable population.

Mayr's needs and ideas came together in 1957, when he was invited to speak to the Anthropological Society of Washington, DC, as one in a series of lectures commemorating the centenary of the publication of Darwin's *Origin*. In preparation Mayr reread Darwin's classic work, and realized that he, no less than Jordan, could be credited with appreciating the uniqueness of each individual in a population. In this context Darwin's nominalism on the species question was a detail that could be ignored. On this public occasion Mayr had larger fish to fry. In Washington, in October 1957, Mayr declared that besides the theory of evolution by natural selection, another of Darwin's great achievements, "equally important but almost consistently overlooked" was that he had "replaced typological thinking by population thinking" (Mayr, 1959:2).

> Typological thinking no doubt had its roots in the earliest efforts of primitive man to classify the bewildering diversity of nature into categories. The *eidos* of Plato is the formal philosophical codification of this form of thinking. According to it there are a limited number of fixed, unchangeable 'ideas' underlying the observed variability, with the *eidos* (idea) being the only thing that is fixed and real while the observed variability has no more reality that the shadows of an object on a cave wall, as it is stated in Plato's allegory. The discontinuities between these natural 'ideas' (types), it was believed, account for the frequency of gaps in nature. Most of the great philosophers of the 17th, 18th, and 19th centuries were influenced by the idealistic philosophy of Plato, and the thinking of this school dominated the thinking of the period ... . The assumptions of population thinking are diametrically opposed to those of the typologist. (Mayr, 1959:2, 1976:27)

Previously Mayr's bits of history only described taxonomists, or biologists, but here Mayr (1959) seemed to be talking about everyone. From this grand perspective, Darwin was the white knight. From this perspective, the enemy was not a particular sort of biologist, but the philosopher Plato and all later philosophers who had been unable to free themselves from his influence.

Mayr would later (1976:26) identify his Washington lecture as "the first presentation of the contrast between essentialist and population thinking, the first full articulation of this revolutionary change in the philosophy of biology," but he did not use the word *essentialism* in 1957. Accepting this philosophical term as the equivalent of the biologists' word *typology*, which he did in 1968 (Mayr, 1968:430), represented Mayr's cautious acceptance of the arguments of a young philosopher of science named David Hull. As a graduate student Hull had taken a course taught by Karl Popper, already celebrated as the author of *The Open Society and its Enemies* (Popper, 1945) and *The Logic of Scientific Discovery* (Popper, 1959). In 1944 Popper coined the term *essentialism* for the stance opposite to nominalism, because the traditional philosophers' label for the Platonic view, where the *eidos* or universals were held to be real, was *realism*, yet to nonphilosophers this seemed counter-intuitive (Popper, 1944:94). Although there were things about Popper and his thought that Hull did not like, he fully agreed with Popper's powerful argument that "every discipline as long as it used the Aristotelian method of definition has remained arrested in a state of empty verbiage and barren scholasticism" (Popper, 1950:206 quoted in Hull, 1965:314). Hull's student paper, demonstrating Popper's thesis with a fresh example, was published in 1965 as "The Effect of Essentialism on Taxonomy — Two Thousand Years of Stasis." Hull immediately sent Mayr a copy, and they began a substantial correspondence.[8]

Over the subsequent decades, the logical difference between population thinking and essentialism, which philosophers view as belonging to the old, deep, and difficult debate over the existence of natural kinds, grew into a large literature, and it is still growing. By his stimulating presence at meetings and his tireless correspondence, Mayr has encouraged several generations of philosophers to take evolutionary biology seriously (Hull, 1994), and he has also substantially encouraged numerous historians of biology (Burkhardt, 1994). Yet to a remarkable extent these two sorts of humanist academics, from the time of their emergence in the 1960s to the present, have behaved like distinct species displaying character displacement. In other words, there must be a sociological reason, for there is no intellectual justification, for the failure of historians, including my own, to address the historical dimension of the story told by Mayr in Washington in 1957. A few biologists have attempted some history of typology (Sokal, 1962; Van der Hammen, 1981; Panchen, 1992), as have a few philosophers (Sober, 1980; Amundson, 1998), but attempts on this topic by certified historians can be charitably described as subtle (Winsor, 1979; Farber, 1976). The essentialist version of the typology story now reads thus:

> Prior to Darwin, with only a very few exceptions that were as noteworthy as they were rare, nearly everyone viewed species as having eternal, immutable essences. Each species had to be characterized by a set of properties which *all* (but abnormal) members of that species possessed, and *only* members of that species possessed. (Hull, 1988:25)

Yet the exceptions were not so few, as Mayr had already suggested. In his Washington lecture he had gestured broadly to the seventeenth, eighteenth, and nineteenth centuries as a period whose thinking was dominated by Platonism, but a decade later, when he returned to the narrower question of the development of systematics, Mayr described the hundred years between Linnaeus's 1758 *Systema Naturae* and Darwin's *Origin* as a transitional period, the age of empiricism.

> Deductive principles, whether based on essentialism or nominalism, were increasingly rejected, and taxonomists to an increasing extent based their taxa on the totality of characters. This was started by Adanson, but a strongly empirical philosophy characterized virtually all the leading taxonomists of that period. The term "natural" acquired a new meaning during this period, signifying a classification unbiased by *a priori* considerations and based on a consideration of totality of characteristics. (Mayr, 1968:547)

Adanson was doubtless the rare and noteworthy exception Hull was alluding to in 1988. By any definition of typology or of essentialism, Adanson will not fit into that box. He had very clearly distanced himself from the idea that reason could identify the essential feature of any plant. Arthur Cain, in order to press the unruly Linnaeus into the square hole of a scholastic Aristotelian, had maintained, by means of a contorted argument, that what Linnaeus meant by *nature* was near enough to count as essential (Cain, 1958:154–156), but he could not pull this off with Adanson and so was left with declaring him "far ahead of his time" (Cain, 1959a:205).

---

[8] Mayr Papers, Harvard University Archives.

Peter Sneath had very rightly emphasized that Adanson's argument about the lesson to be derived from his 65 artificial systems was that the description of all of his natural families would be polythetic (Sneath, 1965:477). Linnaeus and Bernard de Jussieu had both felt, before Adanson, the desirability of discovering nature's own orders or families, but the result of their labors was only to point to those higher taxa, that is, to list the names of genera they believed belonged together, without listing characters definitive of the groups. (Jussieu taught his students about the taxa that seemed natural to him, and planted the garden he supervised accordingly, but published no descriptions; Linnaeus of course published the characters of his sexual, *artificial* classes and orders, but he indicated his *natural* orders simply by naming the genera belonging to each [Stearn, 1959:97—101]).

Adanson went far beyond this, attempting to fully describe each of his 58 natural families by listing the genera belonging to it and their characters. He did so with descriptions, rather than definitions, and his descriptions were certainly not essentialist, for in almost every family, Adanson said that most genera have one feature but some have another. He presents the character-states found in each genus in a table, but the states of those characters can vary among the genera in one family and even among the species in one genus. Nothing could be further from the logic attributed to Plato or Aristotle than Adanson's descriptions of the groups he perceived. It must have been from Stafleu's contribution to the 1963 Pittsburgh conference (cited in Mayr, 1969) that Mayr learned about Adanson's empiricism.

But was Hull right to call Adanson the rare exception, or was Mayr right to say that almost all taxonomists of the generation before Darwin resembled Adanson? We still do not know. Mayr repeated his picture of a hundred years of empiricism in his *Growth of Biological Thought,* but he admitted he was handicapped by the absence of good research (1982:864–865). Since then, although a few biologists have continued Stafleu's tradition of scholarly historical research (Stevens, 1994; Lindberg, 1998) and a few historians have taken an interest in taxonomic practice (Winsor, 1991; Ritvo, 1997; Buhs, 2000; McOuat, 2001; Müller-Wille, 2001), we are still largely ignorant about the history of systematics. The trouble is that few taxonomists, including those most respected by their peers, wrote anything reflective, and those who did usually represented their principles as more rational than they actually were. What we are most in need of are detailed examinations of past taxonomic practice, and no one is better qualified than taxonomists to undertake that sort of history.

Yet is not the moral of this story that scientists ought to leave the history of their subject to professional historians? Absolutely not. First, because in that case, precious little of it would be written at all, and second, because the significance of past events for understanding the present and foreseeing the future would be given short shrift. Surely scientists are just as capable as historians are of telling the difference between rhetorical allusions based on slight evidence and sound history based on a critical evaluation of sources. What we may hope for is a more sophisticated understanding by historians, philosophers, and scientists of how their distinct motivations and skills might contribute to making histories that are more accurate and more useful.

Acknowledgments

I am indebted to the Social Sciences and Humanities Research Council of Canada for financial support of my excursion into metahistory and to the Systematics Association for their generous invitation to the September 2001 meeting at Imperial College, London, which occasioned my study of Adanson. Victoria University has gently administered my funding and provided me with working space, and the University of Toronto librarians have patiently supplied me with reading matter. Sara Scharf, Brigit Ramsingh, and Danai Dalaka were dependable research assistants. Ernst Mayr kindly allowed me to read his correspondence, accessible at the Harvard University Archives, where the staff smoothed my way. Peter

Sneath graciously welcomed my interest in this episode. For advice on technical matters I am indebted to Roger Hansell, George Estabrook, Norm Platnick, and Nadia Talent. I particularly thank Gary Nelson for his penetrating 1979 article and his many stimulating emails. I am grateful to Peter Stevens for his profound research and selfless encouragement. Special thanks are due to the Hunt Botanical Library for creating the Adanson volumes and sending me a copy 36 years ago, even though I have been slow to appreciate their value.

References

Adanson, M., *Familles des plantes*, 1763–1764, facsimile reprint. J. Cramer, Lehre, 1966.

Amundson, R., Typology reconsidered: two doctrines on the history of evolutionary biology, *Biol. Philos.*, 13, 153–177, 1998.

Bather, F.A., Biological classification, past and future, *Q. J. Geol. Soc. London*, 83, lxii–civ, 1927.

Bernier, R., Système et méthode en taxonomie: Adanson, A.L. de Jussieu et A.-P. de Candolle, *Nat. Can.*, 111, 3–12, 1984.

Buhs, J.B., Building on bedrock: William Steel Creighton and the reformation of ant systematics, 1925-1970, *J. Hist. Biol.*, 33, 27–70, 2000.

Burkhardt, R.W., Jr., Ernst Mayr: biologist-historian, *Biol. Philos.*, 9, 359–371, 1994.

Burtt, B.L., Adanson and modern taxonomy, *Notes R. Bot. Gard. Edinburgh*, 27, 427–431, 1966.

Cain, A.J., Logic and memory in Linnaeus' system of taxonomy, *Proc. Linn. Soc. London*, 169, 144–163, 1958.

Cain, A.J., Deductive and inductive methods in post-Linnaean Taxonomy, *Proc. Linn. Soc. London*, 170, 185–217, 1959a.

Cain, A.J., The post-Linnaean development of taxonomy, *Proc. Linn. Soc. London*, 170, 234–244, 1959b.

Cuvier, G., Éloge historique de Michel Adanson, *Memoires de l'Institut National de France. Classe Sci. Math. Phys.*, 7(1), 159–188, 1807. (Reprinted in *Recueil des Éloges Historique lus dans les Séances Publiques de l'Insitute Royal de France*, Paris, 1819, 3 vols. Facsimile reprint, 1969, Vol. 1, pp. 267–308, Brussels : Culture et Civilisation)

de Candolle, A.-P., *Théorie Élémentaire de la Botanique*, Paris, 1813.

Farber, P.L., The type concept in zoology during the first half of the nineteenth century, *J. Hist. Biol.*, 15, 145–152, 1976.

Gerson, E.M., Scientific work and social worlds, *Knowl.: Creation, Diffus., Util.*, 4, 356–477, 1983.

Ghiselin, M.T., *Metaphysics and the Origin of Species*, State University of New York Press, Albany, NY, 1997.

Gilmour, J.S.L., Taxonomy and philosophy, *The New Systematics*, Huxley, J.S., Ed., Clarendon Press, Oxford, 1940, pp. 361–374.

Guédès, M., La méthode taxonomique d'Adanson, *Rev. Hist. Sci.*, 20, 361–386, 1967.

Hull, D.L., The effect of essentialism on taxonomy: two thousand years of stasis, *Br. J. Philos. Sci.*, 15, 314–326; 16, 1–18, 1965.

Hull, D.L., *Science as a Process: An Evolutionary Account of the Social and Conceptual Development of Science*, University of Chicago Press, Chicago, 1988.

Hull, D.L., Ernst Mayr's influence on the history and philosophy of biology: a personal memoir, *Biol. Philos.*, 9, 375–386, 1994.

Hull, D.L., History and philosophy of science: a time now passed?, in *Proceedings of the 21st Congress of History of Science*, Gayon, J., Ed., Universidad Nacional Autónoma de México, in press.

Huxley, J., Ed., *The New Systematics*, Clarendon Press, Oxford, 1940.

Jacobs, M., Adanson: the first neo-Adansonian?, *Taxon*, 15, 51–55, 1966.

Junker, T., Factors shaping Ernst Mayr's concepts in the history of biology, *J. Hist. Biol.*, 29, 29–77, 1996.

Lindberg, D.R., William Healey Dall: a neo-Lamarckian view of molluscan evolution, *Veliger,* 41, 227–238, 1998.

Lovejoy, A.O., *The Great Chain of Being: A study of the history of an idea,* Harper Torchbooks, New York, 1960.

Mayr, E., *Systematics and the Origin of Species from the Viewpoint of a Zoologist,* Columbia University Press, New York, 1942.

Mayr, E., Karl Jordan's contributions to current concepts in systematics and evolution, *Trans. R. Entomol. Soc. London,* 107, 45–66, 1955.

Mayr, E., Species concepts and definitions, in *The Species Problem: A Symposium Presented at the Atlanta Meeting of the American Association for the Advancement of Science, December 28-29, 1955,* Mayr, E., Ed., American Association for the Advancement of Science, Washington, DC, 1957, pp. 1–22.

Mayr, E., Darwin and the evolutionary theory in biology, in *Evolution and Anthropology: A Centennial Appraisal,* The Anthropological Society of Washington, Washington, DC, 1959, pp. 1–10.

Mayr, E., Introduction, in *On the Origin of Species by Charles Darwin: A Facsimile of the First Edition,* Harvard University Press, Cambridge, MA, 1964.

Mayr, E., Theory of biological classification, *Nature,* 220, 545—548, 1968 (reprinted in Mayr, 1976).

Mayr, E., *Principles of Systematic Zoology,* McGraw-Hill, New York, 1969.

Mayr, E., *Evolution and the Diversity of Life: Selected Essays,* Harvard University Press, Cambridge, MA, 1976.

Mayr, E., *The Growth of Biological Thought: Diversity, Evolution, and Inheritance,* Harvard University Press, Cambridge, MA, 1982.

Mayr, E., When is historiography Whiggish?, *J. Hist. Ideas,* 51, 301–309, 1990.

Mayr, E., Linsley, E.G., and Usinger, R.L., *Methods and Principles of Systematic Zoology,* McGraw-Hill, New York, 1953.

McOuat, G., Cataloguing power: delineating "competent naturalists" and the meaning of species in the British Museum, *Br. J. Hist. Sci.,* 43, 1–28, 2001.

Müller-Wille, S., *Botanik und weltweiter Handel: zur Begründung eines natürlichen Systems der Planzen durch Carl von Linné (1707-1778),*Verlag für Wissenschaft und Bildung, Berlin, 2001.

Nelson, G., Cladistic analysis and synthesis: principles and definitions, with a historical note on Adanson's *Familles des plantes* (1763-1764), *Syst. Zool.,* 28, 1–21, 1979.

Nicholas, J.-P., Adanson, the Man, in *Adanson: The Bicentennial of Michel Adanson's <<Familles des plantes>> Part 1,* Hunt Botanical Library, Pittsburgh, PA, 1963, pp. 1–21.

Outram, D., The language of natural power: the "Eloges" of Georges Cuvier and the public language of nineteenth century science, *Hist. Sci.,* 16, 153–178, 1978.

Panchen, A.L., *Classification, Evolution, and the Nature of Biology,* Cambridge University Press, Cambridge, 1992.

Popper, K., The poverty of historicism. I, *Economica,* 11, 86–103, 1944.

Popper, K., *The Open Society and its Enemies,* Routledge, London, 2 vols. 1945.

Popper, K., *The Open Society and its Enemies,* rev. ed., Princeton University Press, Princeton, NJ, 1950.

Popper, K., *The Logic of Scientific Discovery,* Hutchinson, London, 1959.

Ritvo, H., Zoological nomenclature and the empire of Victorian science, in *Victorian Science in Context,* Lightman, B., Ed., University of Chicago Press, Chicago, 1997, pp. 334–353.

Robson, G.C., *The Species Problem: An Introduction to the Study of Evolutionary Divergence in Natural Populations,* Oliver and Boyd, Edinburgh, 1928.

Sneath, P.H.A., Some thoughts on bacterial classification, *J. Gen. Microbiol.,* 17, 184–200, 1957.

Sneath, P.H.A., Some aspects of Adansonian classification and of the taxonomic theory of correlated features, *Annali. Microbiol. Enzimol.,* 8, 261–268, 1958.

Sneath, P.H.A., Mathematics and classification, from Adanson to the present, in *Adanson: The Bicentennial of Michel Adanson's "Familles des plantes" Part 2*, Hunt Botanical Library, Pittsburgh, PA, 1965, pp. 471–498.

Sneath, P.H.A. and Sokal, R.R., *Numerical Taxonomy: The Principles and Practice of Numerical Classification*, W. H. Freeman, San Francisco, 1973.

Sober, E., Evolution, population thinking, and essentialism, *Philos. Sci.,* 47, 350–383, 1980.

Sokal, R.R., Typology and empiricism in taxonomy, *J. Theor. Biol.,* 3, 230–267, 1962.

Sokal, R.R. and Sneath, P.H.A., *Principles of Numerical Taxonomy*, W. H. Freeman, San Francisco, 1963.

Stafleu, F.A., Adanson and his "Familles des plantes," in *Adanson: The Bicentennial of Michel Adanson's "Familles des plantes" Part 1*, The Hunt Botanical Library, Pittsburgh, PA, 1963, pp. 123–264.

Stafleu, F.A., Introduction, in *Familles des Plantes*, Michel Adanson, facsimile reprint, J. Cramer, Lehre, 1966, pp. v–xv.

Stearn, W., Four supplementary Linnaean publications: *Methodus* (1736), *Demonstrationes Plantarum* (1753), *Genera Plantarum* (1754), *Ordines Naturales* (1764), in *Carl Linnaeus Species Plantarum: A Facsimile of the First Edition 1753*, Vol. 2, Ray Society, London, 1959, pp. 73–104.

Stevens, P.F., *The Development of Biological Systematics: Antoine-Laurent de Jussieu, Nature, and the Natural System*, Columbia University Press, New York, 1994.

Tort, P., *La Raison Classificatoire: Quinze Études*, Aubier, Paris, 1989.

Van der Hammen, L., Type-concept, higher classification and evolution, *Acta Biotheor.,* 30, 3–48, 1981.

Vernon, K., The founding of numerical taxonomy, *Br. J. Hist. Sci.,* 21, 143–159, 1988.

Vernon, K., A truly taxonomic revolution? Numerical taxonomy 1957-1970, *Stud. Hist. Philos. Biol. Biomed. Sci.,* 32, 315–341, 2001.

Winsor, M.P., Louis Agassiz and the species question, *Stud. Hist. Biol.,* 3, 89–117, 1979.

Winsor, M.P., *Reading the Shape of Nature,* University of Chicago Press, Chicago, 1991.

Winsor, M.P., Species, demes, and the omega taxonomy: Gilmour and the new systematics, *Biol. Philos.,* 15, 349–388, 2000.

Winsor, M.P., Cain on Linnaeus: the scientist-historian as unanalysed entity, *Stud. Hist. Philos. Biol. Biomed. Sci.,* 32, 239–254, 2001.

Chapter 2

Launching the Society of Systematic Zoology in 1947

Joe Cain

Abstract

Richard Blackwelder's account of launching the Society of Systematic Zoology overemphasizes the role of conflict with evolutionary systematics and the exclusion of defenders of other taxonomic approaches. The momentum behind this society certainly included disagreements over the usefulness of evolutionary theory and phylogenetic reconstructions for taxonomic practice, but this disagreement was only one dimension of a more complex story. For example, organizers took action only after failing to imbed their interests in other groups. More important, the society's launch occurred during a period of frantic postwar reconstruction of scientific communities and professional infrastructure. Previously ignored archives provide glimpses into the many kinds of jostling under way during this complex transition in American science. Blackwelder's historical accounts omit the postwar context and the more involved elements of the original calls for organization. His particular emphases distort events in ways consistent with his position in other flash points of his career-long conflict with Ernst Mayr over the relevance of evolutionary systematics.

Introduction

Thirty years after the fact, American entomologist Richard Blackwelder (1909–2001) was still angry. In an article celebrating 25 years since the launch of *Systematic Zoology*, Blackwelder barely held his tongue. "Our era started in an atmosphere of controversy. There was difference of opinion on basic ways to look at systematics. And when it came time to form an organization to further the science of systematics, there was strong difference of opinion on how this was to be done" (Blackwelder, 1977:109). This split, Blackwelder explained, led to the formation of two societies rather than one. In Blackwelder's account, his side had tried to be civil and flexible. The other side's refusal to cooperate, however, led to a reluctant divorce. Rejected, his side went their own way in 1947, organizing the Society of Systematic Zoology (SSZ).

This paper examines the SSZ's launch. The basic outline of events is recited frequently — normally deriving from Blackwelder (1977:109–112) or Hull (1988:106–107). These accounts lack detail and context. They also fail to draw from archival material now available. This paper aims to fill these omissions. Why did the society form when it did? Who was at its center? How do these efforts fit into the larger context of American life sciences in the aftermath of World War II? Most important, I challenge Blackwelder's account of the society's founding. Blackwelder entered the society rather late in the organizing process. His motivations (genuine though they were for him) were only part of a more complex picture. The conflict that motivated him seems peripheral to the aims and objectives stated by those involved in the first steps of organizing.

Organizing the Society: 1946–1947

The SSZ was organized during the December 1947 meetings of the American Association for the Advancement of Science (AAAS) in Chicago. Leading this project were Waldo LaSalle Schmitt[1] and George Wharton.[2] The decision to organize was not spontaneous. Schmitt and Wharton worked for more than a year to produce the organization. Most participants had been invited to join via a circular distributed four months before. The long process was completed with the approval of the permanent constitution in September 1948.

According to Blackwelder's legend, a misunderstanding provided the crucial spark. A year and a half before the SSZ's formal launch in Chicago, Schmitt and Wharton attended the organization meeting for another group, the Society for the Study of Evolution (SSE).[3] Prominent systematists led this organizing effort (e.g., Ernst Mayr, Alfred Emerson, George Gaylord Simpson). Others strongly backed it (e.g., Carl Hubbs). There seems to have been a general impression before their meeting that the proposed society would lean heavily toward the interests of systematists or that it would work to place systematics at the center of the biological sciences,[4] but organizers of the evolution group disappointed those with this impression. SSE organizers had a fairly clear idea of where they wanted to steer the new society. First, they wanted it to focus on the analysis of evolution: its patterns, processes, and mechanisms. Second, they wanted to use it as a lever for elevating the status of evolutionary studies in their home disciplines. These relatively narrow interests would come at the expense of other developmental issues within those home disciplines.[5]

[1] Waldo LaSalle Schmitt (1887–1977) was one of America's most respected taxonomists (Blackwelder, 1979; Cox, 1983). A head curator at the United States National Museum (USNM), Schmitt's expertise was Crustacea. He had risen through the institution's ranks since 1910. He also played important roles in fisheries research and studies of faunas around oceanic islands.

[2] George Willard Wharton, Jr. (1914–1990) built his career at Duke University, remaining in the zoology department after his PhD (1939) until 1953, when he moved to the University of Maryland (1953–1961), and then to Ohio State University (from 1961, becoming emeritus in 1976). After developing the Acarology Laboratory at Maryland and taking it to Ohio State, Wharton served as its official director (1969–1976) as well as its guiding force (Mitchell et al., 1969). Wharton's expertise was mites and ticks, with an emphasis on their systematics, ecology and parasitology.

[3] Documents verify that Schmitt and Wharton attended the organization meeting of the SSE that took place on March 30 1946 at the St. Louis AAAS convention. Cain (1994:401, Table 1) lists them in attendance, and their signatures appear on the attendance sheet (Smocovitis, 1994: Figure 1). Blackwelder was not present.

[4] The impression that this new society might center around principles of systematics was not unreasonable given the reputations of those prominent in the organization for defending the values and virtue of systematics within biology (e.g., Mayr [1942], Emerson [1937, 1941] and later Allee et al. [1949], Kinsey [1936] and Simpson [1944, 1945]). Huxley's (1940, 1942) conception of the "new systematics" and a "modern synthesis" of evolution similarly highlighted this emphasis. This impression also would have been bolstered by those following the advertised connections between the new evolution society and the short-lived Society for the Study of Speciation, organized in 1939 but functioning only for about a year. In the speciation society, the concerns of taxonomists formed a large part of the discussion (Cain, 2000b). Schmitt, Wharton, and Blackwelder were members of the speciation society.

[5] Mayr's mission at this particular time was to raise the status and value of systematics within the life sciences, not to develop — as Schmitt wanted — professional infrastructure within systematics. As he explained to Schmitt in March 1948, Mayr saw a noble purpose in this outward-looking effort: "The trouble with us taxonomists is that we are so busy with all sorts of jobs we don't get around to doing research, and when we do we usually don't have enough leisure to produce a good, all-around biological paper but have to be satisfied more or less with taxonomic revision" (Mayr to Schmitt, March 9 1948, Mayr/*Evolution* Papers, folder "Schmitt").

As Blackwelder's legend continues, Schmitt rose at the organizing meeting of the SSE during their discussion of terms of reference. He asked for a broadening of the new society's domain to cover all of systematics (i.e., rules of nomenclature, collections management, etc.) and not simply mechanisms of evolution. This supposedly provoked an extended argument with the organizers, who complained such an expansion in goals would dilute their more specific interest in evolutionary processes. As Schmitt's argument stalled, Wharton supposedly tugged on his jacket. He asked Schmitt to sit down and not to worry. They could start their own group along the lines Schmitt was suggesting.[6]

The first record of Schmitt and Wharton taking action toward a society for systematists dates to April 1947, 12 months after the SSE's launch. While looking ahead to a set of meetings in the summer that would bring Wharton and other sympathetic colleagues to Washington, DC, Schmitt returned to the question of an organization for zoological taxonomy.[7] For reasons he never explained, Schmitt found an organizational model in the American Society of Plant Taxonomists, founded in 1935 (Schofield, 1998). He adapted their corporate documents and asked Wharton to consider if this approach provided a reasonable starting point. Wharton replied on April 1 1947:

> The constitution of the Plant Taxonomists seems to be readily adaptable to a society of taxonomic zoologists. I would suggest the title for our society "Union of Taxonomic Zoologists." The purpose of the society might be expressed simply as follows: Purpose — To advance the study of taxonomic zoology and the welfare of taxonomic zoologists by providing a medium for the expression of the needs, ideals, achievements, and purposes of its memberships. ... In preparation for this meeting it would be desirable to contact taxonomists who are interested in the project at as many museums and universities as possible in order to line up membership in their respective institutions.[8]

---

[6] This legendary moment most likely did not happen as was repeated so often by Blackwelder. The minutes of the March 1946 SSE meeting record no such conversation between Schmitt and the organizers, while noting several other discussions and identifying numerous speakers making comments. Schmitt is identified in this meeting's minutes at two points, each to make motions: (1) that the society be named the "Society for the Study of Evolution" instead of several alternatives and (2) that the society affiliate with the AAAS. Elsewhere in the minutes, Edgar Anderson is identified objecting to the purposes of the society as too narrow, preferring instead to advocate "studying systematics and species vigorously" under the general study of speciation. Mayr and Simpson objected that this would exclude palaeontologists. This is a different point from what the SSZ legend suggested, though it may have provoked a sidebar involving Schmitt and Wharton (either at the meeting or later) that was passed on to Blackwelder after the fact (which he may also have embellished). These minutes are in "Materials Relating to the Society for the Study of Evolution," SSE Papers, box "Council Meetings," folder "Minutes." Schmitt became a charter member of the SSE and was consulted in the initial organizing (e.g., Mayr to Schmitt, May 7 1946 Schmitt Papers, box 24, folder 9). Neither Wharton nor Blackwelder joined the SSE (see the July 1947 membership directory in Schmitt Papers, box 52, folder 2). Minutes of subsequent meetings of both society and council likewise record no conversation as Blackwelder describes. This conversation may have taken place in another setting or it might be a conflation of memories. Blackwelder (1979:157), for example, places Schmitt's confrontation at Boston in December 1946, but this cannot be true because Wharton was not recorded present at that meeting and the minutes show no discussion of the topic (while recording much discussion on other issues). In the same presentation, Blackwelder (1979:157) states Schmitt's comments lead to Robert L. Usinger's volunteering to help. This could have happened only at the March 1946 meeting, where Usinger was present. In a circa 1951 printed brochure about the SSZ, Blackwelder states a group met "shortly afterward to organize another society." (Brochure in Simpson Papers, folder "Society of Systematic Zoology," p. 3.)

[7] These meetings probably were for the Biological Society of Washington. In 1947, Schmitt served as the BSW's president.

[8] Wharton to Schmitt, April 1 1947, Schmitt Papers, box 60, folder "[SSZ]: Correspondence, T–Z."

Wharton cautioned against moving too quickly. "At the meeting in Washington this summer a nucleus of an organization could certainly be started," he suggested. However, they could not yet take any formal action (such as naming charter members, adopting a constitution, or electing officers) lest colleagues elsewhere "identify the society too closely with Washington," i.e., with the USNM. Wharton feared this might hinder their intentions for a "truly national" character to the group.[9] Schmitt quickly agreed on the slow pace. "We should do as the Society for the Study of Evolution did, forming only a tentative organization and leaving the formal organization for the AAAS meeting in December."[10] The kernel of a plan agreed upon, these two busy men set their organizing aside for other, more pressing, work.

A 4-month break in correspondence separates this basic arrangement with the next substantive action recorded.[11] Schmitt picked up his correspondence with Wharton again in early August 1947. Apparently, the "tentative organization" had not come to fruition over the summer, so Schmitt's focus moved to the upcoming AAAS meetings in December. On August 7, Schmitt sent Wharton the draft for a circular "we might mail out to interested parties" together with a draft constitution.[12] As previously agreed, the constitution was co-opted largely from the plant taxonomists. Schmitt asked Wharton to be critical of the text, however, and to think hard about the circular because he wanted to appear certain of their intentions once they went public. Schmitt also passed on to Wharton informal comments from Loren Woods (Chicago Museum of Natural History), who suggested the circular be "more emphatic about the fact that there is no other society to serve the interests and viewpoints of the taxonomic workers as such." This might have been an effort to distance their plans from the evolutionist society; however, if Schmitt meant this, he left references to the evolutionists in the subtext.

In his August 1947 correspondence with Wharton, Schmitt also worried about securing an enthusiastic and competent secretary-treasurer for the administration of their society. With many years experience watching people in such roles, Schmitt knew the importance of this position for a new organization. He also knew the burdens it would impose on the volunteer.[13] "If we are to launch the society, and I am anxious to do so," Schmitt told Wharton, "someone will have to volunteer and I will do that now, unless we can find someone more willing to take over."[14]

Considering who to invite for their organizing meeting, Schmitt proposed going through several of the most recent issues of *Zoological Record* and *Biological Abstracts* for names, as "this would reach the more active workers and in that way get spread around by word of mouth ..." Needing to keep tight control of costs, Schmitt was forced to rely on informal transmission of their news: "... we will scarcely be able to circularize the full membership of each society that may be made up largely of taxonomists as desirable as this might be." Schmitt concluded his letter to Wharton by affirming Wharton's concern about signatories on the circular. Wharton had suggested these signatories be recruited with a careful eye toward representing various specialties and regions. Still anxious to avoid impressions of a Washington bias to the group, Schmitt agreed.[15]

[9] Wharton to Schmitt, April 1 1947, Schmitt Papers, box 60, folder "[SSZ]: Correspondence, T–Z." The interest in inclusiveness might be a subtle comment against the leaders of the evolution society, who strongly clustered around only a few locations, especially New York and Berkeley (Cain, 2002).

[10] Schmitt to Wharton April 3, 1947, Schmitt Papers, box 60, folder "[SSZ]: Correspondence, T–Z."

[11] No records exist for what transpired in Washington during the summer meetings of Schmitt, Wharton and others. Schmitt certainly talked up the idea of a society, but to whom and in what detail is not preserved. Important for context, the first issue of Mayr's journal, *Evolution*, appeared in mid-July 1947.

[12] Schmitt to Wharton, August 7 1947, Schmitt Papers, box 60, folder "[SSZ]: Correspondence, T–Z."

[13] As illustrations of his point, Schmitt could look at Mayr's success with the SSE (Cain, 1994) and at Alfred Emerson as one overwhelmed in the Society for the Study of Speciation, to the society's detriment (Cain, 2000b).

[14] Schmitt to Wharton, August 7 1947, Schmitt Papers, box 60, folder "[SSZ]: Correspondence, T–Z."

[15] Schmitt to Wharton, August 7 1947, Schmitt Papers, box 60, folder "[SSZ]: Correspondence, T–Z."

Wharton's reply on August 18 was enthusiastic and filled with suggestions on how best to "spread the word." He added rhetorical flourishes to their circular. He also affirmed again Schmitt's suggestion about copying the evolutionists, proposing "we form an organization committee this Christmas and have the formal adoption of a constitution and election of officers a year from now."[16] The two men agreed, then set to work recruiting signatories for the circular.[17] Schmitt approached candidates himself.

Enthusiastic responses by possible signatories encouraged these men on. Alexander Petrunkevitch (Osborn Zoological Laboratory, Yale University) must have been overwhelmingly supportive in his reply. His letter brought Schmitt to tell Wharton, "Doesn't [his] response make you feel like cheering?" Replies sometimes also proposed additional scope for the future society. Petrunkevitch's letter must have included the suggestion of publishing a journal. "I like the sound of a 'Journal of Systematic Zoology,'" Schmitt told Wharton. "That is something to dream about and without question for the future."[18]

Their circular was ready in early September 1947. Schmitt and Wharton chose five signatories, themselves included. The others were Carl Frederick William Muesebeck (U.S. Department of Agriculture, economic entomologist), Karl Patterson Schmidt (Chicago Museum of Natural History, herpetologist), and Arthur Sperry Pearse (Duke University, parasitologist). Schmitt distributed 500 copies of their circular according to their previously agreed upon strategy; a draft constitution accompanied each circular.[19] The title of their circular left no doubt as to their goal.[20] Because of its importance, their two-page circular is quoted here in full.

Proposal for a Society of Systematic Zoologists [September 8 1947]

Dear Colleague:

The world is still in a ferment; times are still troublesome. We are all of us busy beyond words, and perhaps do not wish to be bothered even with things of very direct concern to us as taxonomic workers.

Where in all the welter of proposals for the advancement of science and of science legislation, passed or pending, has the importance of the knowledge of the kinds

[16] Wharton to Schmitt, August 18 1947, Schmitt Papers, box 60, folder "[SSZ]: Correspondence, T–Z." Schmitt clearly felt the stronger sense of urgency for organizing. Wharton's caution seems motivated by a desire for procedures that maximized inclusion.

[17] For example, signatories are discussed in Schmitt to Wharton, August 29 1947, Schmitt Papers, box 60, folder "[SSZ]: Correspondence, T–Z."

[18] Schmitt to Wharton, Aug 29 1947, Schmitt Papers, box 60, folder "[SSZ]: Correspondence, T–Z." Petrunkevitch's original letter has not been located. Schmitt always credited Petrunkevitch with suggesting the idea of a journal, e.g., Schmitt to Petrunkrevitch, 21 September 1948, Schmitt Papers, box 60, folder "[SSZ]: Correspondence, M–P." Also Schmitt to Pitelka, 14 July 1950, in same folder, where Schmitt added, "He said we needed it to be able to maintain our place and position in the scientific world. Aside from all other considerations, it has a prestige value that cannot be gainsaid."

[19] The number of copies is stated in Schmitt to WH Camp, September 29 1947, Schmitt Papers, box 60, folder "[SSZ]: Correspondence, A–E." No directory of recipients has been located.

[20] Zoologists was the preferred reference, rather than zoology. This point was discussed explicitly at the organization meeting in December 1947, when "Dr [E. Raymond] Hall moved that the name remain the Society of Systematic Zoologists for the present, and Dr [Thomas] Park amended the motion to the effect that the name be taken up by the committee who would prepare a constitution" (minutes of December 29 1947 meeting, in SSZ Records, box 24, folder: "1947 Organiz. Mtg."). In correspondence through August 1948 (e.g., draft constitution in Hubbs Papers, box 34, folder 63), Schmitt and Wharton refer to the "Society of Systematic Zoologists." The approved version of the constitution dated September 13 1948 (copy in Hubbs Papers, box 34, folder 61) refers to the "Society of Systematic Zoology."

of animals received its due recognition? In the Vannevar Bush report, "Science, the Endless Frontier," the "description of species of animals, plants, and minerals" has been grouped with the establishment of standards for hormones, drugs, and x-ray therapy, as background research. Pure research, on the other hand, is defined as "research without specific practical ends" resulting "in general knowledge and understanding of nature and its laws." We may have our own opinions as to which type of research our labors as systematists are to be classed. But are individual personal opinions enough to make our legislators, other scientific colleagues, and the laymen aware of both the needs and the worthwhileness of taxonomic study?

The plant taxonomists saw the light some eleven years ago and founded the American Society of Plant Taxonomists. So successful has the society been in advancing the common interests of the taxonomic botanists that they now have a flourishing society of more than 600 members, American and foreign.

The fact that there is no similar society to serve the interests and viewpoint of the taxonomic workers in zoology has prompted this letter. We, the undersigned, feel that there is a definite need for a Society of Systematic Zoologists and that such an organization can, and will, be of mutual benefit to its members and make a distinct contribution to the welfare and future of systematic zoology.

Its purpose and activities as we envision them would be:

First, to provide recognized official support for the point of view of the systematic zoologists so that their needs, as well as their contributions to zoology, may become better known to workers in other fields of science, and to the public in general.

Second, to serve as a clearing house for workers in systematic zoology, for the exchange of pertinent information and concepts, the identification of specimens, and the advancement of the study of systematic zoology.

Third, to compile and issue a directory of its members, including their special field of study, the groups of animals which they are willing to identify, and the faunal areas or regions in which they may be especially interested. [page 2]

Fourth, to publish a news letter at least once a year giving news: of current taxonomic research work for facilitating cooperation and avoiding duplication of effort; of material desired for the furtherance of specific studies; of expeditions in progress or planned that may be of service to systematists, or may be seeking trained taxonomic personnel; and of such other matters, domestic or foreign, as may be of value to systematic workers.

Fifth, to cooperate with existing societies whose membership may be in part or largely made up of individuals interested in systematic zoology in the support of projects of special interest to systematic zoologists.

Sixth, to provide government agencies and other establishments having need of experts or specialists in systematic zoology with names of those best able to serve them, and

Last, but not least, to hold an annual meeting in conjunction with other zoological societies, perhaps in the nature of a luncheon and business meeting, without the presentation of scientific papers.

If you are interested in the formation of a society along the lines suggested, will you not kindly let either Dr Waldo L. Schmitt [postal address omitted] or Dr. G. W. Wharton [postal address omitted] have your frank reaction to our proposal, enclosing, if you please, a carbon of your letter. An organization meeting will be held during the Chicago meeting of the AAAS. The Association has arranged for us to have the use of Parlor D, Congress Hotel, at 10 am, Monday, December 29, and will announce this fact in their official program. For purposes of discussion at that meeting, we append what is largely a paraphrase of the constitution of the American Society of Plant Taxonomists.

Because of practical limitations we have not been able to send this letter to everyone interested in systematic zoology. We should, therefore, appreciate it if you would "pass the word along" to others so that as many as possible can cooperate in this endeavor.[21]

In their two-page proposed constitution, Schmitt and Wharton identified the new society's object as "to promote the interests of taxonomy and systematic zoology for both invertebrate and vertebrate animals, museum and experimental work, and the scientific welfare of its members." Anyone "sincerely interested" in these objects and who "has done serious work in systematic zoology" was eligible to join. The society's governing Council would have the job of deciding which applicants fit these criteria.

Analysis: Themes Motivating a Sense of Need

Schmitt and Wharton developed four major themes in their September 8 1947 circular. These are key for understanding why the SSZ launched when it did. This section breaks from the narrative about organization to analyze the perceptions of need expressed in this circular.

Serving the Work of Day-to-Day Taxonomy

Though it might be seen now as a rather generic call for fellowship, Schmitt and Wharton's circular points to a general absence of infrastructure within systematics in the immediate postwar years. This lack of support involved several dimensions of day-to-day work.

First, there was the return to peace. Schmitt and Wharton's circular contains a call to reestablish contact. A great deal had changed since the start of the war. Careers had changed as a result of national service and shifting research opportunities to support the war effort. Institutions had shifted their missions. Research programs had grown and absorbed new techniques and new demands. Disruptions to the normal flow of casual communication —

---

[21] Proposal for a Society of Systematic Zoologists' September 8 1947, Schmitt Papers, box 60, folder "SSZ: Memoranda, Minutes of Meetings, Election Ballots, Financial Records, Committee Reports, 1947–1959." Another copy is located in Hubbs Papers, box 34, folder 60. This is the letter mentioned in Blackwelder, 1977:109). The first draft of this proposal, handwritten by Schmitt with modifications by Wharton, accompanies Wharton to Schmitt, August 14 1947, Schmitt Papers box 60, folder "[SSZ]: Correspondence, T–Z."

caused, for instance, by the cancellation of meetings owing to travel restrictions — meant many felt out of contact with other members of their specialties. Some of Schmitt and Wharton's efforts to organize no doubt simply sought a mechanism for reestablishing contact with old friends, catching up on professional and technical developments, collectively assessing developments over the past half decade and relaunching the work of their particular specialties.

Moreover, the return to peace meant a return to work. In terms of absolute load, curatorial work in the postwar period was rather more busy than usual. Collections in many museums were being returned, unpacked and resorted. (This was especially true of large institutions along the American eastern seaboard where collections had been evacuated.) To this, a general shortage of technical labor slowed progress, returning curation to prewar levels. Collections and research also had been expanding rapidly in areas with potential military and medical value or where service personnel had been stationed overseas. Systematists associated with teaching institutions also saw large increases in student populations and training demands. At the very least, systematic workers in the immediate postwar years were extremely busy professionals.

One subtext in Schmitt and Wharton's call to organize was a concern over the rapidly expanding scale of science in the previous decade. A national commitment for science and technology facilitated a change in its scale. AAAS membership rose sharply following the war: from 27 thousand in 1945 to 47 thousand in 1950 (Wolfe, 1989; Lewenstein, 1999). The number of new societies and publications also ballooned (Price, 1961; Bates, 1965). In the life sciences, this had the effect that experts in one specialty now stood little chance of using informal networks to monitor relevant activity in both their own and related specialties.[22] In this context, calls for directories and annual meetings, newsletters and notices of activities sought to provide workers with tools for benign surveillance. Through these mechanisms, "duplication of effort" could be avoided, "matters ... of value" could be traced, and a "clearing house" approach could work to bring disconnected activities or untapped collections to the service of interested specialists. Schmitt and Wharton's efforts were by no means unusual in the immediate postwar years. Scientists in every discipline were struggling to keep up with the torrent of activity and change.

Another subtext in Schmitt and Wharton's call involved concerns about nomenclatural practice. It is hard to recreate the sense of uncertainty over zoological nomenclature during the immediate postwar years. In 1947 the most recent International Congress of Zoology (XII, 1935 in Lisbon) was more than a decade in the past. Another was expected (Paris, July 1948), but details were vague during most of 1947.[23] With European institutions far from recovered and East-West cooperation increasingly curtailed, Americans felt a keen nervousness about the current state of affairs. This is a complicated subject, but the key point to remember is that it was by no means clear in the mid-1940s how nomenclatural disputes could be adjudicated. Local bodies, such as the USNM's Committee on Zoological Nomenclature, also tried to assert national and international influence as a way to fill this vacuum (e.g., Blackwelder et al., 1947). Moreover, arguments regarding nomenclature required considerable skill as from litigators. International conventions existed in a nether-

---

[22] In the early 1960s, Price (1963) identified this problem, noting the formation of "invisible colleges" and proposing their use to effectively manage disciplines.

[23] The situation had stabilized considerably by the end of 1948. The XIIIth International Congress took place in Paris (July 1948). A good deal of revision was done to Commission procedure and structure, but the central goal was to reconstitute the international framework and stabilize working practices (Hemming, 1948). The whole *Bulletin of Zoological Nomenclature* 1950, Number 4 was devoted to changes from the Paris meetings of the Commission (also see Melville, 1995:52–55). Intensive work in the 1950s led to the first (1961) edition of the *International Code of Zoological Nomenclature*, the first complete and official revision to nomenclature rules since the *Règles* (Rudeval, 1905).

world between the 1905 *Règles* (Rudeval, 1905) and a long series of opinions issued by the International Commission on Zoological Nomenclature.[24] Numerous efforts to summarize the current state of the rules appeared following the Lisbon congress. This reflects the degree of confusion around the bundle of poorly digested case law.

Promoting an official platform for American expertise was one step toward stabilizing the situation. Schmitt's plans for the SSZ included committees to do precisely this. As an aside, this paralleled events in Britain, where the revival of the Association for the Study of Systematics in Relation to General Biology and its consolidation into the Systematics Association in 1946 can be understood as another effort to get the systematics community back on its feet (Anonymous, 1948).

Advancing the Principles of Systematic Zoology

Schmitt and Wharton's emphasis on principles underscored two themes important to the postwar context. One involved a fundamental transition in American life sciences.

As professional groups reassembled in the immediate postwar years, some undertook fundamental reassessments of priorities, methods and principles. In the life sciences, this furthered a process of disciplinary realignment under way since the start of the century involving a transition from object- to process-focused work, or from zoology to biology.[25] In this realignment, new specialties were represented as transcending traditional disciplines grounded in the comparative affinities of organisms (such as between mammalogy, ornithology, ichthyology and so on) and as building on functional or process affinities (such as genetics, developmental biology, ecology, endocrinology, behavior and evolution). The extraction of a discipline-transcending, principle-based biological specialty of systematics (sometimes systematic zoology or systematic biology) from organism-focused taxonomic specialties came rather late in the overall sequence of realignments in the life sciences. Nevertheless, it represented one stage in this much broader shift in interests and emphasis. The SSZ provided a community infrastructure for promoters of this realignment. It helped them draw attention not only to extracted principles within systematics but also to the extracting process itself.

On this point, when *Systematic Zoology* first appeared in 1952, contributions were solicited especially with reference to "principles and the application of principles of wide implication and general interest in any phase of systematics, such as comparative anatomy, zoogeography, paleontology, taxonomy, classification, evolution, or genetics ..."[26] Blackwelder later emphasized:

> ... *Systematic zoology* is not intended for the publication of descriptive papers. The Council sees its field as the philosophic aspects of systematics, its principles

---

[24] The official statutes for zoological nomenclature in 1947 remained the 1905 *Règles Internationales de la Nomenclature Zoölogique* (Rudeval, 1905), with the French original superseding the German and English translations. Amendments to the *Règles* were adopted at nearly every international congress. News of subsequent opinions rendered by the International Commission on Zoological Nomenclature peppered the specialist literature and were summarized in various compilations, such as Schenk and McMasters (1936) — revised by Keen and Muller (1948) — and Van Cleave (1943). During 1910–1936, the Commission issued 133 opinions. Between 1939 and 1948, they issued 61 more (Melville, 1995). The *Bulletin of Zoological Nomenclature* began as part of an effort to systematically record this information. As with all systems based on case law, differences in interpretation accumulated.

[25] Allen (1975) called this the "revolt from morphology." The precise nature of this process has been debated extensively (e.g., Pauly, 1987; Rainger et al., 1988; Maienschein, 1991; Cittadino, 1990; Benson et al., 1991). Holton (1972) and Thackery (1992) cover postwar changes to science generally; also see Heims (1991) for cybernetics; Hagen (1992) for ecology and Allen (1975:187–228) and Kay (1993) for molecular biology.

[26] This information is printed on the inside back cover of Volume 1, Number 1.

and problems, as well as news of systematists, their institutions, courses, and publications. Systematic material is not completely ruled out, but it must be subordinate to the discussion of principles. (Blackwelder, 1952:92)

A second theme in Schmitt and Wharton's emphasis on principles focused on the consideration of particular principles in themselves. Blackwelder's many retellings of the SSZ launch place conflict over principles at the very center of Schmitt and Wharton's motives. In his founding narrative, these three lined up on one side of the conflict. On the other side were evolutionists of the SSE with their phylogenetic approach, especially Mayr and Simpson.[27]

Despite Blackwelder's later representations, Schmitt and Wharton's original call for the SSZ did not express a desire to promote any particular systematic principles or philosophy over any other. Interestingly, they neither identified these differences in their circular nor constructed a sense of conflict in which their side needs organization. Schmitt was not openly hostile to the evolutionists.[28] He maintained good relations with Mayr throughout his career. He also encouraged the evolutionist society during its organizing phase, even writing a strong letter in support of Mayr and Simpson's proposal to support the journal *Evolution*.[29] He was asked to assist with a 1949 membership campaign for the evolutionist society.[30] Even in deep disagreements with Simpson, Schmitt was cordial. When discussing his intellectual differences with the evolutionists, Schmitt presented them as a matter having little to do with him. He regularly excluded himself from assessments of quality — apologizing for "not being an evolutionist ..." when asked to consider issues of evolutionary taxonomy. Ignoring the presence of a conflict allowed Schmitt to disagree with their approach and maintain his own emphasis on other methodologies without having to strain his working relationships with increasingly prominent colleagues.[31]

Some proponents of different approaches certainly made their conflicts known (e.g., Winsor, 1995), and considerable discussion took place among systematists about the relative strengths of various approaches and how their results could be best reconciled (e.g., Huxley, 1940). Moreover, polarization and affiliation into tribal groups played an important role in SSZ activities after the mid-1950s (Hull, 1988), but conflict and polarization were not Schmitt and Wharton's style. Their calls for the "exchange of pertinent information and concepts" and for "the advancement of the study of systematic zoology" represent an interest in discussing all aspects of systematics within an overarching pluralism. Though each man had strong views on these matters, they chose to step back from tribalism to consider larger disciplinary issues. They promoted this approach in their circular.

[27] Cain (1994, 2000a, 2000b, 2002) examines the evolutionists' program in depth. Evolutionary taxonomy underwent considerable growth in popularity during the 1940s. It focused attention on intraspecific variation and mechanisms for speciation. It challenged taxonomy to reflect degrees of evolutionary relatedness and used nomenclature to imbed theories about the dynamics of divergence or isolation. Mayr (1942) and Simpson (1937, 1945) are the best remembered American advocates of this approach. Other prominent exemplars at the time include Kinsey's (1936) studies of wasps, Emerson's (1941, 1937) work on the phylogeny of termite nests, and Lack's work on robins and finches (1943). Hull (1983) shows a strong presence of evolutionary systematists in the earliest volumes of *Systematic Zoology*.
[28] Wharton's views are difficult to assess, as very little correspondence in his voice survives.
[29] Schmitt to Mayr, June 4 1946, Schmitt Papers, box 24, folder "Mayr, Ernst," in which he explained to potential funders, "For us systematists evolution has been, throughout, a guiding principle and basic to all our ideas of classification and species. That there may yet be a journal in which the dynamics of evolution and of species will be freely discussed, where pertinent observations and the results of special investigations can be brought together for all who are interested sounds almost too good to be true."
[30] Schmitt to Theodor Just, May 11 1949, Schmitt Papers, box 52, folder "Society for the Study of Evolution." Schmitt told Just, "The last few numbers have been particularly good and, from my humble systematic point of view, most pertinent."
[31] See, for example, Schmitt to Mayr, June 7 1965, Schmitt Papers, box 24, folder "Mayr, Ernst."

Too much analytical focus on the conflict between supporters and detractors of evolutionary taxonomy obscures developments in systematic principles that occurred during the 1930s that cut across this divide and that were just beginning to assert wide influence as the international situation deteriorated. Most important, experimental taxonomy brought important changes to systematics in inter-war America. Overall, its appeal involved the promise of greater objectivity and consistency in determinations, plus the hope that criteria could be chosen for keys that somehow were more fundamental to understanding the kind of organism on hand. As experimental techniques became fashionable, researchers used this label rather loosely. By the late 1930s, experimental practice in taxonomy included several general areas of activity:

1. Use of characters accessible through the laboratory — such as biochemistry, serology and karyology (chromosome banding patterns, chromosome numbers) — and quantitative methods
2. Use of characters whose extent of phenotypic variability can be tested through experimental manipulation or crossing
3. Use of new categories of characters that seemed more biological and functional than morphological and structural — such as those involving behavior, physiology or ecology
4. Use of criteria derived from experimental tests (such as hybridization trials or crosses) and measurements of process (such as chromosome/gene flow)
5. Use of tools to express confidence about determinations (such as inferential statistics)

Hagen (1984) provided a general introduction to some of these dimensions; most are rather poorly studied by historians. Some widely recognized exemplars of these methods included work by Babcock's team on *Crepis* (Babcock and Stebbins, 1938; Babcock et al., 1942), Kinsey's (1936) work on *Cynips*, Miller's (1941) studies of *Junco*, Anderson's (1949) hybrid work on *Tradescantia* and *Iris*, Patterson's (1941) field work with *Drosophila* and Hubbs' (1940) hybridization experiments in fishes. Simpson and Roe's (1939) *Quantitative Zoology* pressed the utility of inferential statistics.

Evolutionary taxonomists — anxious in the 1930s to create more solid epistemic foundations — absorbed many of these experimental techniques as a way to solve persistent complaints from other biologists about the objectivity and testability of their work. This was expressed most obviously in efforts to promote biological species concepts (Dobzhansky, 1937; Mayr, 1942) and in attempts to document evolution in action through the identification of subspecies, *Rassenkries* and clines (Mayr, 1941; Huxley, 1939a, 1939b; Rensch, 1929). Huxley's (1940) excitement about a "new systematics" derived largely from a sense of elation over the union between experimental and evolutionary themes in systematics.

Evolutionary systematists were not the only ones appropriating developments in experimental taxonomy. Blackwelder and Boyden (1952) provide a competing example.[32] These experts represented themselves as practical, atheoretical taxonomists also anxious to promote experimental techniques as part of a broad program to set systematics on more secure epistemic foundations. Methods developed by experimental taxonomists held great promise for improving the quality of all classifications. Experimental data played increasingly prominent — then dominant — roles in taxonomic determinations (just as DNA sequencing has done since the 1990s). However, this work needed to take place, Blackwelder and Boyden argued, independently from the work of establishing evolutionary connections or studying evolutionary processes. Taxonomic categories should only reflect degrees of similarities and differences, groupings should not build on phylogenetic connections, and nomenclature need

[32] Boyden's considerable expertise in serology served as background (see e.g., Boyden, 1953; Cole, 1958).

not reflect the degree of evolutionary association. Breaking the connection between phylogeny and taxonomy removed one layer of hypotheses and put systematic work on the best possible evidential grounds.

With experimental techniques on the rise, more researchers coming into the subject and the prospect of improved senses of objectivity and testability, this was an exciting time for systematists — opportunities for considerable advancement seemed on the horizon. In calling for an organization to consider systematic principles, Schmitt and Wharton might have been far more interested in considering the value and implications of new experimental techniques for systematic work. All taxonomists would be worrying about these issues in the postwar years and considering how to balance their relative value for day-to-day work.

Offering a Service Role to Those Needing Taxonomic Expertise

Schmitt and Wharton's emphasis on service to those requiring the special skills of taxonomists tapped into two themes regarding human resources and expertise.

First, a role channeling queries about expertise reflects an approach to securing personnel that became common during the war and continued into the 1950s and 1960s.[33] These efforts to rationalize and centralize procurement of expertise offered efficiency gains while also breaking away from other likely gatekeepers, such as the National Academy of Sciences or the AAAS. This was an easy service to offer. In correspondence about SSZ committees, Schmitt later suggested directories — listing the groups each member felt qualified to consult about — as the most suitable means for developing this service. Blackwelder produced the first directory in 1949.[34]

Second, this emphasis on promoting service roles spoke directly to many potential members. Taxonomists rarely lack a service role in their employment.[35] More specifically, Wharton assisted the U.S. Navy in pest identification and control during the war.[36] Schmitt watched innumerable consultancies under way at the U.S. National Museum.[37] In postwar America, professional societies of biologists with service roles dwarfed those with predominantly academic interests.[38] The interest in mixing practical diagnostic taxonomy with discussions of systematic principles also arose in the only newsletter produced by the Society for the Study of Speciation in 1940.[39]

---

[33] By the middle 1950s this would lead to the Scientific Manpower Commission (Bates, 1965:195). The American Institute of Biological Sciences also became involved in manpower issues for the life sciences.

[34] "Systematic directory of members" (no date, nine pages, by Blackwelder), Schmitt Papers, box 60, folder "SSE: Memoranda, Minutes of Meetings, Election Ballots, Financial Records, Committee Reports, 1947–1959." A copy in Hubbs Papers, box 34, folder 61, is stamped as received March 17 1949. A directory of members also appears in the first *News Letter*, dated December 7 1949.

[35] Dupree (1957, 1963, 1972) discussed service at the federal level in America. Greenberg (1967) and Appel (2000) considered the political dimensions. This relation between science and the state was hardly new either to American systematics or to the twentieth century, e.g., Koerner (1999) or Brockway (1979). Lewenstein (1999:105) argued that the language of service reflected an increasingly outward-looking language in American science during the postwar years as it became increasingly focused on improving human welfare. With Appel (2000) in mind, this language seems more to do with shifting patterns of patronage and the rhetoric or justifications those patrons found politically acceptable.

[36] The Ohio State University's "Faculty Information Service" October 5 1968 biography of Wharton emphasized his service to practical problems in parasitology. Thanks to Michelle Drobik (University Archives at Ohio State University) for this item and other biographical material.

[37] Blackwelder (1979) described Schmitt's activities during the war.

[38] The American Institute of Biological Sciences compiled membership data on U.S. biological societies, e.g., *AIBS Bulletin*, Vol. 1 (January 1951), p. 10. The largest societies were in applied subjects: the Society of American Foresters (6500 members), the Society of American Bacteriologists (4800), the American Dairy Science Association (2700) and the American Association of Economic Entomologists (2500). Of the 43 organizations listed, 24 reported memberships higher than the SSZ.

[39] The newsletter is reproduced by Cain (1999) and discussed in Cain (2000b).

A service role became increasingly important for naturalists in their relations to other life sciences. Colleagues trained less in subjects defined by particular organisms (such as mammalogy, entomology or comparative anatomy) and more in biological subjects. They lacked taxonomic expertise and subsequently earned scathing comments. "My blood boils," Horton Hobbs, Jr. (University of Virginia, Crustacean taxonomist and ecologist) complained to Schmitt, "at the way some of the physiologists and anatomists disregard (I feel certain through ignorance on their part) even the basic rules of Zoological Nomenclature ... and even worse, slave away on the intricacies of some organism about which they haven't the slightest idea as to its identity, let alone its affinities. Don't let me get started on that."[40] Whether or not the service was wanted by these colleagues, supporters of taxonomic work increasingly promoted consultancy as a way to add value to their own role in the life sciences. Thus, when taxonomists were unable to compete institutionally with other biologists they could tie themselves squarely to programs supporting work in the service of medicine, public health, genetics, ecology, oceanography and so on. This use of service rhetoric was also a common tactic used by museum directors. It allowed them to claim high value for their institutions' collections and personnel while also offering research frameworks designed for modern science and the public interest (e.g., Parr, 1939, 1959).

Representing Systematics within the Sciences

A great deal of competition and jostling for position took place between specialties in the life sciences during the postwar years. In part, this continued disputes arising from ongoing disciplinary realignments beginning before the war. It also reflected growing unease with the increasing influence of disciplines such as genetics, biochemistry, cellular biology and molecular biology. These communities not only succeeded in acquiring growing percentages of the overall resources available to life scientists, but they also increasingly displaced other specialties in universities and research institutions. This meant shifts in the distribution of staff, space, money, students and opportunities. Along with many others in the naturalist tradition, systematists were acutely nervous about their status within the life sciences, and a great deal of discussion took place regarding marginalization.[41]

Schmitt and Wharton's call for organization certainly would have been understood as an attempt to stake a claim within this context. Their claim to represent implied an effort to fight back. Calls to unify and to project a common front were extremely common in the life sciences during the postwar years.[42] Thus, Dobzhansky quipped that "nothing in biology makes sense except in the light of evolution," and James Watson asserted that there was "only one biology ... and it is molecular biology." Schmitt and Wharton's call was simply a bit more verbose: "It is a major aim of [this] Society to bring about a recognition of the importance of systematics in zoology. This does not require justification ... but ignorance of [its] contributions and the extent of the service has prevented some

---

[40] "Horton" to Schmitt, October 4 1947, Schmitt Papers, box 60, folder "[SSZ]: Correspondence, F–L." Such resentment was common across the life sciences. For example, it underlies Mayr's often quoted initial approval of Dobzhansky as "here is finally a geneticist who talks like a naturalist" (Mayr, 1992:2). In the 1940s, the rise of ecosystem ecology provoked a similar complaint about ignorance of specific organisms (Hagen, 1992), and these resembled complaints by stratigraphers in palaeontology as well as experimenters in high-energy theoretical physics (Traweek, 1988).

[41] For example, Ball (1946). Vernon (1993) discusses this point generally. The concern over status was repeated in the 1950s with respect to funding within the U.S. National Science Foundation (Appel, 2000) and in the 1960s in conflicts between cell and molecular biologists and evolutionary, organismal and ecological biologists (Dietrich, 1998; Beatty, 1994). Wilson (1994:218–237) described this tension as it manifested at Harvard University. Larson (2001) and Kleinman (1999) examine cases where naturalists adapted to the postwar changes in priorities and thrived.

zoologists ... from recognizing the fundamental nature of systematics and its indispens-
ableness to many other fields."

Much later, Blackwelder recalled that the SSZ leadership considered a more overarching
role:

> At the beginning the S.S.Z. was looked upon in part as an association of societies.
> This was because the field of taxonomy never had been united, and there were
> already two dozen specialty societies ... A program to invite affiliation of these
> societies was planned but never really got started. At about this time, the Amer-
> ican Institute of Biological Sciences [AIBS] was formed ... and it appeared that
> his function of affiliation might be better handled there. In time, it became obvious
> that the [AIBS] could not enlist the taxonomic societies of zoology, but by this
> time the plan had been dropped by the SSZ.[43] (Blackwelder, 1979:158–159)

Mayr was involved in a long-running parallel campaign to improve the status of systematics. "After all," he asserted in his 1942 *Systematics and the Origin of Species*,

> it must not be forgotten that the average taxonomist is more than a mere caretaker
> of a collection. In most cases he collects his own material, he studies it in the
> field, and develops thereby the techniques and point of view of the ecologist.
> Furthermore, most of the younger systematists have had a thorough training in
> various branches of biology, including genetics. This experience, both in field
> and laboratory, gives the well-trained systematist an excellent background for
> more ambitious studies. It is only natural that he cannot and will not content
> himself with being merely a servant to some other branch of biology. Systematics
> is for him a full-fledged science. (Mayr, 1942:9)

At times Mayr's focus concerned the viewpoint of the systematist within evolutionary
studies. At other times it ranged across many other elements of the life sciences and combined
promoting both systematics and evolutionary studies within the life sciences. In fact, Mayr
had taken steps toward organizing systematists for these goals several years before Schmitt
and Wharton. Early in January 1944 he made discrete inquiries on the matter. Writing to
Carl Hubbs, for instance, Mayr explained,

> I have been asked by a number of colleagues to organize a society of animal
> taxonomists. It seems doubtful whether it is wise to start another society. At
> the same time it is true that the taxonomists are about the most disorganized
> lot of scientists.

---

[42] In evolution, calls for unity are discussed by Cain (1994). In genetics, Ralph Cleland to "colleague," July 20
1946, Dunn Papers, folder "National Academy of Sciences, #1." Cleland was working to build support for a
July 10 1946 petition to create a new section for genetics in the National Academy membership as a way to
build "unity of spirit and purpose." Parr's unifying schemes centered on ecology and oceanography (Parr, 1939).
Parallel calls came to unify academic biology as a whole (e.g., via the American Institute of Biological Sciences,
the American Society of Zoologists, and *The American Naturalist*) or through the medical sciences (e.g., the
National Society for Medical Research). This paralleled calls to unify science as a whole (e.g., Lee, 1951;
Shapley, 1950).

[43] Blackwelder (1979:158–159). Given his sense of rivalry among groups of systematists, this organizing tactic
represents an attempt to develop positional superiority (Said, 1978) in which his camp would stand to speak
for all systematists.

Such a society might do a great deal to increase the prestige of taxonomy, particularly if it can manage to publish a journal which embodies the most valuable generalizations resulting from taxonomic work. I have been corresponding with [Julian] Huxley, and he has encouraged me in this matter.

The time is extremely unfavorable to start an active organization. On the other hand it might be worth while to discuss these plans in an informal way and then go ahead with a well crystallized plan after the completion of the war. What do you think of this matter? Don't you think it would be a good idea to get the entomologists, ornithologists, etc. together to discuss taxonomic principles?[44]

During the postwar jockeying for position, Mayr expressed concerns similar to Schmitt and Wharton's. The "time has come," Mayr confided to a colleague, "to bring taxonomists back to the fold of general biology. The truth is that the experimental biologists have had for many years the opinion that taxonomists were not biologists."[45] With Schmitt, Mayr was sufficiently comfortable to be blunt:

... the production of papers of general interest by taxonomists is practically a survival necessity. We must impress on general biologists that taxonomic work is not merely a clerk's job but real, genuine biology. Furthermore, we must impress on them that it yields results that are inaccessible to any other branch of biology. I think we have made a good beginning and taxonomy enjoys a greater prestige now than it had for a generation or more. We have to keep at it to improve our position, otherwise we will slide back again.[46]

Ever zealous, Mayr went so far as to complain to Schmitt about how "regrettable" it was that the USNM was "not more prominent in this movement of bringing together taxonomy with the other biological sciences."[47] (Given Schmitt and Wharton's work for a systematics society, Mayr's criticism seems misplaced.)

The self-conscious concern over the status of systematics within biology sometimes produced a backlash. For instance, it led Frank Pitelka to doubt the wisdom of declarations about the discipline's importance or place in the scientific community, complaining,

Why should systematists as a group be self-conscious, apologetic, or defensive? Notwithstanding the current extent of antagonism toward systematics, the fact remains that such antagonisms, when expressed, are more a measure of a man's breadth as a biologist than an indictment of systematics. Whatever traumata any one of us individually may have experienced, our prestige and stature will be the

[44] Mayr to Hubbs January 6, 1944, Hubbs Papers, box 35, folder 56. Mayr had complex reasons to suggest this plan. These focused on promoting systematics within the life sciences more than inward-looking concerns of taxonomic practice. Hubbs' reply and a follow-up letter from Mayr are located in the same folder. In Mayr's follow-up, he proposed a *Journal of General Systematics* and a "Society of Animal Taxonomists" (Mayr to Hubbs, January 15 1944, Hubbs Papers, folder 35, folder 56). When Mayr proposed launching the Society for the Study of Evolution several years later, he pressed for an emphasis on "evolutionary systematics" and an organization that sought "to bring the taxonomists and geneticists together" (e.g., Mayr to Hubbs, January 11 1945, Hubbs Papers, box 35, folder 56 and Mayr to Hubbs, September 23 1947, Hubbs Papers, box 34, folder 58). This period is examined from Mayr's perspective in Cain (1994, 2002).
[45] Mayr to Karl Schmidt, April 26 1946, SSE Papers.
[46] Mayr to Waldo LaSalle Schmitt, March 9 1948, Mayr/*Evolution* papers, folder "Schmitt."
[47] Mayr to Schmitt, February 17 1948, Mayr/*Evolution* Papers, folder "Schmitt."

outcome of what we are and do, not what we tell someone we are and are trying to do.[48]

Instead, he predicted, hard work would be recognized in its own right. "I think the professional ego of systematists is safe, and the less said about it publicly, the better."[49]

This jostling for position in the postwar years also was motivated by debates over how best to structure federal support for scientific research. The spark igniting a frantic round of such jostling was the publication of Vannevar Bush's (1945) *Science: The Endless Frontier* (Zachary, 1997; Reingold, 1991). Bush defended the need for liberal funding of basic research in the sciences as the best strategy for long-term results. "It provides scientific capital," he argued, for both foreseen and unforeseeable national problems (Bush, 1945:13). He called for substantial funding of universities, government science agencies and other research institutions. Paradoxically, Bush defended the need for heavy investment in basic research using notions of service to the nation. He argued, for example, that medical progress often was due to discoveries in "remote and unexpected fields of medicine and the underlying sciences." In this approach, Bush arranged underlying basic science along lines feeding into medicine. Thus, much of basic chemistry and physics and all of anatomy, biochemistry, physiology, pharmacology, bacteriology, pathology, parasitology and so on were set into subordinate relations.

Schmitt and Wharton's circular specifically identified Bush's work. Systematics ranked very near the bottom of Bush's priorities. When Bush's report became the focal point for national debates over a possible national science foundation, academic biologists reacted defensively, and their maneuvering grew intense (Appel, 2000). In response, advocates in many specialties argued for two actions: organization and lobbying. Centrality became a claim most specialties asserted. Unity around central themes became the party line. Schmitt and Wharton's circular clearly asserted that systematics possessed both needs and worthwhileness when set against the backdrop of the sciences more generally. Organizing was designed to establish a unified voice and designate spokespeople to represent them as a constituency.

Strong Support for Schmitt and Wharton

The writing campaign to organize systematists worked wonders. Schmitt and Wharton received 96 responses to their September 1947 circular. They reported 72 as favorable. Most replies were shared between the two men and discussed in detail.[50] A few respondents sent extensive comments on possible goals and projects.[51] Overall, the general tone of the replies reinforced their planning decisions. Wharton summarized the results plainly, "From the replies ... I am satisfied that the society can be formed."[52]

Most supporters recognized the potential for inward-looking service for systematists and outward-looking service for wider communities. Concerns about marginalization were common, too, such as Horton Hobbs' letter (quoted above).[53] Henry Bronislaw Stenzel (University of Texas, invertebrate palaeontologist specializing in Crustacea) peppered enthusiasm with reminders for inclusion. "You are entirely right in stressing the needs of such a group and I heartily approve of your proposal. As a minor criticism I wish you would definitely

[48] Frank A. Pitelka to Lee R. Dice, April 19 1950, Hubbs Papers, box 34, folder 62.
[49] Frank A. Pitelka to Lee R. Dice, April 19 1950, Hubbs Papers, box 34, folder 62.
[50] Wharton to Schmitt, December 3 1947, Schmitt Papers, box 60, folder "[SSZ]: Correspondence, T–Z"
[51] For instance, Louis Hutchins sent seven pages of detailed suggestions (Louis Hutchins to Schmitt, October 3 1947, Schmitt Papers, box 60, folder "[SSZ]: Correspondence, F–L"). A short note of approval was written by Hubbs to Schmitt (September 30 1947, Hubbs Papers, box 34, folder 62).
[52] Wharton to Schmitt, October 6 1947, Schmitt Papers, box 60, folder "[SSZ]: Correspondence, T–Z"
[53] "Horton" to Schmitt, October 4 1947, Schmitt Papers, box 60, folder "[SSZ]: Correspondence, F–L"

include systematic palaeontologists in the constitution. Most palaeontologists have to do a great deal of systematic work, perhaps more than those who work with living forms. At the same time a close co-ordination between those who work with fossils and those who work with living forms is desirable."[54]

With an eye on the national funding debates, annelid specialist John Percy Moore strongly supported what Schmitt and Wharton were attempting. "Taxonomy certainly needs to be vivified and justified in the eyes of the scientific world and of the sponsoring foundations, to whom it has been misrepresented." He suggested "weighty arguments" in favor of the plan.

> Taxonomy, the great grandmother (medicine her consort) of biological science and for centuries practically its dominant discipline, has been so largely brushed aside by the modern experimentalists, who would file it among the family genealogical records, that it is being largely deserted by most of the younger biologists, who are unable to bear up under that *unwanted feeling*.

> Of course there has been some recent reversal of that attitude. The ecologists have continued to associate with taxonomists as good and useful pals and the geneticists (perhaps rather shamefacedly) have found that, outside of the purely cytological and hereditary fields, they are themselves microtaxonomists, dealing with mutations and populations and their relation to speciation. That is the significance of [the] formation of the Society for the Study of Evolution, and for the rise of the 'new systematics.'

> Taxonomy itself has broadened. Many systematists regard living organisms as complexes of physico-chemical properties rather than of static characters and at least theoretically look upon classification as an expression of *all* that is known about animals or plants, whether derived from anatomy, physiology, ecology, genetics, chemistry, et al.[55]

By the end of September 1947, Schmitt and Wharton were planning a program for the December AAAS meeting (Chicago).[56] Their objective for that meeting was to appear certain in their goals and specific in the actions they proposed. In Chicago, they would seek a mandate to put this plan into effect. Schmitt was not shy about enlisting advice. For instance, he invited W. H. Camp (New York Botanical Garden and key to organizing the Society of Plant Taxonomists) to speak at the meeting. "Your organization is certainly the prototype of ours and we would love to have a few words from you. Here's hoping!"[57]

The demands of other business took Schmitt away from further organizing until late November 1947. Apologizing to Wharton for the break in his attention, Schmitt presented ambitious plans for society activities from its very start. These plans, he suggested, would give direction. Even if these projects were set aside for the moment, Schmitt said, he wanted to catch the organizing momentum and steer it toward clearly defined goals. Thus, he suggested committees on policy and standards, cooperation with international committees of nomenclature and the *Zoological Record*, cooperation with systematic botanists, membership, publicity and the society's newsletter, directory and journal. He wanted a group to revise the constitution. He also proposed committees to promote systematics in schools and

[54] Stenzel to Schmitt, September 29 1947, Schmitt Papers, box 60, folder "[SSZ]: Correspondence, R–S."
[55] J.Percy Moore to Schmitt, September 28 1947, Schmitt Papers, box 60, folder "[SSZ]: Correspondence, M–P."
[56] Wharton to Schmitt, October 6 1947, Schmitt Papers, box 60, folder "[SSZ]: Correspondence, T–Z."
[57] Schmitt to W.H. Camp, September 29 1947, Schmitt Papers, box 60, folder "[SSZ]: Correspondence, A–E."

universities, to develop a textbook on practice and procedures and to coordinate celebrations for the anniversary of Linnaeus in 1958.[58] Given his early enthusiasm for the idea of a journal and the ambition of his other committee plans, Schmitt now hesitated about prospects for a journal. "Several have suggested that a journal be published by the Society," he reminded Wharton. "I can't yet see eye to eye with them, but the matter should certainly be considered, perhaps only for the future." Instead, Schmitt proposed creating a large committee charged with producing a newsletter. This was to ensure it was "representative of the country and the interests to be served." [59] A journal could come later.

Schmitt put considerable effort into recruiting prospective members for possible committee roles. For example, he invited Norman Stoll (parasitologist at the Rockefeller Institute in Princeton, NJ) to Chicago so he might help in the effort to talk over their proposal. "You will be a very essential man to our organization and we especially want you for a committee that we hope to form for cooperation with the International Committee [sic], of which you are a member."[60]

Though generally supportive of these efforts, Wharton again suggested Schmitt was moving too fast by enlisting committee members and their chairs.[61] This pushed too hard and gave the impression of a one-man job. First, they had only been empowered so far to launch the society, not administer it. Schmitt was initiating activities in the name of the society before its constitution had been formally approved and before the likely membership had been consulted. Second, naming and empanelling committees was the kind of work best left for society presidents. Though Schmitt said he only wanted to ensure things got under way, Wharton warned this approach might cause more problems than it solved.

Sixty-six people attended the organization meeting at the Chicago AAAS meeting on December 29 1947. Following their September circular, an announcement in *Science* read, "The Society of Systematic Zoologists has scheduled a preliminary meeting ... to consider the organization of a society of animal taxonomists" (Anonymous, 1947:560). A morning session would be dedicated to questions of organization. An afternoon session was organized for a presentation by Arthur Francis Hemming about the upcoming meeting of the International Commission on Zoological Nomenclature at the 1948 XIIIth International Congress of Zoology in Paris.[62]

All went pretty much according to plan at the organization meeting.[63] Schmitt was elected president. Though supporters of Schmitt and Wharton meant this as a sign of their support, it caused a technical glitch. It meant Schmitt could not serve in the crucial role of secretary. Thus, Wharton was elected as secretary-treasurer until the council could elect an

---

[58] The year 1958 was one of several plausible bicentenaries for Linnaeus (Koerner, 1999). The first editions of his pamphlets *Systema Naturae* (*System of Nature*, 1735 first edition) and *Genera Plantarum* (*Genera of Plants*, 1737), used the sexual system, but his use of this approach preceded these publications. Linnaeus first used trivial names for species in *Species Plantarum* (*Species of Plants*, 1753) and in the tenth edition of *Systema Naturae* (*System of Nature*, 1758). Use of the year 1758 probably is meant to celebrate the adoption of trivial names and the consistent use of binomial nomenclature.

[59] Schmitt to Wharton, November 28 1947, Schmitt Papers, box 60, folder "[SSZ]: Correspondence, T–Z." A list of committees serving in the first year is located Schmitt Papers, box 60, folder "SSZ: Memoranda, Minutes of Meetings, Election Ballots, Financial Records, Committee Reports, 1947–1959." One person declined nomination to a committee chair. Libbie Hyman was asked to run the Committee on Improving Publication Sources but declined. As Schmitt told her colleague Alexander Petrunkevitch, "It was my hope that she would begin to break ground looking toward a periodical for systematic zoology ..." (Schmitt to Petrunkevitch, September 21 1948, Schmitt Papers, box 60, folder "[SSZ]: Correspondence, M–P." Blackwelder (1977:109–110) mentions some of this work though he attributed its start to 1948 rather than 1947.

[60] Schmitt to Norman Stoll, November 28 1947, Schmitt Papers, box 60, folder "[SSZ]: Correspondence, R–S." The reference is certainly to the International Commission on Zoological Nomenclature.

[61] Wharton to Schmitt, December 3 1947, Schmitt Papers, box 60, folder "[SSZ]: Correspondence, T–Z"

[62] British lepidopterist and civil servant Arthur Francis Hemming (1893–1964) had served as Secretary for the International Commission on Zoological Nomenclature since 1936 (Melville, 1995).

alternative. The basic result of the meeting was a license for Schmitt and Wharton to proceed with the society's organization, create its operating infrastructure and undertake the formalities of its official constitution. The group also agreed to keep charter membership open to any people who paid their dues before the next (i.e., the first annual) meeting.[64]

The SSZ held its first annual meeting in Washington, DC, on September 12 1948.[65] The constitution was officially adopted.[66] Alexander Petrunkevitch (Osborn Zoological Laboratory, Yale University) was elected president, though he did not attend the meeting. Schmitt anticipated Petrunkevitch's election. Before the meeting, Schmitt offered his vision for the next steps forward:

> I wish that we might count on you for our September 12 breakfast meeting of the Society of Systematic Zoology. I have never forgotten your several suggestions on the improvement of the proposed Constitution, and especially the one regarding a journal of systematic zoology. I am certainly going to do my bit to get such a journal subsidized. I am sure that if we finance it for just a few years, two or three, it will make a place for itself. Besides discussions of nomenclatorial problems, systematic practice and proceedings, I should like to see such a journal instituted for the purpose of making the decisions of the International Commission more widely known than at present is possible. Furthermore, I would also like to see the first descriptions of new species on card form, letter size, for convenience in filling, issued as a supplement to this journal. This is something on the order of the card index to types of Devonian fossils put out by the Wagner Free Institute of Science [Philadelphia, PA]. No matter what may be said about a new species in subsequent monographic series, its first appearance should be on a standard reference card, letter size, because

---

[63] Minutes for the December 1947 Chicago meeting were made by Wharton and are located in the Schmitt Papers, box 24, folder "1947 Organiz. Mtg." They do not list attendees. Technically, the Society had not yet launched because they voted only to proceed with the organization, not to approve the constitution. Schmitt and Wharton's draft constitution was approved as a provisional structure for one year. The formalities are documented in extensive correspondence (e.g., Schmitt to Wharton, January 9 1948; Wharton to E.R. Hall. January 11 1948; and Schmitt to Wharton, February 6 1948, all in Schmitt Papers, box 60, folder "[SSZ]: Correspondence, T–Z"). Some of the first complaints about organizational decisions arrived before March. F.G. Walton Smith complained to Schmitt about the lack of a potential councilor as marine invertebrate specialist. Mayr complained about the nominating process for officers. See Schmitt to Wharton, March 9 1948, Schmitt Papers, box 60, folder "[SSZ]: Correspondence, T–Z." A good summary of activity is provided by Wharton to "Dear Colleague," April 1 1948, letter/circular, Schmitt Papers, box 60, folder "SSE: Memoranda, Minutes of Meetings, Election Ballots, Financial Records, Committee Reports, 1947–1959."

[64] Membership lists for the society as of May 10 1948 and December 1 1948 are located in the Schmitt Papers, box 60, folder "SSE: Memoranda, Minutes of Meetings, Election Ballots, Financial Records, Committee Reports, 1947–1959."

[65] In 1948, annual meetings of scientific societies associated with the AAAS broke with the tradition of convocation week at the end of December. This was because the AAAS celebrated its centennial in September and moved its annual convention accordingly: September 13–17 1948 in Washington, DC. The new American Institute of Biological Sciences (AIBS) organized meetings of its charter biological societies in conjunction, held on September 10–13 1948. The SSZ met within the AIBS meetings. Schmitt called the September 12 1948 meeting "our definitive organization meeting" (Schmitt to Walter Necker, September 7 1948, Schmitt Papers, box 60, folder: "[SSZ] Correspondence M–P").

[66] Schmitt Papers, box 60, folder "SSZ-Constitution and By-Laws" contains a constitution marked as the original copy and dated as adopted on September 13 1948. The date may be an error, mistaken for the 12th. Details on the state of society business are summarized in a letter from Schmitt to Petrunkrevitch, September 21 1948, Schmitt Papers, box 60, folder "[SSZ]: Correspondence, M–P." Some time after the meeting, Wharton circulated a mimeographed letter to members of the society describing the organizational process to date, announcing the formation of ten committees, and counting the number of members at 345 (Wharton to Hubbs, no date but stamped as received March 17 1949, Hubbs Papers, box 34, folder 61).

file cabinets for that size are standard, and the sheet is large enough to carry figure descriptions and whatever remarks an author may care to make. The institution of such a publication of new species would obviate the many little scrappy notes describing species scattered through many different and often obscure journals.[67]

After the meeting, Schmitt delivered the news of his election to Petrunkevitch. "I write to tell you how happy I am that this is the case and I hope that you will have no hesitation in accepting the office. We need some one of mature judgment like yourself to guide us at least through our first definitive year."[68]

The SSZ was well and truly under way.

Where Does Blackwelder's Narrative Fit?

By all accounts, Richard Blackwelder[69] played a fundamental role in the SSZ's first decade. He deserves credit not only for running the society as a going concern but also for being a driving force behind many SSZ activities. He was involved in the launch of the *News Letter*, then *Systematic Zoology*. He created the society's book exhibit and represented the SSZ in many settings.[70] Formally, he served as secretary, 1949 to 1960, then president in 1961. Blackwelder poured his heart into the SSZ.

Blackwelder enters the archival record of the SSZ in March 1948, after the launch was well under way.[71] Following the December 1947 organization meeting, Schmitt and Wharton corresponded with many colleagues in their efforts to build on the foundations created in Chicago. In February 1948 Wharton drafted a second circular updating colleagues on progress thus far and calling for charter members. He sent this draft to Schmitt for comments. At some point Schmitt passed this to Blackwelder for comment. He replied in a long letter, rewriting Wharton's draft and raising many technical points about procedure for the two men to note. Blackwelder seems to have thought long and hard about this task. Closing his reply to Schmitt, Blackwelder offered, "I can foresee great possibilities for the society and hope to make some definite suggestions for its activities as soon as it is going."[72] Schmitt

[67] Schmitt to Alexander Petrunkevitch, September 7 1948, Schmitt Papers, box 60, folder "[SSZ] Correspondence M–P." Schmitt's organizational efforts to date were represented in his "List of Committees serving prior to September 12, 1948," Schmitt Papers, box 60, folder "SSE: Memoranda, Minutes of Meetings, Election Ballots, Financial Records, Committee Reports, 1947–1959."

[68] Schmitt to Petrunkevitch, September 21 1948, Schmitt Papers, box 60, folder "[SSZ] Correspondence M–P." Schmitt also offered more advice about appointing committees and putting them to work. He paid particular attention to council work regarding nomenclature. "No matter how uncertain our present dealings with nomenclatorial problems may be, it is better to be somewhat more conservative than radical, as Hemming seems to have become." Petrunkevitch was slow to agree to the post, as is shown in correspondence as late as Schmitt to Petrunkevitch, October 15 1948, Schmitt Papers, box 60, folder "[SSZ] Correspondence M–P."

[69] An expert in Coleoptera, Richard Eliot Blackwelder (1909–2001) earned his Ph.D. at Stanford University (1934), held a postdoctoral position at the U.S. National Museum (1935–1938), was assistant curator at the American Museum of Natural History in New York (1938–1940), then returned to the USNM as assistant (1940–1942) and associate curator (1942–1954). Conflicts with management (largely over the relative balance of his time spent on SSZ business) led to his resignation from the USNM. Blackwelder moved to St. John Fisher College in Rochester, NY (1956–1958), then to Southern Illinois University at Carbondale, rising through the professorial ranks (1958–1977, then emeritus). In his retirement, Blackwelder became a leading collector of Tolkieniana (Blackwelder, 1990, 1993). He donated this substantial collection to Marquette University (which also holds the J.R.R. Tolkien Papers).

[70] Blackwelder summarized the main points of his work in the SSZ in Blackwelder (1977:110–114).

[71] Wharton's minutes of the December 1947 meeting do not identify Blackwelder as speaking and provide no list of attendees. This makes it impossible to say whether Blackwelder was in attendance.

forwarded these comments to Wharton. Though rejecting most of Blackwelder's suggestions, Schmitt did not hesitate to convey a sense of potential value in his new recruit.

> Dick Blackwelder is tremendously interested in all that we are trying to do with the Society for Systematic Zoologists. ... He wants to see the various nomenclature groups throughout the country coordinated in some fashion and feels that our Society is just the place for it, ... Because of his interest, I asked him to work over your draft ...[73]

This seems to mark Blackwelder's entrance into SSZ business. Blackwelder's letter to Schmitt has a formality that suggests a lack of familiarity. While Schmitt consulted Blackwelder further before December 1948, it seems to have been neither extensive nor central to Schmitt's decisions about the SSZ. After December 1947, Schmitt listed Blackwelder as a prospective member for some of the committees he planned for the SSZ. No doubt Blackwelder occasionally interacted with Schmitt at the U.S. National Museum: They worked in the same building. However, Blackwelder seems to have been brought into SSZ business after the launch took place. From there, of course, Blackwelder earned Schmitt's confidence, and their camaraderie developed. By the end of their careers, Blackwelder acknowledged the debt he owed Schmitt with a devotional biography (Blackwelder, 1979).

Blackwelder's narrative of the SSZ's launch placed at its center the tension between evolutionists and taxonomists. For him, this conflict motivated the original need for organization. The earliest version of Blackwelder's narrative appeared in the draft copy of an SSZ recruiting brochure, circa 1951. The spark came at the founding meeting of the evolutionists' group, he wrote. A request was made to broaden the scope of the new society to include "other fundamental aspects of systematics," but this suggestion was rejected. "That decision may have been wise, and undoubtedly resulted in a more unified purpose, but it also had the effect of shutting out those systematists whose primary interests were not in evolution as such."[74] Blackwelder repeated this basic version of events for the rest of his life (e.g., Blackwelder, 1964, 1967:329–372, 1977). He included it in an oral history produced in 1978[75] and in a biography of Schmitt he published the following year (Blackwelder, 1979:157).

Blackwelder's professional writing expressed the same polarization. His contribution to the first issue of *Systematic Zoology*, with Alan Boyden, was a broadside against "phylogenists" within systematics. Their criticisms were aimed specifically at Dobzhansky, Mayr, Simpson, and Huxley (Blackwelder and Boyden, 1952). Blackwelder (1964) claimed the "speciationists have stolen and twisted" his generation's revolution in systematics techniques and called for a return to an "omnispective approach" that he later explained in detail (Blackwelder, 1967). Even his short guide to the classification of animals raised this conflict (Blackwelder, 1963). This polarization continued in private correspondence, where Black-

---

[72] "Memo for Dr. Waldo L. Schmitt" from Blackwelder, March 6 1948 and Schmitt to Wharton, March 9 1948, both in Schmitt Papers, box 60, folder "[SSZ]: Correspondence, T–Z." The original version of Wharton's second circular is lost so it is impossible to learn much about what Blackwelder changed in his revision.

[73] "Memo for Dr. Waldo L. Schmitt" from Blackwelder, March 6 1948 and Schmitt to Wharton, March 9 1948, both in Schmitt Papers, box 60, folder "[SSZ]: Correspondence, T–Z"

[74] The draft text is located in Hubbs Papers, box 34, folder 60. A copy of the printed brochure is located in Simpson Papers, folder "Society of Systematic Zoology" (quote in note 6). Blackwelder published a similar version of events in his "Secretary's Special Issue" of the SSZ *News Letter*, Series 2, Number 7, June 1957 (SSZ Records, box 23).

[75] The oral history is deposited as Record Unit 9517, Smithsonian Institution Archives, Washington, D.C. His discussion of the SSZ's formation is on pp. 34–42.

welder could be scathing. For instance, writing to Schmitt about a supportive paper soon to appear in the journal, Blackwelder confided,

> Don't fail to read the lead article in *Systematic Zoology* this issue - by Borgmeier. I was afraid Brooks [John L. Brooks, who edited the journal, 1952–1957] wouldn't accept it because it's so hard on the New Systematics people. Really rides over Mayr rough shod; he obviously deserves it. It's high time that somebody brought together the larger amount of disagreement with the popular fictions of these people who think that the business of systematics is to detect evolution at work. Incidentally, I can take a bow for Alan Boyden and me in the remarks of Borgmeier [1957].[76]

For a myriad of intellectual reasons Blackwelder disagreed with principles underlying evolutionary taxonomy. He certainly was not alone. However, for some reason — no doubt complex reasons — he chose to represent this disagreement such that he needed to reject their project absolutely. Blackwelder created a self-identity based on this opposition and constructed his relation to evolutionary taxonomy in terms of bitter conflict and rivalry. This construction was more than rhetorical. It was personal.[77]

Blackwelder tied this rivalry to the SSZ. He regularly compared his SSZ with the evolutionists and the SSE. This provided an out-group for defining the SSZ. Defending the "taxonomic side of systematics" gave purpose to his role as secretary. It explains why he chose to emphasize the way his side had been "shut out" from the organizing of the SSE. It also explains why his historical narrative about the SSZ's launch exaggerated this moment and polarized the issues in play far beyond what is supported by the archival evidence. Origin myths reinforce social bonds in important ways (Abir-Am and Elliott, 2000). Blackwelder put history to work for his own purpose.

Blackwelder's emphasis on conflict with evolutionary taxonomy haunted him from the start of his involvement with the SSZ. He suffered his first major losses during efforts to launch the new journal.

When SSZ officers finally decided in 1951 to launch a journal, Blackwelder headed the effort to secure seed money. They first turned to the American Philosophical Society (APS). This seemed an obvious choice at the time because the SSZ could follow the evolutionists who received an APS grant to launch *Evolution*.[78] When Blackwelder approached the APS for support with a similar proposal, he deliberately drew an analogy with the SSE and asked for equal treatment.[79] Blackwelder's plan was a bit more ambitious: The grant would provide for 64-page issues six times a year produced with high-quality illustrations. He arranged support from distinguished colleagues,[80] but the SSZ request was rejected. Blackwelder thought the reasons were suspicious. It did not help that Simpson was an APS member who would have been an influential voice in the relevant committee. Simpson, who had yet to join the SSZ, declined to support Blackwelder's proposal within the APS.[81] Twenty-five years later Blackwelder still expressed anger over the issue when he recalled how the proposal

---

[76] "Dick" to "Waldo," December 2 1957, Schmitt Papers, folder "Blackwelder, R.E., 1949–1959."

[77] In contrast, Schmitt used a different representation, one that minimized the extent of their intellectual overlap and that provided for coexistence.

[78] Securing funds to launch *Evolution* was by no means simple (Cain, 1994). After failing elsewhere, Mayr and Simpson approached the APS in 1946 only because a colleague pointed out they had surplus funds not spent during the war. After some delicate negotiations and a campaign of supporting letters, the money was awarded. This support was unusual for the APS, who worried about the precedent it might set.

[79] A copy of the "Proposal to Establish a New Journal, Systematic Zoology, ..." accompanies Blackwelder to Simpson, July 27 1951, Simpson Papers, folder "Society of Systematic Zoology #1."

had been declined "with the explanation that the APS does not support publications (not ours anyway!)" (Blackwelder, 1977:113).

Far worse, Blackwelder was unable to secure the editorship of *Systematic Zoology*. His role as Editor *pro. tem.* for the first issue resulted from his desire to see the project launched as he saw fit. Schmitt wanted to propose Blackwelder as permanent editor but hesitated, telling Frank Pitelka, "He is a most excellent secretary and a most indefatigable one. I know that he would make also a top-notch editor for the journal, but we cannot give him all the work that the Society needs to have done."[82] No doubt encouraged by Blackwelder after he was passed over, Alan Boyden put himself forward for the editorship as a way to keep their side on top.[83] However, the editorship went to John L. Brooks (Osborn Zoological Laboratory, Yale University). While not an overt speciationist, Brooks followed in the traditions of the Osborn Laboratory and advocated a phylogenetic approach to classification (e.g., Petrunkevitch, 1952b).

Brooks' editorship began with the second issue. Petrunkevitch (1952a) introduced his young colleague, stressing his credentials as a researcher with an "enviable past and a promising future." Blackwelder (1952) added his own polite welcome. Under the surface the transition was fraught with conflict between these two men. Among other issues, Blackwelder insisted on a large space in the journal devoted to SSZ news. He would produce this (e.g., Blackwelder, 1952). This conflict became sufficiently acrimonious that senior colleagues felt the need to intervene.[84] Blackwelder's difficulties with Brooks were not made easier by the preponderance of papers appearing in *Systematic Zoology* defending evolutionary taxonomy and classifications based on phylogenies (such as Huntington, 1952;

[80] For example, Hubbs to Edward Conklin, August 3 1951, Schmitt Papers, box 60, folder: "[SSZ]-Correspondence F–L." Hubbs argued, "That our Society is filling a real need and place is indicated by the success of its programs and by the fact that it has very rapidly grown to a membership of approximately 1000, with prospects for future growth. A real reason for the success of the meetings and of the Society is the feeling that there is need for an integrative association of specialists in different fields of systematic zoology. The trend of development in systematic zoology has been very strongly towards the broad general interpretive or 'philosophical' approach. An increasingly large number of workers, in our Society representing the majority in the dominant group, are concerned very much with the broad general definitely biological aspects of the science. There has been a strong integrative tie-in between systematic zoology and evolution, genetics, modern ecology, embryology, and other fields. My feeling is that it would be very good for biology in general to have established on an adequate basis a journal of 'Systematic Zoology' that would emphasize the broadly applicable and fundamental aspects of systematic work. This would make for a better understanding between zoologists in different fields and would certainly increase integration. The journal would certainly be mighty good for systematic zoologists, as it would tend to lead them further into fundamental thinking and fundamental research." Other correspondence from Hubbs in this folder shows the extent of efforts to obtain this support.

[81] Simpson to Blackwelder, August 21 1951, Simpson Papers, folder: "Society of Systematic Zoology"

[82] Schmitt to Pitelka, July 14 1950, Schmitt Papers, box 60, folder: "[SSZ]: Correspondence, M–P." No doubt, Schmitt was aware of the growing concern in the museum about Blackwelder's workload and the time he spent on SSZ activities.

[83] Boyden to Hubbs, January 22 1951, Hubbs Papers, box 34, folder 64

[84] Alfred Romer (SSZ president in 1952) wrote Blackwelder to intervene on Brooks' behalf in June 1952. "You have pretty much kept the editorial reins in your own hands and left him little leeway." Romer tried to be polite but forceful. "You have done so much for the Society, Dick, that I hate to appear to oppose in any way anything you wish." But in Brooks, Romer saw "another guy that looks to be able to develop an equal enthusiasm and carry a chunk of the burden of Society work." Schmitt rushed to Blackwelder's defense, complaining that Brooks was inexperienced and acted with a rather strong hand. This led to meetings between Romer and Blackwelder, and Blackwelder and Brooks before the end of the month. In the end, Romer found reconciliation, "You know, Dick, I think that the fact we can have a bit of a dither of this rot is something to be pleased over rather than the contrary. It emphasizes the fact that in the Society and its Journal we really have something worth getting excited about." Romer summarized the situation best when writing to Schmitt: "In Dick and Brooks we have two young bucks so eager to do a good job that they were getting in each other's hair." Related correspondence is peppered throughout the Schmitt papers, box 60. For quotes, see Schmitt to Romer, June 6 1952, June 10 1952, and June 23 1952, Schmitt Papers, box 60, folder "[SSZ]: Correspondence, R–S"

Maslin, 1952; Myers, 1952; Stunkard, 1952; Van Cleave, 1952; the relative balance of content in the journal is examined by Hull, 1983).

Through the 1950s Blackwelder's rivalry focused most directly on Mayr, and these two repeatedly clashed over who would create and lead the infrastructure within systematic zoology. This rivalry was unfortunate for Blackwelder. Mayr grew in prominence during the 1950s to become a senior statesman in the discipline. Blackwelder did not. This relative imbalance must have been frustrating for Blackwelder. His involvement with the SSZ — and his sense of the SSZ's historical role in defending his cause — came to mean more and more.

Losing the editorship of *Systematic Zoology* was an early setback in this rivalry. It meant Blackwelder had failed to match Mayr's largely single-handed success with *Evolution* (Cain, 2000a). To compensate, Blackwelder tried to make the most of the news section in the journal, then relaunched the *SSZ News Letter*, partly "because of the desire of the editor to remove this material from the journal and partly because of the desire of the Secretary-Treasurer to have more flexibility ..."[85]

In the early 1950s Mayr became disillusioned with the SSE and *Evolution*. The relative dominance of geneticists in the society frustrated his attempts to secure a balance with the concerns of evolutionary systematics. Moreover, while editor of *Evolution* (1946 to 1949), Mayr largely failed (despite considerable effort) to recruit papers from systematists representing an evolutionary perspective. This situation did not change under the next editor, Edwin Colbert. As Mayr turned away from *Evolution*, he turned toward *Systematic Zoology*. Mayr joined the editorial board in 1953 just as his co-authored textbook on systematic principles appeared (Mayr et al., 1953).[86]

In 1953 Mayr also left his curatorial position at the American Museum of Natural History for an Agassiz professorship at Harvard University. This was an important personal transition for him. It offered the kind of independent, prestigious position he had long sought. As part of this move, Mayr fully committed himself to a statesman's role representing systematics at a national level. Added momentum came with his election to the International Commission of Zoological Nomenclature (1954 to 1976).[87] Plus, in 1954 he was elected to the National Academy of Science. "Taxonomy has had a considerable rejuvenation during recent decades," he wrote his Academy colleagues, "but we must continue to make every effort to broaden its basis even more. If we do not, the best among the students will go into other fields. And there is still so much to be done ..."[88] Mayr quickly involved himself with policy committees of the National Research Council and the American Institute of Biological Sciences, most importantly one focused on funding priorities for the preservation of biological materials.[89] Another focused on core biological concepts, and their report influenced policy decisions within the National Science Foundation (NSF) (Appel, 2000). In 1961, Mayr replaced Romer as director of the Museum of Comparative Zoology.

---

[85] Blackwelder's comment in the 1 March 1953 *News Letter*, which was the first issue of the second series.

[86] This and Simpson (1961) upstaged Blackwelder, who wanted to write a book on general principles. He finally succeeded (Blackwelder, 1967). Mayr replied with a revision to his earlier text (Mayr, 1969). By this time, the systematics community was well and truly polarized into tribal groups and both men were being upstaged by other issues (Hull, 1988).

[87] Also elected to the International Commission in 1954 was another advocate of the new systematics, Alden Miller. He served from 1954 to 1966, and as president from 1963 to 1965.

[88] Mayr to Hubbs, May 5 1954, Hubbs Papers, box 24, folder 60. The 1954 election included Mayr and Petrunkevitch, bringing Hubbs to calculate the presence of seven taxonomists (also including George Simpson, Alfred Romer, Alexander Wetmore, Remington Kellogg and Carl Hubbs himself).

[89] Schmitt Papers, box 40, folder "American Institute of Biological Sciences," covers 1950 to 1972, including "Report of the Committee on Systematic Biology of the [AIBS]," undated 24 pages. Serving on this committee were Alfred Emerson, C.O. Erlanson (USDA), Th. Just (Chicago), Karl Schmidt, George Simpson, W.C. Steere (Stanford), Oswald Tippo (Yale), Victor Twitty (Stanford) and Ernst Mayr (Chair). This project related to the NSF in 1954 and discussed at length the possibility of creating a national center for systematic biology.

In contrast, Blackwelder faired poorly. Though recognized for the considerable energy he spent on SSZ activities, he largely failed to move into leadership roles. Perhaps Blackwelder tied his own sense of identity too strongly to his role in the SSZ. This blurred the distinction between the Society's purposes and his own professional goals and intellectual interests.[90] In 1954, such blurred distinctions led to serious questions about Blackwelder's commitment to his employer, the USNM, and questions about overuse of the institution's largess for the SSZ's gain. Though he adamantly denied any wrongdoing, Blackwelder was forced to resign over this matter.[91] Four years of unrelated employment and failed attempts at fellowships followed before Blackwelder moved back on track, joining the expanding zoology department at Southern Illinois University in Carbondale. Blackwelder sorely needed the SSZ during this period if simply for his continuing professional identity.

The differences in careers were obvious to all. In 1955 Mayr proposed creating an international directory of systematists, using his connections with NSF to secure funding. When he consulted Schmitt on the matter, Mayr received the expected support, "... naturally [the idea] strikes a responsive chord here. There is no question about the need for such a list." However, Schmitt peppered this support with a plea to assist his friend. He asked Mayr to give the project to Blackwelder, who could devote much time to it, use his established connections as SSZ secretary, and see to the frequent updating. "I think that I am aware of Blackwelder's faults as well as anyone. There are many assignments which someone else could do better, but this one seems peculiarly suited to his capabilities."[92] Mayr conceded. Blackwelder and his wife saw the project through (Blackwelder and Blackwelder, 1961).

Conclusion

This paper develops two historical points. First, it describes the launch of the SSZ, drawing primarily on previously unconsidered archival sources. This focuses attention of the efforts of Waldo LaSalle Schmitt and George Wharton to rebuild an infrastructure for American systematists following World War II. Close analysis of their September 1947 circular peels away successive layers of historical context. It provides an appreciation for the complex senses of need that shaped decision making at the time.

Second, this paper separates the launch of the SSZ from the founding narrative presented by Richard Blackwelder. Blackwelder came to SSZ business relatively late in the organizing process. While he played a key role in subsequent activity, Blackwelder's account of the launch was second-hand. In retelling that narrative, Blackwelder was heavily influenced by his own construction of the state of systematics in 1940s America. This emphasized rivalry and polarity. It set the SSZ into competition with evolutionary taxonomists and tied its success to that competition. Rather than complain about authenticity, the purpose of this analysis has been to contextualize Blackwelder's founding narrative. This origins myth says much about the man who promoted it, and this paper is similar to other studies of the role

[90] Blackwelder belonged to no other major professional organization in his career other than AAAS and SSZ. His affiliations are followed in successive editions of *American Men and Women of Science*. This sense of ownership is illustrated in his correspondence with Carl Hubbs during Hubbs' presidency in 1951 (see Hubbs Papers, box 34, folders 60–64).

[91] Blackwelder's resignation letter is Blackwelder to Leonard Carmichael, October 23 1954, Schmitt Papers, box 3, folder "Blackwelder, R.E., 1949–1959."

[92] Schmitt to Mayr, November 10 1955, Schmitt Papers, box 24, folder "Mayr, Ernst." Thinking ahead, Schmitt prepared Blackwelder for the possibility of a visit from Mayr with this idea in mind. "Don't feel as I first did about it — that he was trying to take part of our original project [of a directory through the SSZ] and make something of it. I may be mistaken but, whether or no, I want you to be receptive so that at least you can explore the proposition. I would like you to take hold of it, perhaps insisting on enough money to make it worth your while." (Schmitt to Blackwelder, November 14 1955, Schmitt Papers, box 3, folder "Blackwelder, R.E., 1949–1959")

of origins mythology in promoting social bonds within tribal communities in science (Abir-Am and Elliott, 2000).

No group came to dominate the SSZ in its first decade. Neither did a broad intellectual consensus form. Instead, the SSZ functioned as a trading zone in which members of different camps met and engaged in exchanges (Galison, 1997, 1999). As a symbol of community, the SSZ also provided a boundary object that different groups could define in their own terms and for their own reasons but that all could support jointly in defense of a shared cause: the promotion of systematics as fundamental within the life sciences (Star and Griesemer, 1989). Appel (2000:150) shows the importance of this collective voice for systematics. For example, of all U.S. federal monies supporting basic systematics in 1954, 61% derived from the NSF. This was the agency most targeted by those with a vision for unified action in systematics. Whether or not they were truly unified misses the point. Their efforts sowed the seeds for a new generation of American systematists. Ultimately that new generation pressed ahead to develop fundamental changes in systematic theory and practice.

Acknowledgments

Many thanks to David Williams for his work organizing this symposium. Thanks also to William Cox (Smithsonian Institution Archives) for considerable assistance as well as to the staffs at the American Philosophical Society Library and Scripps Institution of Oceanography. Glen Needham, James Keirans, and Michelle Drobik provided biographical information about George Wharton. This project was undertaken as part of a Research Fellowship from the Leverhulme Trust.

Archival Collections

The following abbreviations are used in this paper: **Dunn Papers**: Leslie C. Dunn Papers, American Philosophical Society Library, Philadelphia, PA; **Hubbs Papers**: Carl Hubbs Papers. Archives of the Scripps Institution of Oceanography, University of California, San Diego, La Jolla, CA; **Mayr/*Evolution***: Ernst Mayr/Society for the Study of Evolution Papers. Collection B/M451s, American Philosophical Society Library, Philadelphia, PA; **Schmitt Papers**: Waldo LaSalle Schmitt Papers, 1907–1978. Record Unit 7231. Smithsonian Institution Archives, Washington, DC; **Simpson Papers**: George Gaylord Simpson Papers. American Philosophical Society Library, Philadelphia, PA. Unless otherwise noted, items are in series 1; **SSE Papers**: Materials Relating to the Society for the Study of Evolution, collection B/M451S 1991 1027ms, American Philosophical Society Library, Philadelphia, PA; **SSZ Records**: Records of the Society of Systematic Zoology, 1947–1975. Record Unit 7226. Smithsonian Institution Archives, Washington, DC.

References

Abir-Am, P.G. and Elliott, C.A., Eds., *Commemorative Practices in Science: Historical Perspectives on the Politics of Collective Memory*, University of Chicago Press for History of Science Society, Cambridge, 2000.

Allee, W.C., Emerson, A.E., Park, O., Park, T., and Schmidt, K.P., *Principles of Animal Ecology*, W. B. Saunders, Philadelphia, PA, 1949.

Allen, G., *Life Sciences in the Twentieth Century*, John Wiley & Sons, New York, 1975.

Anderson, E., *Introgressive Hybridization*, John Wiley & Sons, New York, 1949.

Anonymous, Programs of the AAAS sections and the societies, *Science*, 106, 558–566, 1947.

Anonymous, The Systematics Association Annual Report VI. (1946–1947), *Proc. Linn. Soc., Lond.*, 160, i–iv, 1948.

Appel, T., *Shaping Biology: The National Science Foundation and American Biological Research, 1945–1975*, Johns Hopkins University Press, Baltimore, MD, 2000.

Babcock, E.B. and Stebbins, G.L., *The American Species of Crepis, Their Interrelationships and Distribution as Affected by Polyploidy and Apomixis*, Carnegie Institution of Washington, Washington, DC, 1938.

Babcock, E.B., Stebbins, G.L., and Jenkins, J.A., Genetic evolutionary processes in *Crepis, Am. Nat.*, 76, 337–363, 1942.

Ball, C.R., Why is taxonomy ill-supported?, *Science*, 103, 713–714, 1946.

Bates, R.E., *Scientific Societies in the United States*, 3rd ed., Pergamon Press, Oxford, 1965.

Beatty, J., The proximate/ultimate distinction in the multiple careers of Ernst Mayr, *Biol. Philos.*, 9, 333–356, 1994.

Benson, K.R., Maienschein, J., and Rainger, R., Eds., *The Expansion of American Biology*, Rutgers University Press, New Brunswick, NJ, 1991.

Blackwelder, R.E., SSZ news, *Syst. Zool.*, 1, 92–5, 1952.

Blackwelder, R.E., *Classification of the Animal Kingdom*, Southern Illinois University Press, Carbondale, IL, 1963.

Blackwelder, R.E., Phyletic and phenetic versus omnispective classification, in *Phenetic and Phylogenetic Classification: A Symposium*, Heywood, V.H. and McNeill, J., Eds., The Systematics Association, London, 1964, pp. 17–28.

Blackwelder, R.E., *Taxonomy: A Text and Reference Book*, John Wiley & Sons, New York, 1967.

Blackwelder, R.E., Twenty five years of taxonomy, *Syst. Zool.*, 26, 107–137, 1977.

Blackwelder, R.E., *Zest for Life, or Waldo Had a Pretty Good Run: The Life of Waldo LaSalle Schmitt*, Allen Press, Lawrence, KA, 1979.

Blackwelder, R.E., *A Tolkien Thesaurus*, Garland, New York, 1990.

Blackwelder, R.E., *Tolkien Portraiture* (compiled by Richard E. Blackwelder, assisted by John Ratcliff, Christine Scull, Charles E. Noad), Cape Girardeau, MO, 1993.

Blackwelder, R.E. and Blackwelder, R.M., *Directory of Zoological Taxonomists of the World*, Southern Illinois University Press for the Society of Systematic Zoology, Carbondale, IL, 1961.

Blackwelder, R.E. and Boyden, A., The nature of systematics, *Syst. Zool.*, 1, 26–33, 1952.

Blackwelder, R.E., Knight, J.B., and Sabrosky, C.W., Letter in "Comments by Readers," *Science*, 106, 315–316, 1947.

Borgmeier, T., Basic questions of systematics, *Syst. Zool.*, 6, 53–69, 1957.

Boyden, A., Fifty years of systematic serology, *Syst. Zool.*, 2, 18–30, 1953.

Brockway, L., *Science and Colonial Expansion: The Role of the British Royal Botanic Gardens*, Academic Press, New York, 1979.

Bush, V., *Science: The Endless Frontier. A Report to the President*, U.S. Government Printing Office, Washington, D.C., 1945.

Cain, J., Ernst Mayr as community architect: launching the society for the study of evolution and the journal Evolution, *Biol. Philos.*, 9, 387–427, 1994.

Cain, J., Ed., *Regular Contact with Anyone Interested: Documents of the Society for the Study of Speciation, 1941*, Riverside Press, Winona, MN, 1999.

Cain, J., For the "promotion" and "integration" of various fields: first years of evolution, 1947–1949, *Arch. Nat. Hist.*, 27, 231–259, 2000a.

Cain, J., Towards a "greater degree of integration": the Society for the Study of Speciation, 1939–1941, *Br. J. Hist. Sci.*, 33, 85–108, 2000b.

Cain, J., Epistemic and community transition in American evolutionary studies: the "Committee on Common Problems of Genetics, Paleontology, and Systematics" (1942–1949), *Stud. Hist. Philos. Biol. Biomed. Sci.*, 33, 283–313, 2002.

Cittadino, E., *Nature as Laboratory: Darwinian Plant Ecology in the German Empire, 1880–1900*, Cambridge University Press, Cambridge, 1990.

Cole, W.H., *Serological and Biochemical Comparisons of Proteins*, Rutgers University Press, New Brunswick, NJ, 1958.

Cox, W., *Guide to the Papers of Waldo LaSalle Schmitt, 1887–1977*, Smithsonian Institution Archives, Washington, DC, 1983.

Dietrich, M.R., Paradox and persuasion: negotiating the place of molecular evolution within evolutionary biology, *J. Hist. Biol.*, 31, 85–111, 1998.

Dobzhansky, T., *Genetics and the Origin of Species*, Columbia University Press, New York, 1937.

Dupree, A.H., *Science and the Federal Government: A History of Policies and Activities to 1940*, Harvard University Press, Cambridge, MA, 1957.

Dupree, A.H., Central scientific organization in the United States government, *Minerva*, 1, 453–469, 1963.

Dupree, A.H., The great instauration of 1940: the organization of scientific research for war, in *The Twentieth-Century Sciences: Studies in the Biography of Ideas*, Holton, G., Ed., W.W. Norton, New York, 1972, pp. 443–467.

Emerson, A., Speciation, *Ecology*, 18, 152–154, 1937.

Emerson, A., Taxonomy and ecology, *Ecology*, 22, 213–215, 1941.

Galison, P., *Image and Logic*, University of Chicago Press, Chicago, 1997.

Galison, P., Trading zone: coordinating action and belief, in *Science Studies Reader*, Biagioli, M., Ed., Routledge, New York, 1999, pp. 137–160.

Greenberg, D., *The Politics of Pure Science: An Inquiry into the Relationship Between Science and Government in the United States*, New American Library, New York, 1967.

Hagen, J., Experimentalists and naturalists in twentieth-century botany: experimental taxonomy, 1920–1950, *J. Hist. Biol.*, 17, 249–270, 1984.

Hagen, J., *An Entangled Bank: The Origins of Ecosystem Ecology*, Rutgers University Press, New Brunswick, NJ, 1992.

Heims, S., *Constructing a Social Science for Postwar America: The Cybernetics Group, 1946–1953*, MIT Press, Cambridge, MA, 1991.

Hemming, F., Important advances in zoological nomenclature achieved at 13th International Congress of Zoology, *Science*, 108, 156–157, 1948.

Holton, G., Ed., *The Twentieth-Century Science: Studies in the Biography of Ideas*, W.W. Norton, New York, 1972.

Hubbs, C.L., Speciation of fishes, *Am. Nat.*, 74, 198–211, 1940.

Hull, D., Thirty-one years of systematic zoology, *Syst. Zool.*, 32, 315–342, 1983.

Hull, D., *Science as a Process: An Evolutionary Account of the Social and Conceptual Development of Science*, University of Chicago Press, Chicago, 1988.

Huntington, C.E., Hybridization in the purple grackle, *Quiscalus quiscula*, *Syst. Zool.*, 1, 149–170, 1952.

Huxley, J.S., Clines: an auxiliary method in taxonomy, *Bijdr. Dierkd.*, 27, 491–520, 1939a.

Huxley, J.S., Ecology and taxonomic differentiation, *J. Ecol.*, 27, 408–420, 1939b.

Huxley, J., Ed., *The New Systematics*, Clarendon Press, Oxford, 1940.

Huxley, J., *Evolution: The Modern Synthesis*, Harper and Brothers, New York, 1942.

Kay, L., *The Molecular Vision of Life: Caltech, the Rockefeller Foundation, and the Rise of the New Biology*, Oxford University Press, New York, 1993.

Keen, A.M. and Muller, S.W., *Procedure in Taxonomy*, rev. ed., Stanford University Press, Stanford, CA, 1948.

Kinsey, A., The origin of higher categories in *Cynips*, Indiana University Publications, Science Series, 4, 1–334, 1936.

Kleinman, K., His own synthesis: corn, Edgar Anderson, and evolutionary theory in the 1940s, *J. Hist. Biol.*, 32, 293–320, 1999.

Koerner, L., *Linnaeus: Nature and Nation*, Harvard University Press, Cambridge, MA, 1999.

Lack, D., *The Life of the Robin*, H.F. & G. Witherby, London, 1943.

Larson, E.J., *Evolution's Workshop: God and Science on the Galapagos Islands*, Basic Books, New York, 2001.

Lee, M., Detlev W. Bronk, scientist, *Science*, 113, 143, 1951.

Lewenstein, B.V., Shifting science from people to programs: AAAS in the post war years, in *The Establishment of Science in America: 150 Years of the American Association for the Advancement of Science*, Kohlstedt, S.G., Sokal, M.M., and Lewenstein, B.V., Eds., Rutgers University Press, New Brunswick, NJ, pp. 103–165, 1999.

Maienschein, J., *Transforming Traditions in American Biology, 1880–1915*, Johns Hopkins University Press, Baltimore, MD, 1991.

Maslin, P., Morphological criteria of phyletic relationships, *Syst. Zool.*, 1, 49–70, 1952.

Mayr, E., The origin of gaps between species, *Collect. Net*, 16, 137–43, 1941.

Mayr, E., *Systematics and the Origin of Species*, Columbia University Press, New York, 1942.

Mayr, E., *Principles of Systematic Zoology*, McGraw-Hill, New York, 1969.

Mayr, E., Controversies in retrospect, *Oxford Surv. Evol. Biol.*, 8, 1–34, 1992.

Mayr, E., Linsley, E.G., and Usinger, R.L., *Methods and Principles of Systematic Zoology*, McGraw-Hill, New York, 1953.

Melville, R.V., *Towards Stability in the Names of Animals: A History of the International Commission on Zoological Nomenclature, 1895–1995*, International Trust for Zoological Nomenclature, London, 1995.

Miller, A., Speciation in the avian genus *Junco*, *Univ. Calif. Publ. Zool.*, 44, 173–434, 1941.

Mitchell, R., Horn, D.J., Needham, G.R., and Welbourn, W.C., Dedication, in *Acarology IX* (Symposia, Vol. 2), Needham, G.R., Mitchell, R., Horn, D.J., and Welbourn, W.C., Eds., The Ohio Biological Survey, Columbus, OH, 1969, pp. xv–xvii.

Myers, G.S., The nature of systematic biology and of a species description, *Syst. Zool.*, 1, 106–111, 1952.

Parr, A., On the functions of the natural history museum, *Trans. NY Acad. Sci.* (2nd ser.), 2, 44–58, 1939.

Parr, A., *Mostly About Museums: From the Papers of A. E. Parr*, American Museum of Natural History, New York, 1959.

Patterson, J.T., The *virilis* group of *Drosophila* in Texas, *Am. Nat.*, 75, 523–539, 1941.

Pauly, P., *Controlling Life: Jacques Loeb and the Engineering Ideal in Biology*, Oxford University Press, New York, 1987.

Petrunkevitch, A., Introducing the editor, *Syst. Zool.*, 1, 71, 1952a.

Petrunkevitch, A., Principles of classification as illustrated by studies of Arachnida, *Syst. Zool.*, 1, 1–19, 1952b.

Price, D. de S., *Science Since Babylon*, Yale University Press, New Haven, CT, 1961.

Price, D. de S., *Little Science, Big Science*, Columbia University Press, New York, 1963.

Rainger, R., Benson, K., and Maienschein, J., Eds., *The American Development of Biology*, University of Pennsylvania Press, Philadelphia, 1988.

Reingold, N., Vannevar Bush's new deal for research; or, the triumph of the old order, in *Science American Style*, Reingold, N., Ed., Rutgers University Press, New Brunswick, NJ, pp. 284–333.

Rensch, B., *Das Prinzip Geographischer Rassenkreise und das Problem der Artbildung*, Gebrüder Borntraeger, Berlin, 1929.

Rudeval, F.R., *Règles Internationales de la Nomenclature Zoölogique*, Privately Printed, Paris, 1905.

Said, E.W., *Orientalism: Western Conceptions of the Orient*, Routledge & Kegan Paul, London, 1978.

Schenk, E.T. and McMasters, J.H., *Procedure in Taxonomy*, Stanford University Press, Stanford, CA, 1936.

Schofield, E.K., *A History of the American Society of Plant Taxonomists: The First Sixty Years: 1936 to 1996*, American Society of Plant Taxonomists, Laramie, WY, 1998.

Shapley, H., Kirtley F. Mather: president-elect, AAAS, *Science*, 111, 131–132, 1950.

Simpson, G.G., The Fort Union of the Crazy Mountain field, Montana, and its mammalian faunas, *Bull. U.S. Natl. Mus.*, 169, 1–287, 1937.

Simpson, G.G., *Tempo and Mode in Evolution*, Columbia University Press, New York, 1944.

Simpson, G.G., The principles of classification and a classification of mammals, *Bull. Am. Mus. Nat. Hist.*, 85, i–xvi, 1–350, 1945.

Simpson, G.G., *Principles of Animal Taxonomy*, Columbia University Press, New York, 1961.

Simpson, G.G. and Roe, A., *Quantitative Zoology: Numerical Concepts and Methods in the Study of Recent and Fossil Animals*, McGraw-Hill, New York, 1939.

Smocovitis, V.B., Organizing evolution: founding the society for the study of evolution (1939–1950), *J. Hist. Biol.*, 27, 241–309, 1994.

Star, S.L. and Griesemer, J., Institutional ecology, 'translations' and boundary objects: amateurs and professionals in Berkeley's Museum of Vertebrate Zoology, 1907–1939, *Soc. Stud. Sci.*, 19, 387–420, 1989.

Stunkard, H.W., Life histories and systematics of parasitic worms, *Syst. Zool.*, 2, 7–18, 1952.

Thackery, A., Ed., *Science After '40*, University of Chicago Press for History of Science Society, Chicago, 1992.

Traweek, S., *Beamtimes and Lifetimes: The World of High Energy Physics*, Harvard University Press, Cambridge, MA, 1988.

Van Cleave, H.J., An index to the opinions rendered by the international commission on zoological nomenclature, *Am. Midland Nat.*, 30, 223–240, 1943.

Van Cleave, H.J., Speciation and formation of genera in Acanthocephala, *Syst. Zool.*, 1, 72–83, 1952.

Vernon, K., Desperately seeking status: evolutionary systematics and the taxonomists' search for respectability, *Br. J. Hist. Sci.*, 26, 207–27, 1993.

Wilson, E.O., *Naturalist*, Time Warner, New York, 1994.

Winsor, M.P., The English debate on taxonomy and phylogeny, 1937–1940, *Hist. Philos. Life Sci.*, 17, 227–252, 1995.

Wolfe, D., *Renewing a Scientific Society: The American Association for the Advancement of Science from World War II to 1970*, American Association for the Advancement of Science, Washington, DC, 1989.

Zachary, G.P., *Endless Frontier: Vannevar Bush, Engineer of the American Century*, Free Press, New York, 1997.

Chapter 3

Explanations in Systematics

Walter J. Bock

Abstract

Not all of evolutionary biology is historical, but rather contains both nomological-deductive explanations (N-DEs) and historical-narrative explanations (H-NEs); hence one must be careful to ascertain the proper type of explanation to use in the different parts of evolutionary studies. Of the five major areas of evolutionary theory outlined by Mayr, four are N-DEs, and only the one dealing with Darwinian evolutionary classification, Haeckelian phylogenies, Hennigian cladifications, evolutionary histories of attributes and groups of organisms and historical biogeographies is historical in nature, with the explanations being historical-narrative. Philosophers of science have largely shied away from consideration of H-NEs and how they are connected to underlying N-DEs. An overview of H-NEs is presented in which it is shown that each such explanation deals only with a singular event and cannot be considered to deal with universals as do N-DEs. Moreover, it is argued that careful attention must be given to the degree of confidence of each H-NE and that most of these explanations have a poor to medium degree of confidence. Care must be taken not to overstate the robustness of the methods and conclusions of H-NEs in evolutionary biology.

Introduction

At the first International Congress of Systematics and Evolution held in Boulder, Colorado in August 1973, I introduced, for better or worse, some of the philosophical ideas of Karl Popper into biological systematic discussions (Bock, 1974). In doing so, I had two major concerns. First, I wished to stress that biological classifications, phylogenies and evolutionary histories of all kinds are theoretical scientific statements, and as such they had to be tested against objective empirical observations, with care taken to clarify the exact nature of the objective empirical observations used to test these different theoretical statements. Second, I was concerned with the demarcation between natural science and nonscience because of the rise of scientific creationism and the arguments advocated by the proponents of scientific creationism against evolutionary biology in its broadest sense, including the classification of organisms. Acknowledging the objections raised by some philosophers against the Popperian demarcation of science as that endeavor of humans in which theoretical statements are potentially able to be disproved by testing them against objective empirical observations, I feel that this criterion provides the best demarcation between evolutionary biology and all forms of antievolution, including scientific creationism.

Nevertheless, when introducing Popper's ideas into evolutionary and systematic thinking, I was greatly worried about certain aspects of theories in evolutionary biology, although I did not express this disquiet in my talk or published paper. I was aware that not all of evolutionary biology was historical in nature, as concluded by many philosophers and biologists (see Bock and von Wahlert, 1963), and that the nature of most empirical tests used for theoretical explanations in systematics differed from those advocated by Popper (Bock, 1977). Popper's central, and perhaps only, interest in science was in the realm of nomological-deductive explanations (N-DEs) and not in any historical areas, such as historical-narrative explanations (H-NEs). The major approach to testing in H-NEs appears to be induction, and for Popper even the slightest whiff of induction in tests of scientific theories was a complete no-no. Yet, I felt that there were certain significant advantages in the introduction of Popperian ideas into systematic biological thinking.

It was nevertheless a complete surprise to me that the Popperian approach to philosophy was almost immediately and strongly, indeed fully, accepted by systematists advocating a cladistic approach to systematics. For cladists, in the 1970s and 1980s, Popper's ideas about science and scientific methods provided the foundation for their entire methodology, including the support for the form of a cladistic classification and the way in which these classifications are tested against empirical observations (see Platnick and Gaffney, 1977, 1978a, 1978b). This acceptance of Popper's ideas was well prior to the rise of transformed cladistics, contrary to Hull's statement (1999:498) that this acceptance came much later. To my knowledge, cladists still accept Popper's approach to the philosophy of science as the foundation for their methods and conclusions in cladistics; I am not aware of any rejection or contrary position proposed by any cladist (but see Rieppel, Chapter 4).

Systematics involves a number of ordering systems, but not all ordering systems are classifications (Mayr and Bock, 2002). Moreover although ordering systems such as Darwinian classifications, Haeckelian phylogenies and Hennigian cladifications are H-NEs, not all biological classifications are H-NEs. Typological or near-typological biological classifications, such as those for cell, tissue and organ types or for ecological communities, are N-DEs (Mayr and Bock, 2002:171) and must be treated quite differently.

I welcome the opportunity of being asked to participate in this symposium and to provide a review of explanations in evolutionary biology, especially in classification and phylogeny, and of the role of Popperian concepts in the philosophy of science for methods and explanations in systematics. I thank the organizers for inviting me to take part in the symposium and contribute a paper to the published volume.

Fields within Evolutionary Biology

A major difficulty in considering types of explanations within evolutionary biology is that most philosophers and biologists do not have a clear understanding of the diversity of fields of study within evolutionary biology, or precisely what is meant by the theory of evolution. This difficulty started in 1859 with the publication of Darwin's *On the Origin of Species* (Darwin, 1859). Darwin spoke of "my theory" — always in the singular. Hence the widespread, but erroneous conception that only a single type of theory exists in evolutionary biology. Since evolution is about the history of life, most biologists and philosophers consider all of evolutionary theory, including classification and phylogeny, to be historical in nature. However, cladists, in accepting the Popperian approach as the basic philosophical foundation for their work have, knowingly or unknowingly, accepted the position that at least this aspect of evolutionary biology is nomological-deductive in nature. After considering the complex nature of Darwin's theory, I will consider types of explanations existing in science and how these explanations apply to the diverse aspects of evolutionary biology.

Darwin's theory of evolution as originally advocated in his *On the Origin of Species* (1859) is actually a bundle of five separate but interrelated theories (Mayr, 1985). These five separate theories found in Darwin's 1859 book can still characterize areas within evolutionary biology today; these are:

a. **Evolution as such** is the theory that states that all populations of organisms are changing over time, with the minimum time period being one generation. The theory is clearly nomological-deductive.

b. **Common descent** implies that all species or populations of organisms have descended with modification from common ancestors; this descent includes both modification and branching. Darwinian common descent is equivalent to Haeckel's (1866) phylogeny. Hennigian phylogeny is equal only to the branching aspect of Haeckelian phylogeny (Mayr and Bock, 2002). Common descent is expressed in Darwinian classifications, Haeckelian dendrograms and Hennigian cladograms, which are all clearly theoretical statements. These theories, as well as all other theoretical statements about the history of organisms, such as conclusions about the evolution of organic features (e.g., the avian wing) and historical biogeography, are H-NEs.

c. **Gradualism** is the idea that evolutionary change takes place in steps of the magnitude seen between parents and offspring and never in large sudden saltations or jumps. Evolutionary jumps do not take place between species or taxa of higher levels such as expressed in the idea that the first bird hatched from a reptilian egg. Again this theory is nomological-deductive.

d. **Multiplication of species** states that there is splitting of phylogenetic lineages as well as transformational change within a lineage. Hence evolutionary change includes two processes — phyletic evolution or transformation and speciation. Some workers would include extinction as a third process and others would include extinction under phyletic evolution. This theory is also nomological-deductive.

e. **Natural selection** is Darwin's mechanism for phyletic transformation. Today this would be expressed as causes or mechanisms of evolutionary change, regardless of the diverse causes that different evolutionists would include; this theory is clearly nomological-deductive.

Thus it is clear that evolutionary theory or evolutionary biology is not just historical (see Bock and von Wahlert, 1963) and indeed that most aspects of evolutionary theory are not historical, but nomological-deductive. Many evolutionary biologists, especially systematists, work in area b (common descent) dealing with historical evolutionary theory. This paper emphasizes historical evolutionary theory and how this is based on nomological evolutionary theory.

Popper and Historical Analyses

Since the publication over a quarter of a century ago of my original paper "Philosophical Foundations of Classical Evolutionary Classification," I have become increasingly interested in systems of explanation in science and especially in evolutionary biology (see Bock, 1978, 1981, 1988, 1991, 1992, 2000; Szalay and Bock, 1991). This was prompted by concerns in 1974 when I introduced into systematics Popper's ideas that the nature of explanation and the methods of testing theoretical statements in classification and phylogeny do not fit easily or at all into the Popperian mold.

There was also the matter of the frequently expressed accusation that Popper claimed that all of evolutionary biology was historical and hence nonscientific and that he had subsequently recanted this claim. This claim appeared to stem from Popper's (1957; see especially pp. 105–119) *The Poverty of Historicism*. If one reads this volume carefully, it is clear that Popper makes the error of regarding all aspects of evolutionary theory or evolutionary biology as historical, which would make it impossible to formulate general laws about evolution similar to those in physics and chemistry. It is equally clear that Popper's position is quite definite (1957:143) that *"history is characterized by its interest in actual, singular, or specific events, rather than its laws or generalizations"* (italics in the original) and that law-like generalizations cannot be derived from historical analyses (1957:143–147).

The presentations in Popper's *The Poverty of Historicism* cannot be taken as meaning that he does not consider evolutionary theory to be nonscientific, but only as a lack of appreciation of the full range of thinking in evolutionary biology. Many other philosophers and biologists are also guilty of this misconception. Moreover, as many other philosophers, Popper did not appreciate the details of H-NEs in science and their relationship to N-DEs. It is clear, however, that because classification, evolution and phylogeny of groups of organisms deal with singular events, Popperian ideas about the philosophy of science cannot be used for their analysis (see also Hull, 1999:383).

Hull (1999) has performed an important service in his analysis of Popperian ideas and their relationship to biology and especially evolutionary biology. Hull's analysis would have been clearer had he included a discussion of Darwin's five theories and thereby provided a clarification of which parts of evolutionary theory are historical and which are not. Nevertheless, Hull points out (1999:498) that it is ironic that cladists have adopted Popperian ideas because of Popper's absolute stance against induction in scientific testing and against the notion that there are theory-free facts.

Explanations in Science

Philosophers of science are especially interested in scientific explanations, but almost all of their analyses have been based on sciences such as physics. Unfortunately, as I have expressed earlier (Bock 2000:33), "... the physical sciences are rather simple compared to the biological sciences and that any philosophy of science based solely on the physical sciences is too simplistic to be realistic." This is because physicists have delimited their science so that it is completely nonhistorical. The standard model for explanations in science is the N-DE, which comes under a number of names such as hypothetical-deductive, covering law, nomological, etc. This is the model advocated by Popper, and hence in accepting Popper's ideas as the foundation for their work, cladists claim that they are advocating N-DEs for their scientific methods and conclusions. However, in some sciences, such as biology and geology, there is a major historical aspect that has to be included in scientific explanations. Philosophers of science have either shied away from historical explanations in science or, if they consider them at all as does Hempel (1965:Ch. 9), they waffle on the details of these explanations.

Two different, but interrelated systems of explanations exist in biology, which are (a) the dichotomy of N-DEs versus H-NEs, and (b) the dichotomy of functional explanations versus evolutionary explanations. The latter system stems from the division of biology into the major areas of functional and evolutionary biology (Mayr, 1982). These two systems of explanations in biology do not have a simple relationship to one another. All functional explanations are N-DEs and all H-NEs are evolutionary, but evolutionary explanations can be either N-DEs or H-NEs. Hence it is essential not only to characterize carefully the properties of N-DEs and H-NEs, but to show which parts of biology, and especially of evolutionary biology, are nomological-deductive and which are historical narrative.

Nomological-Deductive Explanations (N-DEs)

N-DEs are the standard form of explanation in science and take the following form. Given a set of facts (e.g., initial and boundary conditions) and a set of laws (be they causes, processes, or outcomes; see Bock 1993), both of which form the explanatory sentence, or *explanans*, a particular conclusion, or *explanandum*, follows (Hempel and Oppenheim 1948; Hempel 1965:335-338). N-DEs answer the question, how has a particular phenomenon [*explanandum*] occurred? N-DEs apply to universals (nonlimited sets of phenomena) and do not depend on the past history of the objects or the phenomena being explained, and their premises (the nomological statements) are assumed to be always true. Saying that N-DEs apply to universals means these explanations are not temporally-spatially restricted within the proper region of the phenomena, which for biology is the earth, and more specifically the surface (i.e., the upper part of the crust) of the earth.

If an *explanandum*, resulting from the conjunction of the set of facts invoked (initial and boundary conditions) and the set of general laws, disagrees with empirical observations, then the N-DE is not valid (i.e, falsified), and one must search for the reason for the falsification. Falsification means that the *explanandum* does not agree with independent, objective, empirical observations, but falsification does not automatically imply that the general laws used in the explanation are in error, although this is a possibility. Possibly, the set of initial or boundary conditions used in the empirical test was in error, or the empirical observations were in error. Examples of N-DEs include clarification of oceanic tides using gravitational laws and of phyletic evolution evoking natural selection (nonrandom, differential survival and reproduction of organisms).

Historical-Narrative Explanations (H-NEs)

H-NEs provide an understanding of the existing attributes of a particular set of objects or phenomena at a specified time; these explanations depend on the past history of these objects, and they must use pertinent N-DEs. The objects explained by an H-NE are singulars, not universals, and have definite spatial-temporal positions. H-NEs are stated on a nondeductive and probabilistic basis with the hope of reaching the most reasonable and probable explanation for the objects studied. Five aspects of H-NEs are stressed, the first being the most important:

1. These explanations are historical, which means that earlier events affect later events. Consequently in any H-NE, special care must be given to formulating the analysis within the correct chronological order of events and changes.
2. H-NEs are given on a probability basis of being correct (Nagel, 1961:26). This is necessary because these explanations frequently involve a number of often conflicting N-DEs employed in the explanation and because of the uncertainty over initial and boundary conditions involved in the explanation.
3. H-NEs must be based on pertinent N-DEs, and these N-DEs, together with the pertinent empirical observations, form part of the chain of arguments used in testing the H-NE.
4. H-NEs are not general, in that a successful explanation for one phenomenon (e.g., origin of homoiothermy in mammals) need not hold for a similar phenomenon (e.g., avian homoiothermy).
5. Because of their complexity, the possible confusion between conflicting explanations and the difficulty in identifying valid confirming or falsifying tests, H-NEs must be stated clearly and unambiguously. Failure to do this may preclude meaningful tests or appraisal of rival H-NEs.

H-NEs in biology include the evolution, phylogeny and classification of organisms or the evolutionary history of their attributes — that is, anything related to the history of life.

Being theoretical scientific statements, H-NEs are available to tests by falsification, but such tests are often extremely difficult and inconclusive. Generally H-NEs are not tested by falsification (in spite of numerous statements in the literature) but usually by confirmation with the addition of more and more corroborating support. This procedure is closely akin, if not identical, to induction in the strict sense of that concept. Objections cannot be raised to inductive testing of H-NEs because they are theoretical statements about a finite number of objects, in contrast to N-DEs, which cover universals or an unlimited number of objects. Testing of H-NEs depends on argument chains involving pertinent N-DEs and a large number of background assumptions (hypotheses, many being initial and boundary conditions), and they are finally tested against objective empirical observations. One should proceed to the empirical observations as directly as possible, although the argument chain is often complex. The empirical observations and their roles as tests, whether falsifying or confirming, should be designated clearly.

What makes H-NEs scientific is point 3 (above), that "H-NEs must be based on pertinent N-DEs, and these N-D Es, together with the pertinent empirical observations, form part of the chain of arguments used in testing the H-NE." If this requirement is not followed for any particular H-NE, than that explanation is not scientific. This is a serious problem for cladists, who generally claim that cladistic analysis must not be based on (presumably nomological, although this is not definitely stated) evolutionary theory (that is, fields a, c, d, and e of evolutionary theory outlined above). As cladistic analysis is clearly an H-NE, then it must be based on a pertinent N-DE, and if this N-DE is not nomological-deductive evolutionary theory, then there is the question: on what N-DE is cladistic analysis based? The argument that scientific order exists in nature by itself and in the absence of any N-DE is simply not valid.

Within evolutionary biology, descent with modification (field b of Darwin's theories), Haeckelian phylogeny, Darwinian or evolutionary classification, Hennigian cladification, historical biogeography and all analysis of the evolutionary history of any organic attribute or group (e.g., the evolution of the vertebrate limb, the evolution of flight in birds, the evolution of mammals from synapsid reptiles, etc.) are H-NEs. In geology, the sequence of sedimentary rocks, the pattern of movement of tectonic plates and the pattern of movement of the magnetic poles are H-NEs. These H-NEs are all explanations of singulars and not of universals. Hence a successful H-NE of the origin of flight in birds is of a singular event, and this H-NE will not serve as the explanation of the evolution of flight in bats even though both H-NEs are based on the same N-DEs such as aerodynamics.

Methodology in H-NEs must be based on the pertinent underlying N-DE. I have stressed this point in earlier papers on methodology in evolutionary classification (Bock, 1977, 1981, 1992) in which I endeavored to show that these methods are based on nomological-deductive evolutionary theory. The method advocated by cladists has to be examined closely to show whether they are based on appropriate N-DEs. I have stated earlier (Bock, 1978) that the basic cladistic method of out-group comparisons, as well as many other methods used by systematists in classification, is directly circular. Perhaps the problem is that cladistics actually start their analysis by first establishing groups of organisms and then use these hypothesized groups as the foundation for further analyses of apomorphy and plesiomorphy. If this is the case, then, cladists are silent about the methods used for establishing and testing theoretical statements about groups of organisms, leaving out a significant segment of their methodology. Clear statements about all of the methods used in any approach to systematic analysis are a major requirement that must still be detailed for most approaches to systematics.

Degree of Confidence of Historical-Narrative Explanations

Although H-NEs are theoretical scientific statements, they are basically tested inductively. A particular H-NE can never be proven, but it can be tested extensively so that one can have great confidence in its credibility. Hence one accepts and uses diverse H-NEs with very different degrees of confidence that depend on many factors, not only the methods used to test the particular H-NE. It is simply wishful thinking on the part of cladists to state that their modern methods are robust and will provide conclusions with a high degree of reliability. H-NEs vary greatly in the degree of confidence that one can have in their results.

Only a few H-NEs possess an exceedingly high degree of confidence. The monophyly of all known parrots (Psittaciformes) has such a high degree of confidence that very extensive contrary evidence must be marshaled to undermine it. The same is true for the conclusion that humans evolved from the anthropoid apes, and in particular from the chimpanzee-gorilla complex. Other H-NEs have varied in their degree of confidence over time. In the 1800s and early 1900s most ornithologists accepted the monophyly of the palaeognathous (i.e., flightless birds and the still flying tinomous) birds with strong confidence. Subsequently, most ornithologists accepted the independent origin of the several groups of these birds with equal confidence. After 1960, the monophyly of the palaeognathous birds was again accepted with high confidence (Bock, 1963). At the present time, the conclusion that the flightless ratites (see Bock and Bühler, 1990) are polyphyletic within the palaeognaths is accepted or rejected with varying degrees of confidence. Among the New World nine-primaried Oscines, there had never been any doubt about the monophyly of the species of Neotropical flower-piercers (*Diglossa*). However, a study of the corneous tongue and the bony palate of the skull demonstrated with considerable confidence that there are two distinct and unrelated groups of these birds with the very distinctively hooked, flower-piercer bill (Bock 1985).

Yet consideration of a large range of H-NEs of the relationships of organisms shows that most have a low degree of confidence at best. Considering birds, doubt still exists on the monophyly of many accepted orders such as the Piciformes, the Coraciiformes, Pelecaniformes, Falconiformes (Do the Cathartidae belong here or with the Ciconiiformes?) Cuculiiformes, Columbiformes (Do the Pteroclididae belong here or with the Charadriiformes?) and Galliformes (Do the Opisthocomidae belong here?). Furthermore, have birds evolved from dromaeosaurian dinosaurs or from an earlier basal group within the Archosauria? Among mammals, are the Glires a monophyletic group? Are the aquatic carnivores, the Pinnipeda, a monophyletic group or do they consist of two independently derived aquatic groups? The number of such examples could be increased well beyond the limits of anyone's patience to list or to read them.

What is needed is the establishment of a set of reasonable guidelines for the estimation of approximate degrees of confidence in H-NEs, even a simple listing of bad, poor, medium, good and excellent for these explanations. These methods depend largely on the empirical tests of hypotheses about taxonomic properties of features (homologies and directions of evolutionary change) and the establishment of a degree of confidence of the veracity of the conclusion.

Conclusions

A few conclusions can be offered. Most important is that before any analysis of explanations of evolutionary theory can be undertaken, one has to clarify the diverse areas within the overarching term *evolutionary theory*. Not all aspects of evolutionary theory are historical. Classifications, phylogenies, dendrograms, cladifications and all other explanations of the evolutionary history of groups of organisms and of particular attributes are H-NEs, and these explanations must be presented and tested in a very different way than the standard N-DEs in science. As for all H-NEs, evolutionary H-NEs must be presented realistically with a sensible estimate of the degree of confidence that one can have in the particular explanation. Otherwise the overall reputation of the entire field of H-NEs in biology will decrease sharply, as occurred at the end of the nineteenth century when there was a major rejection by biologists of the speculative conclusions being offered by evolutionary biologists. Investigations of historical aspects of biology are difficult enough without adding unreasonable claims about the methods of study, the nature of the explanations and the robustness of the conclusions (see Bock, 1992).

References

Bock, W.J., The cranial evidence for ratite affinities, *Proc. XIII Int.. Ornithol. Congr.*, 1, 39–54, 1963.

Bock, W.J., Philosophical foundations of classical evolutionary classification, *Syst. Zool.*, 22, 375–392, 1974.

Bock, W.J.. Foundations and methods of evolutionary classification, in *Major Patterns in Vertebrate Evolution*, Hecht, M.K., Goody, P.C., and Hecht B.M., Eds., NATO Advanced Study Institute, Series A, Plenum, New York, 1977, pp. 851–895.

Bock, W.J., Comments on classification as historical narratives, *Syst. Zool.*, 27, 362–364, 1978.

Bock, W.J., Functional-adaptive analysis in evolutionary classification, *Am. Zool.*, 21, 5–20, 1981.

Bock, W.J., Is *Diglossa* (?Thraupinae) monophyletic? *Ornith. Monog. (A.O.U.)*, 36, 319–332, 1985.

Bock, W.J., The nature of explanations in morphology, *Am. Zool.*, 28, 205–215, 1988.

Bock, W.J., Explanations in Konstruktionsmorphologie and evolutionary morphology, in *Constructional Morphology and Evolution*, Schmidt, N. and Vogel, K., Eds., Springer-Verlag, Heidelberg, 1991, pp. 9–29.

Bock, W.J.. Methodology in avian macrosystematics, *Bull. Br. Ornithol. Club*, 112(Centenary Suppl.), 53–72, 1992.

Bock, W.J., Selection and fitness; definitions and uses: 1859 and now, *Proc. Zool. Soc. Calcutta*, Haldane Commemorative Volume, 7–26, 1993.

Bock, W.J., Explanations in a historical science, in *Organisms, Genes and Evolution*, Peters, D.S. and Weingarten, M., Eds., Franz Steiner, Stuttgart, 2000, pp. 33–42.

Bock, W.J. and Bühler, P., The evolutionary and biogeographical history of the palaeognathous birds, in *Current Topics in Avian Biology*, van den Elzen, R., Schuchmann K.-L., and Schmidt Koenig, K., Eds., Deutschen Ornithologen-Gesellschaft, Bonn, 1990, pp. 31–36.

Bock, W.J. and von Wahlert, G., Two evolutionary theories: a discussion, *Br. J. Philos. Sci.*, 14, 140–146, 1963.

Darwin, C., *On the Origin of Species by Means of Natural Selection, or, the Preservation of Favoured Races in the Struggle for Life*, fascimilie reprint, 1964. Harvard University Press, Cambridge, MA, 1859.

Haeckel, E., *Generelle Morphologie der Organismem*, 2 volumes, G. Reimer, Berlin, 1866.

Hempel, C.G., *Aspects of Scientific Explanations*, Free Press, New York, 1965.

Hempel, C.G. and Oppenheim, P., Studies in the logic of explanation, *Philos. Sci.*, 15, 135–175, 1948 (reprinted in Hempel, C. G., 1965)

Hull, D.H., The use and abuse of Sir Karl Popper, *Biol. Philos.*, 14, 481–504, 1999.

Mayr, E., *The Growth of Biological Thought*, Harvard University Press, Cambridge, MA, 1982.

Mayr, E., Darwin's five theories of evolution, in *The Darwinian Heritage*, Kohn, D., Ed., Princeton University Press, Princeton, NJ, 1985, pp. 755–762.

Mayr, E and Bock, W.J., Classifications and other ordering systems, *J. Zool. Syst. Evol. Res.*, 40, 169–194, 2002.

Nagel, E., *The Structure of Science: Problems in the Logic of Scientific Explanation*, Harcourt, Brace and World, New York, 1961.

Platnick, N.I. and Gaffney, E.S., Systematics: a Popperian perspective, *Syst. Zool.*, 26, 360–365, 1977.

Platnick, N.I. and Gaffney, E.S., Evolutionary biology: a Popperian perspective, *Syst. Zool.*, 27, 137–141, 1978a.

Platnick, N.I. and Gaffney, E.S., Systematics and the Popperian paradigm, *Syst. Zool.*, 27, 381—388, 1978b.

Popper, K., *The Poverty of Historicism*, Routledge & Kegan Paul, London, 1977. (paperback reprint of 3rd ed., 1961, Harper Torchbooks, New York)

Szalay, F.S. and Bock, W.J., Evolutionary theory and systematics: relationships between process and patterns, *Z. Zool. Syst. Evol.forsch.*, 29, 1–39, 1991.

Chapter 4

What Happens When the Language of Science Threatens to Break Down in Systematics: A Popperian Perspective

Olivier Rieppel

> To study history one must know in advance that one is attempting something fundamentally impossible, yet necessary and highly important. To study history means submitting to chaos and nevertheless retaining faith in order and meaning. It is a very serious task, young man, and possibly a tragic one. (H. Hesse)

Abstract

The central tenets of the falsificationist philosophy of Karl R. Popper are reviewed in detail, and the way they do or do not apply to systematics and phylogeny reconstruction is analyzed. Cladistic analysis, cast in either maximum parsimony or in maximum likelihood approaches, is not compatible with Popperian falsificationism. The main reasons are the absence of a deductive link between a hypothesis of phylogenetic relationships and character distribution on a tree, which translates into the absence of the basic asymmetry of falsification versus verification. This sets Popper's philosophy of science apart from inductive systems. In cladistic analysis, falsification (disconfirmation) is symmetrical to verification (confirmation), which reveals an inductive and hence probabilistic background. The basic problem of systematics as an empirical science resides in character conceptualization and its critical evaluation.

Introduction

In June 2001 *Systematic Biology* published four papers (DeQueiroz and Poe, 2001; Kluge, 2001a; Faith and Trueman, 2001; Farris et al., 2001), each appealing to Popper's philosophy of science in support of distinctly contradictory viewpoints. Evidently, something must have gone wrong, unless it can be shown that Popper's philosophy is ambiguous or contradictory. At the same time, a long list of stagnant debates in systematics (Rieppel and Kearney, 2001, 2002) raises the question of whether Popperian falsificationism is effective in resolving such conflicts.

   This paper is not about parsimony or likelihood analyses of phylogenetic relationships; it is also not about whether Popper's philosophy of science is good, brilliant, adequate, consistent, accurate, practical, universally applicable, or true to history — I am not qualified to make value judgments in the field of philosophy of science. Instead, this paper is about what Popper wrote and whether his philosophy applies to phylogenetic systematics. I believe that the debate about parsimony versus likelihood can take place. However, as I will try to show, it cannot take place in terms of Popperian falsificationism, but rather only with respect to the criterion of simplicity, or parsimony, of scientific hypotheses. I know that in so saying, I will argue against what many phylogenetic systematists believe today, but if we want to be Popperian, we should not shy away from critical discussion. In this case I also argue against some of what I have myself said in the past. As Popper said of such circumstances, I'll have to eat some humble pie.

Popper referred to himself as a "common sense philosopher" (thus setting himself apart from Wittgenstein's scepticism) and argued that common sense should form the starting point for philosophy (e.g., Popper, 1973:Ch. 2). Consequently, Popper insisted on a simple language, and I will try to do the same. In the interest of clarity, I should also point out that I am using the words *test* and *falsification* exclusively in the way Popper used these terms, i.e., in the *modus tollens* form of argument (Popper, 1976:§18). Essentially, a test is based on the logical deduction of a singular (forbidden) statement from a universal proposition; falsification occurs if an instantiation of that forbidden statement is accepted to have occurred. The statistical rejection of a null hypothesis does not constitute a falsification in a Popperian sense, because it merely corresponds to the statistical support for another hypothesis that deviates from the null hypothesis (see also footnote 1). Likewise, a greater degree of likelihood does not test a hypothesis, but supports one hypothesis better than alternative hypotheses. In general, although statistical support for a theory can be equated with the complement of its degree of falsification, or, more accurately, degree of disconfirmation, this symmetry does not do justice to Popper's falsificationism, as will be discussed in detail below.

Popper not only insisted on simple language; he also insisted on rationality as the basis for argumentation. To him "the 'rationalist tradition' ... is the tradition of critical discussion — not for its own sake, but in the interest of the search for truth" (Popper, 1989:101), truth in this context meaning a regulative principle only (see also Appendix II). Popper equated "rationality with the critical attitude ..." (Popper, 1989:248). Given Popper's broad impact on natural sciences and scientists, most systematists probably strive to adopt a critical attitude, a culture of critical discussion, which in a sense represents a weak form of Popper's falsificationism. In that broad sense, modern systematics is — hopefully — Popperian.

The Language of Science

Popper defined c*ognition* (*Erkenntnis*) and *experience* (*Erfahrung*) as follows: "*Cognition* is: the search for natural laws, or, more precisely: the conjecture of natural laws and their methodical testing (irrespective of the question whether such natural laws exist)" (Popper, 1979:78). This conception of cognition echoes that of Schlick who had characterized its significance as relating to the realm of *formal relations*, which for Popper is deductive reasoning (Tscha Hung, 1992).

"*Experience* is: a well-defined method of deciding the status of sentences or systems of sentences" (Popper, 1979:425). Here, Popper transcends Schlick's notion of *Erleben*, which Schlick considered a private (subjective) experience of the outer world rendered by affirmations (Tscha Hung, 1992).

From these definitions it is evident that method was very important to Popper, as well as language, because all scientific discoveries are communicated in language and all natural laws are formulated as statements, or propositions. However, statements, or propositions, and therefore language, have a certain logic to them, and if scientific knowledge can be communicated through language, this knowledge must be structured, too (Popper 1979:64). Popper's method, furthermore, is deductive, not descriptive, because only deductive systems, not descriptions, offer logically necessary entailments (Foucault, 1971). Popper's interest, therefore, is centered on the logic of scientific discovery, that is, on the logical analysis of the language through which scientists communicate their discoveries. It was well known to Popper, and he stressed it repeatedly in his first book (Popper 1979; written in the years 1930 through 1933), of which *Logic of Scientific Discovery* (Popper, 1976) is a summary, that the ultimate, finite and conclusive falsification of a theory cannot be achieved in practice (just as the ultimate verification of a theory likewise remains impossible; Ayer, 1952). However, the practice of empirical science was not what he was primarily interested in. He was mainly interested in the logical problem of induction, and its logical solution, as well as in the demarcation of science from metaphysics.

Those two problems are, indeed, interdependent. Popper considered the issue of how science is done a matter of interpretation (Popper, 1976:§80); the issue of how science should be done he deemed a matter of sociology (Popper, 1974a:1036). Yet Popper stressed repeatedly that a philosophy of science must take into account the way science is actually practiced (Popper 1979:134). Beyond the mere cultivation of a critical attitude, the question thus becomes whether the logic of systematics matches in any way the logic of scientific discovery as analyzed and formalized by Popper (1976).

Since Popper's interest lay in the logical analysis of scientific language, a precise understanding of the meaning of his own language is of great importance. However, Popper's first language was German, and *Logic of Scientific Discovery* was initially written in German, then translated into English (and other languages; for this paper, I used the German edition, and sometimes present my own translation). This may have created problems (see Kuhn, 1974a:268 for a discussion of the difficulties inherent in translations). For example, in the *Logic of Scientific Discovery*, Popper (1976:§14) uses the term *Individualien* for particulars, as opposed to *Universalien*, that is, universals. *Individualien* means (logical) particulars, or terms that refer to an individual (i.e., particular) concept (*Individualbegriff*), which in German is readily distinguished from *Individuum*, an individual (singular) person, for example. English does not allow us to make this distinction that easily, and *Individualien* was consequently translated as *individuals* in the English edition, with possibly important consequences, as will be discussed later.

Therefore, since language is so important, contextual care must be taken to correctly understand the notions that figure in Popper's language. For example, much has been written in the recent systematics literature on a formalism that Popper used to characterize the degree of corroboration (C) of a scientific hypothesis, which increases with severity of test. Thus,

$$C(h, e, b) = p(e, hb) - p(e, b)/p(e, hb) - p(eh, b) + p(e, b)$$

With reference to such formalisms, it has been stated that background knowledge (*b*) must be kept to a minimum, which is purportedly guaranteed by parsimony analysis but not by likelihood approaches, because (poorly corroborated) model assumptions increase background knowledge beyond the necessary and sufficient. Alternatively, severity of test, which depends on the evidence (*e*) relative to the empirical content of a theory (*h*), and hence the degree of corroboration, or explanatory power, increases with the number of taxa or characters included in the analysis (Kluge, 2001a); or the degree of corroboration is based on the general principle of likelihood (DeQueiroz and Poe, 2001:306). The above formula denotes logical relations, and many of its components, such as logical probability (*p*) have an intuitive appeal the way Popper used them, as he readily admitted. Formalisms such as the one given for degree of corroboration do represent measure functions, it is true, but in their "actual application we depend entirely on those rare cases which are comparable on non-metrical and, as it were, qualitative or *general logical* grounds, such as cases of *logically* stronger or weaker competing theories ..." (Popper, 1973:59).

These formalisms therefore are supposed to work in principle as a matter of logic that governs the relations between scientific statements or complexes thereof, not as a matter of probability calculus (Popper, 1973:51; see also Appendix II). In his later writings, Popper dealt extensively with probability arguments in order to deal with the problems that the developing insights in quantum physics caused for his solution to the problem of induction; but his efforts in this direction always concerned "physical effects" in the form of "reproducible regularities" (Popper, 1976:§68), not historical relationships.[1] In a similar sense, the axiom of likelihood applies to statistical populations of data only, not to historical processes (Edwards, 1992).

Therefore, in order to understand such notions, it is important to investigate the context within which Popper used them. A simple definition of terms such as *degree of corroboration* or *logical probability* will not do. Popper never considered definitions very important, for several reasons: Not only are they arbitrary and *a priori*, but the use of undefined words is in fact unavoidable because, in the last resort, all definitions will have to include undefined terms (Popper, 1973:58, 328). From this it follows that the "(essentialist) doctrine that we can 'define' (or 'explicate') a word or term or concept, that we can make its meaning 'definite' or 'precise', is in every way analogous to the mistaken doctrine that we can prove, or establish or justify the truth of a theory; in fact, it is part of the latter ('justificationist') doctrine" (Popper, 1989:402). Instead, the best definition of terms often lies in the fact, or intent, of how these are used (its use determines the meaning of a word). According to Popper it is not the *a priori* definition of a term that should determine its use; rather, it is the intended use of a term that should determine its definition (*Gebrauchsdefinitionen* according to Popper, 1979:366–367). This, however, requires context (the "context concept" follows from Frege's imperative "never to ask for the meaning of a word in isolation, but only in the context of a proposition"; Weiner, 1999:51).

[1] Popper's dealings with probability calculus must be seen in the context of his falsificationism, which involves many possible interpretations of probability calculus, amongst which Popper distinguished subjective interpretations (analog to truth as subjective conviction or belief) and objective interpretations (analog to the logical notion of truth as "correspondence to facts" [Popper, 1989:404]). This is reflected in the notion of probability: (a) "as a theory of the logical probability of a statement relative to any given evidence ..." and (b) "as a theory of the probability of an event relative to any given *ensemble* (or 'collective') of events" (Popper, 1989:59). In the attempt to cover both interpretations under a formal theory of probability, he obtained the *propensity* interpretation of probability, where propensity is the probabilistic disposition of an object to behave in a certain way. Popper acknowledges that although the tests of quantum theory are necessarily statistical, and although that theory may have statistical consequences, "it need not have a statistical *meaning*" (Popper, 1989:60, italics added). Objective propensities or field propensities would be measurable by statistical methods, without themselves being statistical (see Popper, 1989:119). Popper's (1989:60) dealing with probability confirms Lakatos' (1974:92) assessment that Popper objected to the "probabilification" of science.

The subjectivist interpretation of probability calculus was understood by Popper, (1976:§48, and Appendix IX, Third Comment [1958]) as a theory that interprets probability as a measure of lack of knowledge or of partial knowledge.

The objective interpretation of probability calculus (Popper, 1976:§48, and Appendix IX, Third Comment [1958]) is necessary because "no result of statistical sampling is ever inconsistent with a statistical theory unless we *make them* with the help of ... rejection rules" (Lakatos, 1974:179; see also Nagel, 1971:366). It is under these rejection rules that probability calculus and logical probability approach each other; these are also the conditions under which Popper explored the relationship of Fisher's likelihood function to his degree of corroboration, and the conditions arise only if the random sample is large and (*e*) is a statistical report asserting a good fit (Farris et al., 2001). In addition to the above, in order to maintain an objective interpretation of probability calculus, Popper also required that once the specified conditions are obtained, we must proceed to submit (*e*) itself to a critical test, that is, try to find observable states of affairs that falsify (*e*).

The Problem of Induction

Popper took up the problem of induction where Hume had left it: "Are we justified in reasoning from [repeated] instances of which we have experience to other instances [conclusions] of which we have no experience" (Popper, 1973:8), or, more simply put, what allows us to expect that the future will resemble the past? Or yet simpler, is it possible to know more than we know?

Popper identified a dual nature of the problem of induction, recognizing psychological and logical components. And he recognized that in order to solve the logical problem of induction, it had to be reformulated. According to Hume, the principle of uniformity of nature allows the past to inform us about the future on the basis of a habitually conditioned trust in causal relations that seem to underlie the constant conjunction of events (Kulenkanpff, 1981). However, the principle of uniformity of nature merely restates the "assumption that past experience is a reliable guide to the future" (Ayer, 1952:49). Thus if the problem is the conclusion of future similar instances from the observation of past repeated instances, then from a logical point of view, those similar instances must be relative to some universal, or lawful, regularity (Popper, 1973:9). Hence Popper's reformulation of the logical problem of induction in terms of universal laws.

The psychological problem of induction results from the fact that we seek regularity in nature and may even impose regularity on nature where there is none. This claim of Popper's (1973:23–24, 1974a) comes close to Kant's notion of *a priori*, with one important difference, however. Kant claimed that we impose, *a priori*, lawfulness on nature, and that there is nothing we can do about it. In contrast, Popper claims that we try to impose, *a priori*, regularity on nature, and that there is something we can do about it: adopt a critical attitude and try to find out whether our expectation of regularity is wrong in any given case. This is Popper's (1973, 1974a) process of learning through trial and error: We try out our theories, our hypotheses, our expectations of regularity by applying them to the world of experience and we learn as we find out that our expectation was wrong in any given case.

The regularity systematists expect, and the one they recover, is that of a (logically subordinated) hierarchy of groups within groups. Do we impose this order on nature *a priori*, or is this hierarchy an empirical concept? This is an important question, for as Kuhn (1974b:17) has argued, "the entire population of the world can always be divided (though not once and for all) into perceptually discontinuous categories"; this is, after all, "how children first learn about dogs and cats, tables and chairs ..." (see also Piaget, 1972). In order to answer this question it would be necessary to specify the conditions, at least theoretically or logically, in which our expectation of a logically subordinated order in nature could, in principle, be wrong. It would require that we could, in principle, fail in our attempt to apply the concept of a logically subordinated hierarchy to the world of experience. Evidently, failure is never obtained.

The reason that our expectation of a logically subordinated hierarchy cannot fail in the application of parsimony analysis lies in the fact that it is not subject to the fundamental asymmetry of falsification versus verification relative to universal statements that Popper discovered in his solution to the logical problem of induction. This conclusion evidently requires some explanation.

As stated above, the constant conjunction of events suggests causal relations between particulars (i.e., particular events) that can be construed as law-like correlations between their types (Foster, 1992:190). In Sober's (1988) terms, Peirce's *type* versus *token* distinction is inherent in Hume's problem of induction (De Waal, 2001). Popper solved the logical problem of induction by reformulating it in terms of universal laws: "Can a *preference*, with respect to truth or falsity, for some competing universal theories over others ever be justified by empirical reasons?" (Popper, 1973:8). The fundamental asymmetry, which constitutes Popper's solution of the logical problem of induction, lies in the fact that — from a logical point of view — universal laws, theories, or more generally, universal propositions can never be verified, but they can be falsified. Because the possible instantiations of a universal proposition are infinite, it can never be verified. However, a singular statement that logically contradicts a universal proposition, and that is accepted to correctly report on an observable state of affairs, suffices to conclusively falsify the latter. In more general terms, whereas verification is in principle impossible, falsification is in principle possible, but not necessarily achievable in practice (Popper 1979:390). From this insight, Popper developed his criterion of demarcation of science from metaphysics, which is the testability, and potential refutability, of scientific theories.

Lorenz (1973:16) exemplified the asymmetry that characterizes Popper's falsificationism in an intuitively appealing way. If a paramecium encounters an obstacle in its path, it first retreats, then moves in a randomly chosen different direction, thus showing that it has learned something about its world. "For *objiecere* means 'throw against': it is the object that is thrown against our moving forward, it is the impenetrable that we run up against" (Lorenz, 1973:16). From our superior vantage point (playing the Laplacean demon: compare Popper, 1988), we could suggest a different, more successful path to the paramecium than the one it has randomly chosen, but what the paramecium has learned about its world is absolutely correct: There is no way to continue straight forward.

Popper's (1979) initial interest, therefore, focused on the very general problem of when it would be possible to declare a statement to be true or false, or, more accurately, verifiable or falsifiable. Representatives of the Vienna Circle had argued that if somebody makes a statement, then that person must know under which circumstances that statement is true or false, because if she does not, she is talking nonsense (Ayer, 1952). Wittgenstein (1922:§7) concluded, "Whereof one cannot speak, thereof one must be silent." Accordingly, "*The Riddle* does not exist" (rejected by Popper, 1979:100). Popper found that the status of a statement is completely determined if it is, in principle (logically, not in practice), both verifiable and falsifiable. "In principle" means that there is no logical reason preventing verification or falsification. Some statements may only be falsifiable, such as universal statements, while others are only verifiable, such as negations of universal statements. Because if it is accepted that a universal proposition is only falsifiable, it follows logically that its negation is only verifiable. For example, if the universal statement "All ravens are black" can only be falsified, the negation of that statement, namely "It is not the case that all ravens are black," can only be verified, since falsification of that statement would verify the universal statement "All ravens are black" (Popper, 1979:307).

Because the potential instantiations of a strictly universal statement are infinite, the statement is never verifiable. (Some will restrict that conclusion to intensionally defined universals [e.g., defined on the basis of some descriptive criteria], but Popper showed [1976:§14] that universals can also be "pointed at" [i.e., ostensively defined] if a particular [individual] is pointed at in a statement that adds "and other similar things" or "and so on.") However, if a singular statement reporting on an observable state of affairs that logically contradicts a universal statement is accepted, then that universal statement must, on grounds of logic, be considered falsified — irrespective of whether this universal law would also be rejected in practice. Popper was always clear on this point, as he again stressed in his reply to Lakatos' criticism of his philosophy, that a distinction must be made between the logical falsification of a theory and its rejection in practice (Popper, 1974a:1009). For example, if we observed an object such as a stone apparently defying the law of gravity, we would probably first question the validity of our observation, rather than immediately rejecting the law of gravity.

The *Logic of Scientific Discovery* is therefore embedded in a method according to which we pretend that a natural law exists (without ever being able to know for sure whether this law actually exists), and then try to falsify this universal proposition by its application to the world of experience[2]. Popper had to posit theories (i.e., universal propositions) *a priori*, since Hume had shown that no universal causal law could be deduced from any finite number of experiential propositions (Ayer, 1952:54). Popper fully realized that he was treading on treacherous ground here, for he conceded to coming very close to Kant's conundrum (Popper, 1979:315). How could Kant dismiss the thing-in-itself as an object of rational cognition without some implicit knowledge of what this thing-in-itself is?

Popper, by contrast, was not positing things (Popper 1988:116); he was positing theories, hypotheses and ideas. True, he wanted these theories to be applicable to things, but that says nothing about the things-in-themselves; it only says something about how well our theories perform in our own world of experience. By claiming that something can be discovered, Popper naturally had to posit that contents of our world of experience exist ("... if you kick a rock hard enough, you will feel it can kick back"; Popper, 1988:116), but as Kuhn (1974b:2) put it, "... the aim is to invent theories that *explain* observed phenomena and to do so in terms of *real* objects, whatever the latter phrase may mean." Popper readily conceded to "essentialism that much is hidden from us, and that much of what is hidden may be discovered," but he strongly objected to the idea that "*science aims at ultimate explanations*" (Popper, 1989:05). Conversely, a successful scientific explanation reduces an observable state of affairs (the known) to a theory (the unknown) (Popper, 1989:83).

Popper has often been criticized for not being specific enough as to the origin of our primitive hypotheses, theories or expectations of regularity. Sober (1988), for example, argued that all knowledge must ultimately originate from observations, but he conceded that, in order to proceed from an observation to a hypothesis, some background theory is indispensable. This argument threatens to lead to an infinite regress, because it begs the question of what comes first: the observation or the background theory — unless the background theory is allowed to correspond to the primitive expectations or beliefs that Popper alludes to in his discussion of the psychological problem of induction. For, as Wittgenstein (quoted in Pears, 1992:398) emphasized, even simple ostensions (pointing gestures or the use of a demonstrative such as "this") remain empty in the absence of any background (context) or repetition.

Popper, who knew Konrad Lorenz as a boy in Altenberg (Popper, 1974b:34), turned to an evolutionary epistemology in order to explain the origin of theories. However, before Popper (1974b:34) focused his attention on the "marvellous theory" of imprinting developed by his childhood acquaintance (Lorenz, 1973: 37; "information translates literally into imprinting"), he was encouraged to pursue evolutionary epistemology by Schrödinger's remarks on "feigned Lamarckism" (Popper, 1974b:133, n278). Reading the work of the evolutionary theorist Julian Huxley, Schrödinger (1969) had recognized the importance of behavioral changes in triggering somatic changes (an aspect also emphasized by Lorenz [1973]) and found this to be an alternative to the "gloomy and discouraging view" propagated by the popular expositions of Darwin's theory that he had encountered. Evolutionary epistemology is what lies at the heart of Popper's slogan, "All life is problem solving."

Universal Propositions and Singular Statements

To apply a universal proposition to the world of experience means to use it as a basis for the deduction of a singular statement or test statement, which in principle is both verifiable

---

[2] Popper did not mean to imply that universality is a characteristic of the real world out there. Instead, universality is a characteristic of our theories, of our theoretical language. Universality is something that theories assert, and that is tested when theories are used as a net to capture the world of experience. There is no ontology implied here, and the fact that our theories only approximate the real world out there is an important part of Popper's argument against determinism (Popper, 1988:45).

and falsifiable. "In principle verifiable" means that there are no logical reasons that would prevent verification (Popper, 1979:301). This may sound easy and straightforward, but it is not; nor is it attainable. From the universal statement "All ravens are black" we cannot logically deduce the statement "There is a black raven here now," but we can deduce from it the forbidden singular statement: "There is no white raven here now" (Popper, 1989:385). The reason is that the existence of black ravens, that is, the instantiation of the universal proposition "All ravens are black," does not follow logically from that universal proposition but must be established by an observation that is logically independent from the universal proposition (we cannot deduce from the proposition "all unicorns have a horn on their forehead" the statement "there is a unicorn here now").

By contrast, the deduction of a forbidden singular statement from the universal proposition is possible because it contradicts the universal proposition in logical space (Popper, 1976:§15).[3] Therefore, if its negation occurs (or rather, if the fact of its occurrence is accepted), it logically falsifies the universal statement. In order to make this logic work, though, that is, in order to solve the problem of induction and thus to separate science from metaphysics, Popper had to show that an unequivocal distinction can be drawn between a universal proposition or statement (e.g., a natural law) and a singular statement (e.g., a specific instantiation of that law). This distinction is not easy to draw, for all singular statements can also be formulated as general implications, that is, as statements that cover the class of all of their occurrences, or instantiations (Popper, 1979:230). However, the distinction required was also known not to be possible for a class and its elements: A class can always become an element of a class of higher order, just as an element of a class can itself become a class of lower order. Moreover, the falsification of a universal statement requires that the universal statement can be deductively related to singular statements, or to particulars, particular instances, or to individuals in a logical sense.

This is not trivial, because no description of any particular, no matter how precisely it is made, will ever fully and unequivocally identify that particular if it is rendered exclusively in (intensionally defined) universal terms. The reason for this is simply that the class of particulars referred to by any (intensionally defined) universal term is, logically, infinite. For example, I can describe my dog as precisely as possible, using universal terms such as dog, 1 year old, blue eyes, brown fur and so on. Practically, this description can be made so precise that it will not fit any other dog than mine; logically, this description does not unequivocally identify my dog,[4] because there is a potentially infinite class of dogs that are 1 year old, have blue eyes and brown fur, etc. However, if I say "my dog Lux," this individual is unequivocally identified, namely by its proper name (Popper, 1979:235).

Popper defined universal statements as those that do not contain a proper name, and cannot be reduced to a proper name. Conversely, singular statements do contain a proper name or they are reducible to a proper name, that is, to a particular or individual that is denoted by a proper name (the notion of a proper name itself must remain undefined [Popper, 1979:234]).

Ultimately, therefore, all singular statements must be reducible to the birth of Christ (the anchor in time) and to Greenwich, or to the individual pope who declared Greenwich as the anchor of the prime meridian (the anchor in space) (Popper 1976:§14). In contrast to a universal statement, the singular statement is thus revealed to relate to a spatio-temporally restricted state of affairs, but not just in a general sense. In order to be unequiv-

---

[3] Given (x) (Rx→Bx) (all ravens are black), follows the contrapositive (x) (~Bx→~Rx) (all nonblack things are not ravens) follows. From this we obtain by DeMorgan's Law ~(Ex) (~Bx→~Rx) (there is nothing that is not black and a raven), which encompasses the statement "no white thing is a raven." Hence the possibility to paraphrase all universal claims in the form of a negative existential statement (forbidden statement *sensu* Popper 1976) (I owe this insight to Michael Rieppel.).

[4] Hence Popper's concern that a language without universal terms could not work (Popper, 1989:119).

ocal, the state of affairs must be specifically restricted in space and time: *Hic et nunc*, here and now, I make this particular experiment that tests a universal proposition. This accounts for the fact that we always attach any empirical statement to an empirical base only at a specific point in time and space (Körner, 1970). The specific restriction in time and space also reveals that repeatability of instances is a cornerstone in Popper's philosophy, necessary to establish intersubjective objectivity in natural science (Popper 1979 :66–., 1976:§22).

In Popper's (1976:§14, 1974a:988–987, 1979: 230ff) logic, therefore, the statement "It rained in Vienna on March 12th, 1933" is a singular statement. It is tied to a particular point in time and space. Also in Popper's logic, Vienna is an individual concept, denoted by a proper name. The battle of Waterloo is as much an individual concept, as my dog Lux or the beautiful German shepherd that carried the dog tag #17 984 in Vienna in the year 1930. This latter statement reveals how Popper wanted his use of proper names to be understood. They serve as a tag attached to something and provide the sense of what they refer to (the dog tag is attached to a beautiful German shepherd).[5]

The battle of Waterloo is also a member in a class, namely the class of all battles fought with firearms, which is a universal concept. In the same sense, my dog Lux, an individual concept, is an element in the classes of dogs, or mammals, which can be seen as universal concepts. The concept of all dogs that live in Vienna today evidently is a class concept, but in this case a class concept that is tied to the proper name Vienna. It is, therefore, an individual class concept (Popper, 1979:239, 1976:§14; i.e., a concept of a particular, instead of a universal, class).

Finally, individuals such as my dog Lux can be a member of an individual (particular) class such as the dogs that live in Vienna today, as well as of a potentially universal class such as mammals. Such intersections show that universal statements can logically be applied to singular statements, for they show that individual concepts may "stand to universal concepts not only in a relation corresponding to that of an element to a class, but also in a relation corresponding to that of a sub-class to a class" (Popper, 1976:§14). Thus, my dog Lux is not only an element in the class of dogs that live in Vienna today, which is an individual concept or an individual class, but it is also an element of the class of mammals, which can be seen as a universal concept or a universal class (Popper, 1976:§14), while the class of dogs that live in Vienna today is a subclass of the class dogs, or mammals.

Individuals

There are sciences whose main interest is the description of particulars or particular instances, rendered in singular statements that either contain, or can be reduced to, a proper name. Popper identified geography and history as such individualizing sciences (Popper, 1979:237; the neo-Kantian Heidelberg School characterized such sciences as ideographic, as opposed to the nomothetic or generalizing sciences). Popper's notion of individuals emphasized that the (logical) problem of induction does not apply to those, nor does his solution of the (logical) problem of induction.

Popper's notion of individuals is firmly rooted in logic: spatio-temporally restricted particulars as opposed to universals. Some authors used a similar notion of individuals in the sense of natural individuals (Ghiselin, 1974). The battle of Waterloo or Vienna thus became natural individuals, as did, indeed, biological species or taxa in general (see Rieppel, 1986 for a review and references; the notion of natural individual is discussed in Boyd, 1999). In this context it is interesting to follow Popper (1976:§14) in his exploration of

---

[5] Class names, by contrast, can be said to have both an intension (such as class membership criteria) and an extension (all the things to which the name refers). Mammals as an individual or universal class name would have intension such as fur and single lower jaw bone and extension such as "this dog," "that squirrel," etc.

the fact that everyday language does not unequivocally identify words such as *dogs* or *mammals* either as individual class names or as universal class names. Intending the word *dog* to refer exclusively to organisms that live on planet Earth, would be designating an ultimately spatio-temporally restricted individual class; hence it would be an individual class name (as it would be in evolutionary studies). If, by contrast, the word *dog* referred to physical bodies that have certain spatio-temporally unrestricted properties, which can be described in universal terms, such as hair, this would designate a spatio-temporally unrestricted class and hence figure as a universal class name (as it would be in classifications). It should be noted that Popper's use of the word *individual* refers to individual concepts or individual class names, as Popper tries to establish the unequivocal distinction between singular and universal statements, a distinction that is prerequisite for his solution to the logical problem of induction.

By contrast, species (or taxa in general) have been considered individuals on account of their spatio-temporal restrictedness. Individual in that sense can mean a logical or an ontological status or both. The question relevant for empirical sciences is whether ontology informs scientific theory in normative ways (Popper, 1979:72–73)[6] or whether ontology flows from scientific theory.

No science can ultimately escape from ontological claims if scientific theories are ultimately meant to be statements about what there is in the world (Gibson, 1982; that also means what there is in the world according to one theory is there for only as long as that theory prevails over its competitors). So if the world is to be carved up into two categories, namely, universals and particulars, then according to the theory of evolution, species are spatio-temporally restricted entities; thus they ought to be considered individuals both in a logical as well as in a natural sense (Hull, 1976, 1999a; Rosenberg, 1985). However, the thesis of species *qua* individuals does not solve the problem it was designed for (Rieppel, 1986): the problem of change, metaphorically expressed as Heraklitus' paradox, "you cannot enter the same river twice" (see also Popper, 1989:145–146 on the problem of change). Furthermore, although admitting fuzzy boundaries, the species *qua* individual thesis is very much built on discontinuity between species, as emphasized in the work of Theodosius Dobzhansky and Ernst Mayr, which portrays a species concept that is rapidly eroding today (e.g., Mallet, 2001:887; Grant and Grant, 2002).

---

6 Lakatos (1974:126) finds no problem in having metaphysical components included in a research program, nor does Popper think that metaphysics can ever be completely excluded from science (Popper, 1974a:971), but there are important distinctions here to be made. First, Lakatos' emphasis is on syntactically metaphysical theories, not on ontology, and he allows those as long as they can cause a progressive problem shift, that is, generate predictions of new facts (in Popperian terms, predictions that transcend current background knowledge). Second, since Lakatos' (1974:132, 137, 155) approach to the philosophy of science is eminently historical, this does not constitute any contrast to Popper's philosophy of science, for Popper always maintained that historically, science grew out of metaphysical ideas and concepts (e.g., Popper, 1976:§85; see also Körner [1959] on the role of metaphysical directives during early stages of the development of science). Third, Popper described himself as a metaphysical realist, since positing natural laws while admitting that their existence could not be known is a metaphysical claim. Fourth, Popper agreed "that non-testable (i.e., irrefutable) metaphysical theories may be rationally arguable ..." (Popper, 1973:40, note 9; see also Popper, 1989:Ch. 8). Finally, Popper believed "that whenever it is possible to find a metaphysical element in science which *can* be eliminated, the elimination will be all to the good. For the elimination of a non-testable element from science removes a means of avoiding refutations; and this will tend to increase the testability, or refutability, of the remaining theory" (Popper, 1983:179–180). Nowhere in relation to his logic of scientific discovery does Popper invoke ontology: "... I wish to make clear ... that it is not my intention to raise any 'what is' questions ... That is I am not offering what is sometimes called an 'ontology'" (Popper, 1977:4–5; see also Popper, 1983:136). Although Popper (1976:Appendix X) did admit that in applying a universal law to our world of experience we do expect this law to correspond to structural properties of this world, we will never know whether this law is true and hence corresponds to reality.

However, let us accept the species *qua* individual thesis in both its logical and natural sense for the sake of the argument and consider species names as proper names. It is evident that treating species, or taxa, as individuals renders systematics, indeed most of organismic biology, an individualizing science. This is so because all statements in systematics would be singular statements, statements that can ultimately be reduced to an individual species denoted by a proper name. The consequence would be that the logical problem of induction would not apply to systematics, nor would Popper's solution to it. So why argue Popperian falsificationism in systematics?

Testability in Systematics

Statements of systematics are singular statements (or conjunctions thereof; Kitts, 1977), not just because of some authors' belief that species are natural individuals, but also because of the historical nature of systematic statements. Statements in systematics are, after all, statements that refer to organisms that live on planet Earth. Whereas it is true, in logic, that the problem of induction as reformulated by Popper does not apply to systematics if all statements in systematics are singular statements, it does not mean that there could not be an element of testability in systematics — as is, indeed, argued by contemporary systematists. In fact, Platnick and Gaffney (1978:140) believed they had identified universal statements in systematics, as they concluded somewhat paradoxically, "'All spiders in the New World have eight legs' is clearly not a spatiotemporally unrestricted law of nature, but it is universal, does forbid the existence of New World spiders with other than eight legs, and is therefore a testable and scientific theory." It is true that the statement "All spiders in the New World have eight legs" is testable, because it forbids the occurrence of spiders with legs of a number other than eight in the New World. Yet it is not a universal statement but rather a singular statement because it includes a proper name, New World. It is testable because it can be connected to the universal statement "All spiders have eight legs." Indeed, the observation of a spider with seven legs in the South of France does not falsify the statement "All spiders in the New World have eight legs," but it does falsify the universal statement "All spiders have eight legs" (if it is accepted that the animal in question is a spider and has seven legs).

This example shows that singular statements can stand to universal statements in the relation of a hypothesis of a lower level of universality to a hypothesis of a higher level of universality. With respect to this example, it must be kept in mind that in order to be testable, and potentially falsifiable, in a Popperian sense, the possession of eight legs must be a universal property of spiders (Popper, 1976:§14), eight legs must therefore be used as a universal term, and spiders as a universal class name (the theory, or universal proposition, must be that nothing can be a spider and not have eight legs).

These problems indicate that in order to further investigate the nature of testability in systematics, we have to expand our discussion of the basic problems of empiricism to the problem of basic statements.

Basic Statements

In his *Logic of Scientific Discovery*, Popper (1976:§28) refers to singular statements that report on observations as basic statements. Logically, the falsification of a universal theory needs the verification of a falsifying basic statement (i.e., the positive instantiation of the negative existential statement that is logically deduced from a universal proposition), which is possible in principle but not in practice. For how would we know that it is the theory that is wrong, rather than the basic statement that appears to falsify it? The basic statement "Tom has brown hair" appears to be verifiable in principle, but only if we agree intersubjectively about what brown hair is; someone who refers to Tom's hair as dark-blond will not accept our verification. The verification that there is, indeed, a white raven here and now will not hold if someone rejects the bird we are looking at as being a raven, precisely because it is white, or that the bird is not really white, but a shade of gray that can still be subsumed under black. The logical falsification of a theory therefore needs to be distinguished from the practical rejection of a theory. Popper maintained that "A refutation can [only] be logically conclusive in the sense that, if we accept [rather than verify!] a refuting instance, we are logically bound to reject the hypothesis" (Popper, 1974a:1034-1-35).

However, we seldom deal with one theory in practice; we usually deal with multiple alternative and competing theories. Indeed, Popper (1976:§37) characterized the universe of possible theories, or hypotheses, that explain the world of experience as infinite — hence the importance of falsificationism and the irrelevance of verificationism: "the world of each of our theories may be explained, in its turn, by further worlds which are described by further theories — theories at a higher level of abstraction, of universality, and of testability" (Popper, 1989:115). Lakatos in particular has capitalized on the fact that Popper portrayed his falsificationism as a naïve "duel between theory and observation, without another, better theory *necessarily being involved*" (Lakatos, 1974:181). This critique is not much more than propaganda (Popper, 1974a :999ff). In his first book, Popper (1979:391–392) had already drawn attention to the intricacies that the layered world of theories poses for falsificationism, but he insisted that in order to get anywhere in practice, empirical science needs to partition its problems.[7]

Yet, even if it partitions its problems, the falsification of a prediction occurs not only relative to the particular theory from which the specific prediction was derived, but also relative to the entire system of theories within which the original theory is embedded (background knowledge, initial conditions, etc.). In other words, if a prediction that is derived from a theory that itself is part of a system of theories is found to be incorrect, then we cannot say with precision and unequivocally that this falsifying instance reflects back only on the specific theory from which the prediction was derived. Instead, we have to admit that this falsifying instance reflects back on the conjunction of all the theories that are part of the relevant context (Popper, 1979:391; Popper [1979:390, note 1] credits Duhem for the priority of this insight). This is also the reason Popper recommends a healthy but limited dose of dogmatism in the defense of an up-until-now well-corroborated theory that is confronted by a first falsifying instance.[8]

So in practice, there can be no talk of a precise, unequivocal falsification of any theory. However, says Popper (1979:391), it is not in his interest to deal with the problem of science in such a practical, naturalistic perspective. Popper considered his own theory of falsifiability, and of scientific method in general, as methodological or philosophical, not as empirical and hence not as falsifiable in itself (analogous to the verificationists' conundrum that the principle of verification is not itself verifiable); to consider the scientific method as empirical is to adopt a naturalistic perspective (Popper, 1974a:1010). Although it is possible to study science from a purely descriptive perspective, recording all the facts, difficulties and failures,

---

7   Even Lakatos (1974:131) had to concede: "... we cannot articulate and include *all* 'background knowledge' (or 'background ignorance'?) into our critical deductive model. This process is bound to be piecemeal and some *conventional* line must be drawn at any given time" (emphasis added for *conventional*). Indeed, without partitioning problem situations there could be no *competing* research programs as required by sophisticated methodological falsificationism.

8   Lakatos (1974:176) characterized this adherence to a theory — or rather to a theory embedded in a research program — in the face of a falsifying instance as the disappearance of the methodological (as opposed to the logical) asymmetry between universal and singular statements and found his philosophy of science to go beyond Popper on that point. Popper, however, always defended a "legitimate place for dogmatism" in science (Popper, 1974a:948, 1010; see also Popper, 1979:135, 1976:§18, 1989:49, 312).

this would be — for Popper (1979:391) — an exercise in the sociology of science.[9] By contrast, his interest was in the logic of scientific discovery, which is ultimately a methodological issue, a method that may generate rules for practicing scientists (Popper, 1979:393), but rules that the practicing scientist may nevertheless choose to ignore, reject or accept yet violate (Körner, 1959).

The central basic problem of science is therefore not, according to Popper, the precise and unequivocal refutation of a theory in practice, but rather the method by which we choose to accept a falsifying basic statement. The necessity to make the acceptance of a falsifying statement a matter of methodology occurs not only because all observation is theory laden and all basic statements are therefore soaked in theory (Popper, 1989:387), but also because of logic, that is, what Popper called the "transcendence inherent in any description" (Popper, 1976:§25, Appendix X). The statement "Here stands a glass of water" is a singular statement that reports on an observation, but its verification cannot result from any observational experience. My observational report — here and now — cannot constitute a verification, because each time I report this observation, it represents a singular observation. No number of singular observational (or descriptive) reports can logically exhaust the universal properties that characterize glass, and water. In other words, no finite use of the words *glass* and *water* unequivocally fixes their meaning. Absolute, or true, falsification of a theory cannot be had (empirically) unless it were possible to inductively verify something first, which in itself is impossible (Ayer, 1952).

This basic methodological problem remains the same even for the most sophisticated falsificationist, because the evaluation of competing theories (strictly universal propositions for both Popper and Lakatos) in the context of a dynamic research program will always require intersubjective agreement on, and acceptance of, some basic statement that reports on some observation, no matter whether the emphasis is placed on falsification, as with Popper, or on corroboration, as with Lakatos.[10]

Because a basic statement reports on observations, it cannot be verified and therefore must itself be amenable to a test and potential refutation. We therefore have to treat basic statements as theories, or hypotheses, of lower order and of a lower degree of universality. A basic statement is testable because it is connected to a host of other theories that are not questioned as one particular theory (among several competing ones) is put to test. If I have a theory that predicts the expansion of iron as a function of heat, I need a thermometer to test that theory. Yet all the theories that constitute the foundation on which to design and construct a thermometer are not questioned when I am using this instrument to measure the expansion of iron as a function of increasing heat. All these theories are part of what Popper called the "background knowledge" (Popper, 1979:§7). Popper recognized this background knowledge as dogmatic relative to an experiment that takes place here and now, but it is a dogmatism of an unproblematic kind as long as the requirement is maintained that all theories that form part of the background knowledge be themselves testable and potentially refutable.

Universal versus Singular (Basic) Statements in Systematics

Systematics is formulated in singular statements. A lot of misunderstanding has surrounded, and continues to surround, the issue of whether or not Popper claimed testability and potential falsifiability for strictly universal statements only or for strictly singular statements

---

[9] Or in the history of science; see, for example, Kuhn (1974a,b) and Lakatos (1974).

[10] See Lakatos, 1974:127, and footnote 20 below. Beyond this basic issue, Lakatos added another problem to his philosophy of science, which is not Popper's, namely, what the criteria are by which scientists decide which theories are to constitute the hard core of a research program *sensu* Lakatos, that is, which statements to render unfalsifiable by *fiat* (Kuhn, 1974a:238).

as well. Popper always insisted that singular statements are testable —but only in connection with some universal regularity, with some universal law (e.g., Popper, 1979:435, 1974a:989) — as must follow from his solution to the logical problem of induction. Thus, the statement "It rained in Vienna on March 12th, 1933" is a singular statement, indeed a historical statement, and it is perfectly testable, but only in connection, for example, with lawful planetary motions that determine the calendar. The statement "Here stands a glass of water" is a singular statement, and it is testable only because glass and water behave according to some lawfulness. If they did not, we could not recognize glass as glass, and water as water. As Popper stated, "... every objective, i.e., intersubjectively testable singular statement contains theoretical, hypothetical, or lawful elements and therefore claims a certain lawful connection of other singular statements (or else it would not be testable) ..." (Popper, 1979:435; see also Popper, 1976:§28, Appendix X).

This contrasts with isolated, purely existential statements, which are not testable and not falsifiable and hence are (syntactically) metaphysical (Popper, 1983:161, 1989:258). We can debate the statement "There exists at least one Loch Ness Monster" until the end of our days without ever reaching a conclusion. The situation would change, however, if there was a museum specimen of such a monster; in fact, even a footprint would be of great help, as it would allow us to connect the debate to the law of gravity ("*Tatzelwurmfrage*"; Popper, 1979:122). What, then, are those universal statements in connection with which we might test the singular statements that constitute systematics? Could it be Darwin's theory of descent, with modification?

If Kluge's (2001a, 2001b) notion of descent, with modification (Darwin's two principles) is being used as a universal of some kind in connection with which singular statements could be tested, the universal descent, with modification would have to be applicable to the world of experience. The problem here is that "descent, with modification" is not a full sentence, but rather a conjunction of two concepts. Because it is not a full sentence, it cannot be applied unequivocally to the world of experience. In order to do justice to the meaning of descent, with modification, and in order to render it a potential universal statement, we must reformulate it. For example, "All new species originate by a process of descent, with modification" (see also Appendix I).

This statement may look like a universal statement, but can it logically clash with the positive instantiation of a forbidden statement, as Popper requests for strictly universal statements? Evidently, it cannot, for it is historical in nature. Its application to the world of experience would require its use as a basis from which to deduce singular forbidden statements that, given identical initial conditions, describe repeatable instances. However, even if we agree to relax the logical constraints and dismiss the need for repeatability of its predictions in view of its historical nature, the universal statement would still have to serve as the basis for a prognosis that is specifically restricted in time and space, for example to a particular speciation event, and that prediction would have to be able to fail (a forbidden event could potentially occur), at least in principle. Furthermore, if it is used in conjunction with a cladogram, the logically subordinated hierarchy would have to be an empirical concept of systematics, successful in its application to the world of experience and hence highly corroborated (see discussion below), and the universal statement "All new species originate by a process of descent, with modification" would have to predict a lawfully dichotomous process of speciation (for an exercise in logic that illustrates the necessity of this prediction see Appendix I). We do not have a universal model of speciation, let alone a law of speciation. In fact, we do not even have a universal species concept (e.g., Vane-Wright, 1992).

Popper acknowledged that "Darwin's Theory of Evolution proposed no ... universal laws" (Popper, 1973:267; see also Popper 1989:145–146). For if Darwin's theory of evolution generated strictly universal propositions (which is what universal laws are), then we would have to be able to express those as negations of strictly existential statements (Popper, 1976:§15). The theory of the origin of new species by means of variation and natural selection does not generate such statements or classes thereof. Instead, historical theories, such as Darwin's theory of evolution, are "as a rule, not universal but singular statements about one individual event, or a series of such events," that is, conjunctions of singular statements (Popper, 1961:107).

Hull (1999b:487) cites a number of potential falsifiers taken from Darwin that do specify conditions that would pose serious problems for Darwin's theory, but none of those can be connected to a strictly universal statement. In other words, the acceptance of instantiations of these potential falsifiers would indeed create serious problems for Darwin's theory, but not in the sense of the Popperian asymmetry inherent in falsificationism. However, this does not mean, that "paleontology, or the history of life on earth" do not contain testable elements, as Popper (1980) emphasized in his letter to *The New Scientist*. We can, in principle, test the historical statement "Mammals originated after fishes had originated," because that statement would be falsified if we found a fossil mammal predating the earliest fossil fish. (*In principle* means logically, i.e., assuming completeness of the fossil record; alternatively, the new theory "fishes originated after mammals had originated" becomes testable on the same grounds). Again, this historical, that is, singular statement, is testable only in connection with universal laws, in this case in connection with the theory of solid bodies and with the theory of gravity, as documented by Niels Steno's (1667) book, *De Solido intra Solidum Naturaliter Contento*, plus, of course, in connection with a host of additional theories.

However, Kluge (2001a) uses descent, with modification not as a universal law, but as the necessary and sufficient background knowledge for systematics[11] (but see Sober, 1988). Kluge uses Popper's formula for severity of test to cast cladistics in a falsificationist framework:

$$S\ (e,\ h,\ b) = p(e,\ hb) - p(e,\ b)/p(e,\ hb) + p\ (e,b)$$

He then seeks to keep background knowledge (b) at a minimum, because — according to this formula — the smaller the background knowledge, the greater the severity of test, but the entries that are critical in this formalism are the nature of the evidence *qua* supporting evidence (e) relative to the empirical content of the hypothesis (h), not the background knowledge (b), which is a given. Popper's definition of severity of test starts from the measure of empirical content (Popper, 1989:390), and the severity of test (as well as the explanatory power) of a theory becomes greater with an increasing (informative or empirical) content of the theory (Popper, 1989:391).

If the hypothesis under scrutiny does not transcend background knowledge at least to some degree (i.e., if (h) = (b)) then its relative empirical content, relative to the unproblematic, (i.e., accepted background knowledge) is zero (Popper, 1973:49),[12] which must affect severity

---

[11] The qualifier "necessary and sufficient" is mine, not necessarily Kluge's (Kluge, 2001a; see also Farris, et al. 2001). The idea, however, is not to admit to background knowledge model assumptions that are not well corroborated, or simply, to defend parsimony as superior to likelihood because parsimony analysis does not require any model assumptions (well corroborated or not) that are required by likelihood analysis. Since parsimony analysis therefore keeps background knowledge to a necessary minimum (compared to likelihood analysis), severity of test and degree of corroboration increase. Since Popper's formalism for degree of corroboration is the same as that for explanatory power ([E(h, e, b) = S(e, h, b)]; Popper, 1989:391), the hypothesis with the highest degree of corroboration also has the highest explanatory power, that is, best explains observed similarities as the result of inheritance and common ancestry (Farris et al., 2001). This approach to Popper's formalism treats background knowledge as an arbitrarily (or conventionally) quantifiable entity, and its application to parsimony analysis rests on a symmetry of falsification and corroboration (see also footnote 22, and text below), where corroboration becomes support or confirmation (*sensu* Sober, 1988).

of test. A test that has been repeated many times, and the results of which have become assimilated to the background knowledge, will at some point no longer be viewed as severe or significant. "There is something like a law of diminishing returns from repeated tests (as opposed to tests which, in the light of our background knowledge, are of a *new kind* ...)" (Popper, 1989:240; see also Nagel [1971:414], who uses numbers and kinds of positive instances to determine the degree of confirmation of a hypothesis). In the formalism above, the evidence (*e*) relative to the hypothesis (*h*) figures as supportive evidence (for a counter-example), because the severity of test for the hypothesis (*h*) is greater the less likely it will be to obtain the supportive evidence (*e*). The supportive evidence (*e*) therefore answers to the riskiest prediction the hypothesis generates, and the risk of prediction increases, the further the hypothesis transcends current background knowledge (*hb*).[13]

What is of interest to Popper is a new hypothesis that transcends background knowledge, and because severity of test increases with an increasing relative empirical content, the most interesting hypotheses are also those that are the riskiest given current background knowledge, that is, those that have the highest risk of being found false (Popper, 1989:240). The same idea can be expressed differently (Popper, 1989:240), if it is understood that a serious test of a theory will attempt its refutation and hence will consist of a search for a counterexample.

However, because the most severe test will address the most risky or most unlikely predictions that are generated by the theory under scrutiny (Faith and Trueman, 2001), background knowledge becomes indispensable in the search for the most severe counter-examples, because the degree of risk, or unlikeliness, of a prediction can only be established relative to the given background knowledge. Therefore, if the test is to be severe, the best effort must be made to find such counterexamples, and this is done by using the background knowledge in order to determine where to look for the most probable counter-example to the most unlikely prediction the theory generates. If the theory passes such tests, these passing incidences provide evidence in support of the theory and are relegated to background knowledge. If the theory continues to survive severe testing, the point may be reached at which, in the light of our newly accumulated background knowledge, there are no places left where we can expect to find counter-examples with a high probability. At this point, the severity of test declines (Popper, 1989:240).

Evidently, Kluge's (2001a; see also Farris et al., 2001) notion of background knowledge differs from Popper's understanding of that term. Background knowledge cannot be held to a conventionally determined minimum (see also footnote 13). "For each experiment that we make to test a theory there are so many theoretical preconditions that it is nearly impossible to test them all: each time when we take a reading from an instrument we rely on the hypotheses of geometrical optics, on the theory of solid bodies, on the correctness of Euclidean Geometry in small space, on the hypothesis of the existence of things, and innumerable other hypotheses" (Popper, 1979: 391). In Popperian terms, background knowledge constitutes the entirety of the scientific knowledge that is not questioned at any particular time when a particular theory, or set of theories, is being put to test. Popper

---

[12] See also comments below on background knowledge being tautological.

[13] Farris et al. (2001:439) point out "that corroboration C(*h, e, b*) for different trees *h* would vary ... according to the other term p(*e, hb*), and as is easily seen from the formula [i.e., Popper's formalism for degree of corroboration] the trees with greatest p(*e, hb*) would have the strongest corroboration C." They go on to point out that "p(*e, hb*) may be high for evidence *e* that strongly favors hypothesis *h*," whereas "p(*e, hb*) is ... low for evidence that undercuts the hypothesis." In Popperian terms, (Popper, 1983), p(*e, hb*) points to the degree to which evidence *e* supports hypothesis *h*, as the latter relates to background knowledge *b*, and the degree of support evidence *e* will provide depends on how unlikely it is, that is, it depends on the degree by which *h* transcends current background knowledge (*b*). This shows that (*b*) cannot be held to an artificial minimum in order to maximize (*hb*). The content, or relative content, of hypothesis (*h*) is defined by its relation to the given background knowledge (*b*), not by fiat: "p(*eh, b*) is close to 0 for hypotheses with high content ... a category that would surely include phylogenetic hypotheses" (Farris et al., 2001:439).

characterized background knowledge as tautological (Popper, 1976: Appendix IX, note 5) if applied to his formalisms — tautological in the sense, and only in that sense, in that it is taken as *a priori* true[14] and hence does not provide any critical information. For Popper, critical information is new information that may result from testing new hypotheses, or new predictions deduced from old hypotheses, but because background knowledge is taken as *a priori* true when any specific theory comes under test, it cannot provide new information in itself. Background knowledge is not tautological in the sense of an analytic statement, because the negation of empirical theories relegated to background knowledge is not self-contradictory, as the negation of a tautology is, such as "Not all bachelors are unmarried men" (or "Some bachelors are not unmarried men"). Note, though, that the tautology "All bachelors are unmarried men," although true, does not prevent a bachelor from getting married — yet if he does, he can no longer be described as a bachelor. This exactly corresponds to Popper's notion of background knowledge: Taken for *a priori* true, here and now, nothing prevents it from changing tomorrow, if parts of it come under critical discussion at that time.

Parsimony as a Method of Systematics

The claim could be made that systematics can proceed without underlying universal laws, for what is required is nothing but a method that allows us to choose a preferred hypothesis from a set of competing hypotheses of relationships relative to some theory such as evolutionary theory. Indeed, we do have a method at our disposal that allows us to do just this, but is it Popperian in its logic? The conformity of parsimony analysis with Popper's falsificationism has been asserted in terms of Popper's concepts of logical probability, explanatory power, degrees of corroboration and severity of test. Let us look at these concepts in more detail.

One of the more complex notions in Popper's philosophy is his concept of logical probability. In a general sense, logical probability is based on the view that probability is a logical relation between propositions analogous to the relation of deducibility or entailment; it is used as a measure for the logical distance between a conclusion and its premises (Nagel, 1971:390). Logical probability, therefore, is not a probability statement about the actual world (which would be a statistical probability rendered in the object language of science), but rather it is a probability statement about the logical relation between two other statements (hence a metalinguistic statement about statements rendered in the object language of science [see also Appendix II]). In Carnap's (1995:35) rendition, it is an analytic, and hence meta-scientific, statement about the degree of entailment of stated evidence (*qua* premises) by a stated hypothesis (*qua* conclusion), in short, a measure for degree of confirmation.

Popper exemplified his concept of logical probability in relation to the concept of the logical playground of a theory. A theory with a larger logical playground has a lesser chance of being found false, and hence it has a higher logical probability than a theory with a small logical playground, i.e., a theory with a smaller logical probability. The logical probability for a theory to be found false (that is, its logical improbability) increases with the increasing class of forbidden statements that the theory generates, for if natural laws are of the form of strictly universal statements, then they can also be expressed as negations of strictly existential statements, that is, in the form of nonexistence statements or "there is not" statements (Popper, 1976:§15). "For example, the law of the conservation of energy can be expressed in the form 'There is no *perpetuum mobile* machine'" (Popper, 1976:§15). In

---

[14] For example, Kluge, 2001a:325: "… I consider Popper's concept of background knowledge *b* to comprise only currently accepted (well-corroborated) theories and experimental results that can be taken to be true …" Compare this to footnote 13 and Appendix II.

other words, natural laws do not predict that something exists (the existence of something does not follow logically from a universal proposition but from an independent observation), but they predict that something does not exist.

Falsification occurs when something is observed that should not have happened. The larger a theory's class of forbidden statements, the smaller is its logical playground and the greater is its risk of being found false, because of the need to generate more precise predictions. It is therefore the increasing preciseness of the predictions it generates that restricts the logical playground of a theory that decreases its logical probability. However, it is also the theory that makes the most precise predictions that has the greatest empirical content that lends itself to increasing severity of test, hence to increasing degrees of corroboration. In sum, the theory that makes the most precise predictions has the highest explanatory power.

Logical probability has nothing to do with probability calculus, as is most easily shown relative to predictability. Logical probability is related to the capacity of a universal statement to predict a particular event (i.e., an observable state of affairs), or rather the nonoccurrence thereof, specifically restricted in time and space. Probability calculus cannot make such predictions — it can only make a prediction that covers a series of events (Popper, 1979:141). The probability of obtaining a six throwing a fair die is one in six. This mathematical probability is maintained whether or not I do, indeed, obtain a six with my next throw. It is also maintained if I say, "*Hic et nunc*, here and now, I will throw the die and obtain a six," yet I get a five.

The same is by no means true for logical probability. Assume a theory with a small logical playground, hence a high logical improbability, and the use of this theory as the basis for the deduction of a singular statement that would constitute a severe test if it were found instantiated. Assume, further, that the experimental results would be accepted as falsifying the theory in question. Then on logical — not on practical — grounds the logical probability of the original theory immediately rises approaching 1 as its degree of corroboration drops toward 0, until further testing.

Consider the infinite number of possible trajectories for a stone, if thrown, to travel through space. The theory that predicts that the stone will travel along no other path but a parabola is, from a logical point of view, infinitely improbable. Given the infinite number of theoretically possible trajectories, the class of forbidden statements generated by that theory is very large, its logical playground consequently very small, and its risk of being found wrong very high. In other words, the logical probability of the theory would be very small, its logical improbability very high. If a stone is thrown, and if it were found not to travel along the trajectory of a parabola, and if that experiment were in principle repeatable, the logical probability of that hypothesis would rise dramatically and approach 1 as its degree of corroboration dropped toward 0 (assuming, of course, the observation was correct, accepted by the scientific community, and no *ad hoc* auxiliary hypotheses are admissible).[15]

In summary, the comparison of competing theories in terms of their logical playground and its equivalent, the logical probability, results in the following insights: The smaller the logical playground, the higher the potential severity of test; the higher the severity of test, the higher the degree of corroboration. The degree of corroboration therefore corresponds to the inverse of the logical probability or it corresponds directly to the logical improbability. The explanatory power of a theory increases as a function of its degree of corroboration, which in turn increases with increasing severity of test, which in turn depends on its empirical content, that is, its logical playground.

---

[15] Note at this junction how Bayesian probability and logical probability approach each other under the boundary conditions of near falsification or near verification.

Popper meant those relations to be logical, not numerical, except under certain limiting conditions in which the logical probability approaches 0 or 1. If Theory A answers questions that Theory B does not answer, it may be concluded, on logical grounds, that the explanatory power of Theory A is greater than that of Theory B, but we cannot put a number on that difference, for in order to do so, we would have to know the truth (Popper, 1983:61) and the degree to which the theories A and B approach it, that is, we would have to know their verisimilitude in metric terms (see Appendix II for further explanation). The same can be said relative to a numerical expression of degree of corroboration, because Popper's notion of verisimilitude can be traced back to degree of corroboration using the concept of logical probability as a bridge (Popper, 1979:151, note 4). A theory with a large logical playground can be put to test innumerable times, but it will never achieve the degree of corroboration of a theory with a very small logical playground that has so far survived one test only. The reason is that the theory with a large logical playground is a relatively trivial, uninteresting theory, whereas a small logical playground renders a theory much more interesting, as it dramatically increases its empirical content. Nevertheless, whether the degree of corroboration is relatively high or low, in every case it constitutes an historical account only. It only specifies the past performance of a theory, the degree to which it has survived testing up to the point of its current critical discussion, with no implications as to the future performance of that theory.

Parsimony, Hierarchy and *ad hoc* Auxiliary Hypotheses

With respect to parsimony analysis, it has been claimed that parsimony maximizes the explanatory power of hierarchically ordered hypotheses of relationships by minimizing the number of *ad hoc* auxiliary hypotheses (i.e., hypotheses of homoplasy) that are necessary to explain the distribution of characters across a number of taxa (Farris, 1983; Farris et al., 2001). This idea is derived from the view that the finite universe of N cladograms possible for n terminal taxa constitutes the closed set of competing hypotheses (Kluge, 2001a) and that characters serve as falsifiers — in the ideal world — of all hypotheses except the most parsimonious one. In turn, characters that contradict the most parsimonious cladogram (i.e., incongruent characters) are dismissed as potential evidence for conflicting (or contradictory) hypotheses of grouping and are instead declared *ad hoc*, explained as instances of homoplasy. This procedure is rooted in Hennig's auxiliary principle: Always assume homology in the absence of contrary evidence. The use of parsimony and ensuing *ad hoc* auxiliary hypotheses of homoplasy in the context outlined (Farris, 1983; Farris et al., 2001) is to find the putatively most highly corroborated hypothesis of phylogenetic relationships within a closed set of competing alternatives. At the same time, the acceptance of *ad hoc* auxiliary hypotheses will immunize the concept (or structure) of a strictly dichotomous hierarchical order of nature in itself from failure.

The use of parsimony thus results in three separate yet closely inter-related issues: (1) the general nature of parsimony and its status in Popper's philosophy of science, (2) the role that parsimony and ensuing *ad hoc* auxiliary hypotheses of homoplasy play in safeguarding the (logically subordinated) hierarchy from failing as it is applied to the world of experience, and (3) the role of parsimony in the putative falsification of alternative hierarchies that stand for alternative hypotheses of phylogenetic relationships. This section will deal with the first two of these issues; the third will be dealt with in the next section.

Whereas parsimony certainly allows us to find the cladogram that minimizes incongruence, it is based on the very general scientific principle of economy of hypotheses, which can be traced back to William Ockham. Popper (1979:214ff) certainly agrees that scientific explanations should in every case minimize *ad hoc* auxiliary hypotheses. Indeed, it is exclusively with respect to this principle of minimizing auxiliary hypotheses — whether *ad hoc* or *a priori* — that the debate of parsimony *contra* likelihood can be had. As noted by Edwards (1992:200), "How much simplicity are we prepared to lose for an increase in the excellence of the fit" of observed data to the model? However, this principle of simplicity is firmly rooted in the inductivist tradition (Popper, 1976:§42) and it has been characterized by Popper (Popper, 1976:§43, 1979), as well as by many others, as the principle that requires that the shortest line must be selected from the infinite number of lines that could be used to connect a given number of (data) points. This principle evidently underlies the logic of a Wagner network (Kluge and Farris, 1969). Inductivists and logical positivists used this illustration of parsimony to refer to the simplicity of a theory, stating that the simplest theory has to be preferred.

From a falsificationist point of view, however, this principle of simplicity can only be a matter of aesthetics, of convention or of pragmatism (Popper, 1979:214–218). For this reason, Popper equated the notion of the simplicity of a theory, that is, its degree of simplicity, to its logical improbability (Popper, 1979:218), because in his view, the theory with the smallest logical probability is also the simplest among competing theories. In Popper's terms, the notion of the logical playground of a theory provides a bridge between the simplicity appealed to by inductivists and his own notion of simplicity (Popper, 1979:217), for if there is no other logical means to choose among competing hypotheses, such as degree of falsifiability, one might just as well choose the simplest one, perhaps because this is the most pragmatic way to proceed (i.e., to use parsimony as a methodological tool) or else because this is the aesthetically most appealing choice.

Yet Popper asked for a different kind of choice to be made. For example, the experienced universe deviates less from Einstein's theory than from Newton's. Hence, Einstein's theory requires fewer auxiliary hypotheses than Newton's and is the simpler theory, but it is also the theory with the smaller logical playground, hence the theory with the higher logical improbability, hence the theory that can be more severely tested. It thus becomes apparent that in Popper's philosophy, the simplicity of a theory increases with an increasing level of universality, that is, with a decreasing logical playground (Popper, 1976:§43; Appendix VIII). The goal of science is to eventually make the logical playground of theories so small that an *experimentum crucis* becomes possible.[16] As Sober noted (1988:49), Popper does appeal to simplicity, but it is a specific kind of simplicity, expressed in an ever-higher degree of universality of theories. However, none of this relates to the role that parsimony plays in systematics. The reason is that no strictly universal statements are involved in systematics.

Again, Popper (1976:§20) was well aware that natural sciences cannot proceed without *ad hoc* auxiliary hypotheses; he was also aware that they need to be kept to a minimum. More is involved, though. For example, Popper wanted to admit *ad hoc* auxiliary hypotheses only if they increase the degree of falsifiability of a theory, that is, if they allow predictions of a new kind that transcend current background knowledge. This evidently cannot be applied to parsimony as it is used in systematics. Popper also stated that *ad hoc* auxiliary hypotheses may be admissible if it can be shown that their generalizations are falsified (Popper, 1979:380). For example, it can be claimed for a particular experiment that appears to falsify a cherished hypothesis, that instrument X, here and now, yielded a mistaken reading. This *ad hoc* hypothesis of a mistaken reading is admissible if the generalization of the *ad hoc* hypothesis is falsified. That means, the statement "All instruments X yield mistaken readings under such and such conditions" must be known to be false. In systematics, we could similarly dismiss incongruent characters *ad hoc* as erroneous observations, but this would be admissible only if the statement "all well-trained systematists make erroneous observations under such and such conditions" were known to be false (Popper, 1979:381). Alas, incongruent characters keep coming.

---

[16] Lakatos (1974) gives an insightful account of how crucial experiments play out in the history of science.

In one sense, however, Popper specifically rejected the use of *ad hoc* auxiliary hypotheses, by adopting a strategy used to immunize a theory from falsification. Popper (1979:185–186) illustrated this strategy using a simple syllogism.

Let us say Socrates is human and all humans are mortal. From this must follow that Socrates is mortal, a logically valid (and factually sound) statement. This syllogism can be generalized: if $x = y$ and if $y = z$ then $x = z$. Rendered in this general form, we can now substitute into this deductive system anything we like. For example, all humans have (naturally) green hair, and Socrates is human; therefore Socrates has (naturally) green hair. Although this conclusion is evidently factually false (unsound), it is logically valid nonetheless. This means that logically valid inferences can be obtained no matter how factually wrong the premises are. Conversely, if we do not like the conclusion, we can change the premises. This reveals the circularity of syllogisms (in the sense that the corollary must be contained in the premises; Popper, 1976, Appendix X), and it can be shown by the following logically valid inference: No human has (naturally) green hair, and Socrates is human; hence Socrates does not have (naturally) green hair. In doing so, though, the meaning of the term *human*, that is, the intended use of the term *human* in its application to the world of experience ("*Zuordnungsdefinition*" *sensu* Popper, 1979:185; see also Popper 1976:§17), has changed.

This shows that if we are allowed to continuously change the meaning (the intended use) of the terms relative to their application to the world of experience, we can immunize every theory against falsification. For example, if my theory can be expressed in the following universal statement, "No human has (naturally) green hair" and there comes along an individual with (naturally) green hair, I can simply dismiss this falsifying singular occurrence by declaring that this individual is not a human. Conversely, I can expand the notion of humans *ad hoc* to include individuals with (naturally) green hair. As Kuhn (1974b:18) specified, "Only if you have previously committed to a full definition of 'swan' ... can you be logically *forced* to rescind your generalization" (that all swans are white). Because Popper (1989:402 and above) showed that a full definition of anything cannot be had, Kuhn's full definition must be read as the intended use of the name *swan* (Popper, 1979:366–367).

This is one reason why Popper placed so much emphasis on methodology: "All admissible means to prevent falsification" of a theory must be used (Popper, 1979:378), but this implies that all means that may be used to immunize a theory against falsification must not be used. Popper insisted on the methodological rule for empirical sciences that "*criteria of refutation* have to be laid down beforehand [before the test, that is]: it must be agreed which observable situations, if actually observed, mean that the theory is refuted" (Popper, 1989:38; the same claim underlies appeals to verification; Ayer, 1952).

This, in turn, implies that "anyone who advocates the empirical-scientific character of a theory must be able to specify under what conditions [s]he would be prepared to regard it as falsified; i.e., [s]he should be able to describe at least some potential falsifiers," and these must, in turn, be falsifiable (Popper, 1989:xxi, xxiii). Hence, if an empiricist uses a universal statement as a basis for the deduction of a singular statement, she must stick to the original meaning (the originally intended use) of the terms used in the singular statement that describes a potential falsifier (Popper, 1979:366–367).

Whereas it is true that no finite number of uses (or applications) of a word can unequivocally establish its meaning, it is also true that the strategy to change the meanings (in the sense of intended use) of words *ad hoc* can be used to immunize a theory from falsification. It is again true that in Quine's "naturalistic language" the meaning of words always remains relative and historically mutable (Gibson, 1982), but if a falsifying singular statement is to stick to the universal proposition from which it was deduced, it must be both necessary and contradictive, yet it can only be logically necessary if its rejection would entail a change of its meaning (Ayer, 1992). If the proposition is a hypothesis to be tested,

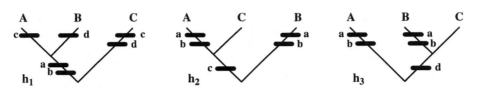

FIGURE 4.1   The application of the *modus tollens* form of argument to a three-taxon statement. H$_{1-3}$ = Hypotheses 1–3. For further explanation see text. (After Cartmill, M., *Z. Zool. Syst. Evol.,* 19, 73–96, 1981.)

or a prediction that is meant to test a hypothesis, immunization from falsification can be achieved by juggling with the meanings of words (Ayer, 1952:96).

However, this is precisely the strategy that is used in cladistic parsimony analysis, which is itself based on Hennig's auxiliary principle: Always assume homology in the absence of contrary evidence. We always assume homology (synapomorphy in Kluge's [2001a, 2001b] terminology) as the meaning of observed similarity, but if the character is incongruent, we swiftly dismiss it as conflicting (falsifying) evidence for alternative hypotheses of grouping and reinterpret it *ad hoc* as homoplasy. Cartmill (1981:73–96) explored the consequences of not allowing constantly shifting interpretations of characters as homologies or homoplasies (Figure 4.1). He did so using the form of the *modus tollens* argument, which states that if the minor premise (the implicate) in deductive reasoning is found false, it falsifies the major premise retrospectively (the implicans) (if *p* implies *q*, then not-*q* must imply not-*p*; quasi-induction according to Popper, 1979:259; see also Popper, 1976:§18). It is thus evident that in Popper's use of the *modus tollens* form of argument, the major premise corresponds to the universal proposition from which the minor premise, the singular (forbidden) statement is deduced.

Since the assumption of homology implies common ancestry (i.e., the *modus tollens* is used under the covering law of evolution and its generally accepted corollaries;[17] Ball, 1982:188), the three-taxon statement ([A, B] C) (Figure 4.1, hypothesis 1) implies the (forbidden) statement that no homology must be shared by B and C. This implication is falsified —if the assumption of homology is not allowed to shift to homoplasy — by character d. Hypothesis 1 also implies that no homology must be shared by A and C, and it is falsified by character c. Again according to the *modus tollens* argument, hypotheses 2 and 3 are falsified for the same reason by characters a and b (Figure 4.1). The conclusion is that all of the three hypotheses are false. Yet the assumption of homology, that is, common ancestry, must imply that at least one of the three hypotheses is true. Following this conclusion, Ball (1982) suggests that we deal with this kind of character conflict by provisionally preserving the most parsimonious hypothesis (in the sense of Popper's "healthy dose of dogmatism") and critically evaluating the potential falsifiers (c and d in Figure 4.1) as possible analogies. However, that requires "criteria for [the] recognition" of adaptation, convergence and parallelism (Ball, 1982:189). Alas, this "bizarre" (Ball, 1982:189) use of parsimony resulted from the insight that no such criteria of recognition can be had outside character congruence (Patterson, 1982).

The liberty to shift from assumed homology to assumed homoplasy and vice versa amounts to the use of *ad hoc* auxiliary hypotheses in an immunizing strategy that protects

---

[17] Ball's (1982) inference of phylogenetic relationships from shared similarity (homology) relative to a background theory (evolution) corresponds to Sober's (1988) "three-place-relation" rather than to a hypothetico-deductive form of argument. Ball is correct in pointing out that Popper used a causal conditional for the *modus tollens*, but the causality implied by Popper is one of lawful regularity, not historical contingency. Ball is also correct that Popper did allow for tests of singular statements, but only in connection with universal laws (Popper, 1974a:989).

the logically subordinated hierarchy *per se* from failure. It might be claimed that transferring the logical structure and properties of syllogisms to that of a cladogram is unfair. However, this claim has no basis, because the cladogram can be represented as a Venn diagram, and Venn diagrams were John Venn's method of representing and evaluating the validity of syllogisms. Venn diagrams also figure prominently in set theory. As shown by Gregg (1954), the syllogism exploited by Popper is easily obtained in set theory if set identity is construed as mutual inclusion. Thus, if the major premise is "All animals with legs are tetrapods," and the minor premise is "Snakes are not animals with legs,"[18] then the logically valid conclusion must be "Snakes are not tetrapods" (Cuvier's conclusion [Rieppel, 1987]). If, in contrast, the first premise is "All tetrapods have legs," and the second premise is "Snakes are tetrapods," the logically valid conclusion will have to be "Snakes have legs" (Platnick's conclusion [Platnick, 1979:543]). This conclusion involves a shift of meaning, because our intended description of snakes has shifted from organisms without legs to organisms with modified legs, that is, lost legs. The name *snake* does not "attach to nature" (Kuhn, 1974a:267) in the same way anymore, or in Carnap's terminology (Mayhall, 2002:69), a different meaning of the word *snake* has been fixed by deduction from different premises.

From a Popperian perspective we cannot decide empirically (i.e., by its application to the world of experience using parsimony) whether the structure of the logically subordinated hierarchy as such (not a specific hypothesis of relationship) is *a priori*, that is, empirically unfalsifiable, or whether and when it may fail in the world of experience. The ultimate reason for this is simply that there exists no deductive connection between phylogenetic hypotheses and character distributions, or conversely, that any phylogenetic hypothesis will always be compatible with any character distribution — at least to some minimal degree (Sober, 1988:135).

However, ordering nature in a logically subordinated hierarchy is not an analytical judgment because its negation is not self-contradictory. There is no contradiction in the statement "Nature is not hierarchically ordered," as there is in the statement "Not all bachelors are unmarried men." We even infer that we have positive knowledge of cases where nature is not hierarchically ordered, such as speciation through hybridization, for example. The cladogram must therefore be a synthetic, or empirical, concept, but the way it is applied when using parsimony does not allow the concept to fail, which means that we do not allow experience to pass judgment on the validity of a hierarchical order of nature.

However, if the logically subordinated hierarchy as a structure (Gregg, 1954; not as a specific hypothesis of relationship) cannot fail as it is applied to the world of experience, it is not really empirical, but rather is synthetic *a priori* relative to the theory (phylogeny) it mediates. The use of the concept of synthetic *a priori* in this context is not meant in a Kantian sense, that is, in the sense of certain (true) *a priori* knowledge about the actual world. Instead, it is meant to be synthetic in the sense that it claims to describe or picture the actual world, and it is *a priori* in the sense that there is no room for alternative descriptions of the actual world under the paradigm of cladistics. For example, there is no methodological rule — and there cannot be any — that would specify the amount of homoplasy that would constitute a crisis for the paradigm that requires nature to be ordered in terms of a logically subordinated hierarchy. In fact, most practicing systematists do not even consider this question, because the general concern is to find at least some degree of

---

[18] Snakes are used here in a paradigmatic sense, not in an actual sense, applied to fossil and extant snakes. The example is also used only in order to illustrate the immunization of the logically subordinated hierarchy *per se*. As far as the classification (or relationships) of snakes are concerned, the old theory (snakes are not tetrapods and have no legs) has been replaced by a new theory (snakes are tetrapods with modified legs), for which an independent test can be postulated, but see below for the nature of test of specific hypotheses of relationships based on the application of parsimony.

congruence; the greater the degree of congruence, the greater the support for a cladogram, the more comfortable systematists feel with their hypothesis.

Yet the appeal to maximal support is certainly not a falsificationist approach; it is because we do indeed find some degree of congruence that we can hang on to the logical structure of the cladogram. It is also this degree of congruence, or regularity of character distribution, that ultimately requires an explanation, the latter representing the *explanandum* (Brady, 1985; in this case, the explanandum is independent from the explanans). Beyond that, bacteriologists raise the question of whether the tree of life is, indeed, an adequate picture, or whether a web or a net might not be a more appropriate picture for genetic, and hence genealogical, relations within their discipline (Dupré, 1999).

One might argue that although the cladogram is a synthetic concept (*qua* hierarchy; Brady, 1983, 1985), it need not be able to clash with the world of experience and therefore can be taken *a priori* because it can be justified with reference to some theory such as Darwin's theory of evolution (Lee, 2002).[19] However, that theory does not qualify as a universal law *sensu* Popper and it does not entail a theory of lawful dichotomous speciation (see also Appendix I for further analysis). Justification of the hierarchy in terms of evolutionary theory furthermore begs the question of how this theory could, in principle, be known to be false. Alternatively, what could the statement "Theory is not known to be false," mean if no theory can unequivocally be falsified in practice? According to Popper (1983:132), a theory that is "not known to be false" can provisionally be accepted only after it has survived the most severe tests, and only if there is independent evidence in its favor, that is, if it is independently testable. Independently testable, however, requires that "apart from explaining all the *explicanda* which the theory was designed to explain, it must have new and testable consequences (preferably consequences of a *new kind*)" (Popper, 1989:241).

However, it could be argued that Darwin's theory of evolution need not justify the logically subordinated hierarchy *a priori*, but is only used as (provisionally accepted) background knowledge to allow the inference from observation (of shared similarity) to a hypothesis (of grouping). In defense of his "three-place-relation," Sober (1988) makes the point that background theories must necessarily be involved in any inference that proceeds from observation to hypothesis — but this is a form of inductive reasoning and hence cannot be used in a Popperian context. As Sober (1988:66) states, "If hypothesis H and an observation statement O are *not* deductively related ([i.e., if] they are logically independent), then O confirms or disconfirms H only relative to a background theory." By contrast, Popper's falsificationism requires that natural laws, that is, universal propositions, serve as basis for the deduction of singular forbidden statements.

The insight that the structure (Gregg, 1954) of a logically subordinated hierarchy represents a synthetic concept *a priori* relative to the theory (phylogeny) it mediates is a general conclusion, which throws us back on Popper's psychological problem of induction (see discussion above). (Indeed, Panchen [1992], already realized the fact that the logically subordinated hierarchy is an *a priori* of systematics.)

---

[19] Kluge (2001a, 2001b) uses descent, with modification in justification of parsimony as a method of phylogeny reconstruction. If, in this context, the cladogram means the logically subordinated hierarchy *per se* (or all the possible cladograms for a given set of terminals), then it is *a priori* and cannot serve as the initial condition. If the most parsimonious cladogram relative to some data set is meant, then the cladogram represents the result of parsimony analysis. *Ergo,* the need to use descent, with modification as justification for the use of parsimony. This, however, requires that speciation is lawfully dichotomous and that character transformation occurs parsimoniously across those speciation events (see above and Appendix I). Such justification cannot be had without recourse to special knowledge (see below).

The Test of Congruence

In this more specific context, characters and their distribution do not test the logically subordinated hierarchy as a general structure, but rather are believed to act as falsifiers of alternative hypotheses of relationships in favor of the most parsimonious one. Characters in systematics are relational concepts. The descriptive statement "This bird has a blue tail" is of no interest to systematists, if indeed it could be made in isolation, that is, out of context. Interest is generated by the statement 'These two birds share a blue tail that is not also shared by this other bird that has a red tail."

A character in systematics is a hypothesis of shared similarity (which implies dissimilarity relative to a third place), and on the basis of Hennig's auxiliary principle, we assume that shared similarity indicates common ancestry in the absence of contrary evidence. If the assumption of homology (synapomorphy) is the original meaning of observed similarity, then the evidence that causes a shift to an *ad hoc* reinterpretation of shared similarity as homoplasy (i.e., independently acquired similarity) does not originate from a test of the character hypothesis in itself but from the relationship among all the characters relative to a subordinated hierarchy. This is because every conjecture of shared similarity is in itself also a hypothesis of grouping (Patterson, 1982). Therefore, some characters, namely the more numerous congruent ones, are believed to falsify alternative hypotheses of relationships, which themselves may be supported by less numerous characters. But in so doing, they falsify the initial assumption of homology (synapomorphy) for those characters on which the now falsified alternative hypotheses of relationship were based. In other words, in the context of parsimony analysis, *ad hoc* hypotheses of homoplasy cannot be tested independently, that is, independent from the problem to be explained, and for this reason are — again — to be rejected according to Popper (1973:15–16, 1989:247).

Let us assume we have what we accept as the most parsimonious hypothesis of relationships for a set of species, or taxa, and that this hypothesis entails some degree of homoplasy, as is usually the case in the world of experience. Let us further accept that upon the introduction of new evidence, that is, new and additional characters, the hypothesis of relationship changes. This must mean that today's homologies (synapomorphies) may become tomorrow's homoplasies, and vice versa. From a philosophical point of view, homoplasies can be characterized as errors, that is, false conjectures of homology (synapomorphy), or — more sophisticated — as falsified but not rejected conjectures of homology (synapomorphy). If homoplasies are false conjectures of homology (synapomorphy) (Platnick, 1979), they cannot become true homologies (synapomorphies) tomorrow for purely logical reasons, since what is false today cannot be true tomorrow, and vice versa. However, if homoplasies are considered as falsified, but not necessarily false (i.e., not rejected) conjectures of homology (synapomorphy), one has to admit that today's homologies (synapomorphies) are the falsifiers of conjectures of homology (synapomorphy) that are incongruent today. If the cladogram changes tomorrow upon the addition of new characters, today's falsified conjectures of homology (synapomorphy) may become in themselves tomorrow's falsifiers of those conjectures of homology (synapomorphy) that were corroborated today. Hence conjectures of homology (synapomorphy) cannot be tested independently.

What becomes apparent here is that parsimony analysis is rooted in Hennig's principle of reciprocal illumination (Wiley, 1981; this understanding of the principle of reciprocal illumination differs from Hennig's [1966]), which robs parsimony analysis of the asymmetry that characterizes Popper's falsificationism. The procedure of parsimony is one whereby the weight of the critical evidence (critical meaning: relative to the explanatory power of the hypothesis) and hence the severity of test, therefore the degree of corroboration, will emerge not from a critical, independent discussion of individual character hypotheses, but from the sheer number of (congruent) characters. In that process, characters symmetrically falsify, or support, alternative hypotheses of grouping (Gaffney, 1979). From this insight, the conclusion can be drawn that "the cladograms chosen and attributes thereby specified as homologous ... are probability statements" (Vane-Wright, 2001:600).

However, Farris (1983) identified an asymmetry in parsimony analysis in the sense that a phylogenetic hypothesis does not differentiate between homology or homoplasy of congruent characters (that support or confirm it; Sober, 1988), but that it speaks to the incongruent characters (that contradict or disconfirm it; Sober, 1988) by unequivocally declaring them to be homoplasies. Because there is no room for unequivocal verification of any observational statement, there is no way to know whether the congruent (i.e., supporting) characters are homologies or whether they could be homoplasies (parallelisms). The incongruent characters, however, are refuted as homologies by the accepted phylogenetic hypothesis. This argument can only be made in conjunction with a given or true tree (Farris et al., 2001; Kluge, 2001b:206).

In the world of experience, though, we do not have a given phylogeny; we do not know the true phylogeny. Farris' (1983) asymmetry does not hold for the method that allows us to accept one preferred, that is, most parsimonious, hypothesis over several alternative and competing hypotheses. This is because in that process of choice, some characters, namely the congruent ones, falsify the assumption of homology that was initially made for characters that turn out to be incongruent relative to the most parsimonious hypothesis. Indeed, taking Farris' argument to its logical conclusion, potential homoplasy (parallelism) would be allowed to potentially falsify conjectured homology.

Kluge (2001a:326) finds that an increasing number of taxa (terminals) increases the degree of corroboration, because the number of possible cladograms increases dramatically with an increasing number of terminals. Kluge (2001a:324) interprets Popper's analysis of logical probability of a theory in the sense that the larger the number of possible cladograms, the greater the content of the closed hypothesis set, and hence the greater the degree of falsifiability; as the degree of fasifiability increases, the degree of logical probability decreases. Content (ct) can thus be viewed as the complement of logical probability: $ct(x) = 1 - p(x)$. "Phylogenetic hypotheses have a high empirical content $1-p(h, b)$, with $p(h, b)$ close to zero" (Kluge, 2001a:326; note that $[h, b]$ expresses the empirical content of the hypothesis $[h]$ in the light of the given background knowledge $[b]$). This means that if the logical probability of hypothesis (h), given background knowledge (b), is maximally low, the content of hypothesis h is maximally large (close to 1). Hence, according to Kluge (2001a), degree of corroboration $C(h, e, b) = 1 - p(h, b)$. From this, Kluge concludes that increasing the number of terminals increases the degree of falsifiability and, given critical evidence (e), the degree of corroboration.

The critical evidence (e) is the falsifiers (which determine the severity of test), that is, "one or more synapomorphies" (Kluge, 2001a:326). Popper (1973:49) defined the relative content of a hypothesis as the degree by which it (h) transcends background knowledge (b), in this case descent, with modification. Straining Popperian logic, one could perhaps argue that because descent, with modification is a unique process, an increasing number of possible cladograms increasingly transcends the one single (true; see Appendix II) historical pattern of relationships. In Popperian logic this would also have to mean that the closed hypothesis set with the smallest logical probability, that is, with the smallest logical playground, carries the largest class of forbidden statements.

If one or more synapomorphies are falsifiers, if they qualify as forbidden statements, would this mean that the closed hypothesis set with the smallest logical probability would forbid the largest number of synapomorphies? Evidently, this conclusion needs to be reconsidered in light of the idea that the hypothesis ([A, B] C) generates the forbidden statement that neither (A, C), nor (B, C) must share a synapomorphy (homology). As the closed hypothesis set of possible cladograms increases with an increasing number of terminals, the class of forbidden statements in that latter sense increases. However, an increasing number of terminals is also correlated with an increasing number of instantiations of these forbidden statements (Sanderson and Donoghue, 1989, 1996), yet no falsification occurs because the shift of meaning that occurs as conjectured homology is interpreted in terms of homoplasy (see the discussion above): The observed white raven is either not white or not a raven.

By minimizing instantiations of forbidden statements, that is, by minimizing *ad hoc* hypotheses of homoplasy, one could argue, as did Kluge (2001a:327), that the most parsimonious hypothesis is the least refuted and hence the most highly corroborated. However, corroboration in Popperian terms results from the absence of an instantiation of a forbidden statement, and the degree of corroboration is a historical account of how many times exactly which forbidden statements — all of them preferably independent from one another — were not found to be instantiated. In the absence of Popper's asymmetry of falsification versus verification, the reasoning that the least falsified hypothesis is the most highly corroborated makes little sense; at least it is not Popperian. For that reason, Sober (1988) specifically replaced the terms *verification* and *falsification* with *confirmation* and *disconfirmation* to mark the distinction of his understanding of phylogenetic inference, based on nondeductive inference, from Popper's asymmetrical falsificationism. On that basis it is possible to identify the most parsimonious hypothesis as the least disconfirmed one, but that means that homologies (synapomorphies) symmetrically confirm a given hypothesis of relationship, rather than falsify alternative hypotheses.

There is one more point to discuss in relation to Kluge's (2001a) idea that an increasing number of terminals increases the empirical content of a closed hypothesis set. The notion of a closed hypothesis set indicates that no matter how many terminals are considered, the number of alternatives is finite. This means that although the content of a hypothesis (*h*) may be immensely large, it can logically, in principle, be extensionally defined, that is, by enumeration. This is certainly not what Popper had in mind when talking about the empirical content of a theory, because if a hypothesis has an exhaustible content (e.g., numerically universal statements; Popper, 1976:§13; see also Kitts, 1977), it could, in principle, be verified. Here again it is important to follow Popper in his distinction of "statement" or "proposition" from "observable state of affairs" (Popper 1979:376, 1974a:1007).

Because Popper views theories as universal statements, these universal statements imply an infinite number of instantiations (i.e., observable states of affairs) or conversely, no number of observable states of affairs will unequivocally, or exhaustively, capture the universal statement. If this were possible, the universal statements would in principle be verifiable, and the logical problem of induction would dissolve, as would Popper's asymmetry of falsification versus verification. This, in turn, shows that an increasing number of terminals does not increase the empirical content of a theory or hypothesis in a Popperian sense. What it does is increase the number of possible alternative (historical) hypotheses (of relationship) against which characters are played out. Because the taxa described in the literature are part of our background knowledge, the inclusion of an increasing number of (known) taxa in an analysis does not increase the relative content of (*hb*), but rather represents a problematization of the given background knowledge to a larger degree. Placing a larger segment of background knowledge under critical discussion may be argued to increase the empirical content of a theory (of relationships) in this particular sense, because it results in a "proliferation of [alternative] hypotheses," a slogan found to be important by sophisticated falsificationists (Lakatos, 1974:121). However, as the number of alternative hypotheses of relationships increases, so does the number of alternatives that the characters be distributed across the taxa, from which results an increasing level of homoplasy (Sanderson and Donoghue, 1989, 1996).

What happens in parsimony analysis, in principle, is that all conjectures of synapomorphies (homologies) represent observable states of affairs or data points, which are then matched against all possible alternative hypotheses of relationships. The choice is based on the best fit in terms of parsimony (i.e., minimalizing incongruence or conflict with exclusion; see below, for further discussion).

Likelihood

Fit of data to a hypothesis invokes methods of maximum-likelihood inference. Indeed, Popper's philosophy of science has recently been appealed to in support of likelihood approaches to phylogeny reconstruction (DeQueiroz and Poe, 2001), but this is possible only if Popper is taken out of context (Kluge, 2001a).

The philosophy of likelihood was succinctly outlined in Anthony Edwards' (1992) book on the subject. From that text I conclude that likelihood methods are firmly rooted in nondeductive inference and hence radically opposed to Popper's deductivism. Fisher, the "father of likelihood," is quoted (p. 100) as saying that likelihood must play the same role in inductive science as probability does in deductive science. The axiom of likelihood applies only within the framework of a statistical model (p. 31). The likelihood method starts with some model, the model being defined as "that part of the description which is *not at present in question*, and may be regarded as given" (p.3), a statement that comes dangerously close to Popper's notion of background knowledge. Likelihood analysis prefers the hypothesis that "gives the highest chance to the fact we observe" (p. 199), given stated assumptions of what the data should look like in order to be relevant for the hypothesis. This again recalls Popper's (1989:38) methodological request that falsifying conditions must be specified in advance. Likelihood divorces itself from parsimony as Edwards raises the question (p. 200), "How much simplicity are we prepared to lose for an increase in the excellence of the fit?" [of observed data to the model], or, the other way around, "What increase in support do we require to justify an increase in complexity of the model?" (p. 200).

These rhetorical questions reveal the nondeductive underpinning of the likelihood approach. Edwards admits (p. 203), "The essence of scientific investigation is, first, an acute sense of curiosity, and second a great power of imagination, for the richer the field of imagined explanations, the greater the chance of a satisfying one." Although Popper would have agreed with the importance of curiosity and creativity in the generation of a bold new hypothesis, he would certainly not have looked for evidence satisfying an explanation or a hypothesis, but for evidence potentially falsifying the explanation. The reason is that there exist, logically, an infinite number of hypothetical explanations — or models — relative to the world of experience, and a different degree of likelihood for one such model over some alternatives does not mean much unless the likelihood is known for all possible models. In that sense, maximum likelihood inference is firmly tied to nondeductive inference.

The ties of likelihood to nondeductive inference are further revealed by the fact that likelihood searches for similarity in different situations, so that from the experience of one situation, something may be learned about another, similar situation (Edwards, 1992:199). Evidently, this entails reasoning from past situations to future similar experiences, which cannot be compatible with Popper's notion of degree of corroboration, which says something only about the past performance of a theory but nothing about its future performance (Popper, 1976:§85).

In Section 72 of Logic of Scientific Discovery (1976), Popper toyed with the idea that a metric might (in principle) be developed for logical probability, and that such a metric would replace the notion of logical probability. However, in Appendix IX to the same book, Popper emphasized, "But there cannot be … a metric of logical probability which is based on purely logical considerations". Also in Appendix IX, Popper recognized limiting boundary conditions in which degree of corroboration can be compared to Fisher's likelihood function. The section deals with degree of corroboration in the context of Popper's objective interpretation of probability calculus (see footnote 1). From this section ("A Third Note on

Degree of Corroboration or Confirmation") DeQueiroz and Poe (2001:305) extracted a sentence that reads, "We can interpret ... our measure of degree of corroboration as a generalization of Fisher's likelihood function." Out of context this sentence suggests that Fisher's likelihood function is a special case of degree of corroboration, or as DeQueiroz and Poe (p. 306) concluded, "We argue that Popper's corroboration is based on the general principle of likelihood." This conclusion neglects the fundamental asymmetry that Popper found to distinguish falsification from verification. The full context reads,

> In this, and only in this case, it will therefore be possible to accept Fisher's likelihood function as an adequate measure of degree of corroboration. We can interpret, *vice versa*, our measure of degree of corroboration as a generalization of Fisher's likelihood function; a generalization which covers cases such as a comparatively large δ, in which Fisher's likelihood function would become clearly inadequate. For the likelihood of h in the light of the statistical evidence e should certainly not reach a value close to its maximum merely because (or partly because) the available statistical evidence e was lacking in precision.

The context of this statement must be sought in the specification of rejection criteria for statistical hypotheses given by Popper (see footnote 1). If h is a statistical hypothesis, and e is statistical evidence or a statistical report bearing on h, then Popper's requirements are that the probability of e be very small (i.e., close to 0; see also Faith and Trueman, 2001), but that of h be very large (close to 1), which is evidently possible only with a very large data set. Let e be the statistical report on "thousands or perhaps millions" of observations of tossing a coin z, for example: "among a million of observed tosses of z, heads occurred in 500,000±20 cases (Popper, 1976: Appendix IX)." Given a deviation of ±20 on a million tosses means that hypothesis h (the probability of heads is one in two) makes a very precise prediction, hence the small δ (which stands for the small deviation; as noted by Nagel [1971:398], δ will often be a function of p and n, but may be a factor of other functions as well).

Therefore, if the probability of e is minimal (close to 0), if the probability of h is maximal (close to 1), if the sample size of statistical data is adequately large and if h makes very precise predictions (hence a very small δ), the measure of degree of corroboration and of likelihood will correspond to each other. This is not surprising, however, because the limiting condition is close to total corroboration: "E [explanatory power] and C [degree of corroboration], can only be large if δ is small and n [sample size] is large; or in other words, if e is a *statistical report asserting a good fit in a large sample*" (Popper, 1976:Appendix IX).

A larger δ indicates lack of precision in e, which is accommodated by degree of corroboration (or explanatory power) but not by the likelihood function. As Popper stated, "It is unsatisfactory, not to say paradoxical, that statistical evidence e, based upon a million tosses and δ = 0.00135, may result in numerically the same likelihood ... as would statistical evidence e', based on only a hundred tosses and δ = 0.135."

Having said all of this, it is important to remember, however (Popper, 1976:Appendix IX), "... that non–statistical theories have as a rule a form totally different from that of the h here described," that is, they are of the form of a universal proposition. The question thus becomes whether systematics, or phylogeny reconstruction, can be construed in terms of a statistical theory that satisfies the rejection criteria formulated by Popper (see footnote 1) and that, in case of favorable evidence, allows the comparison of degree of corroboration versus Fisher's likelihood function. As far as phylogenetic analysis is concerned, I found no indication in Popper's writing that history is subject to the same logic as the test of random samples of statistical data. As far as a metric for degree of corroboration relative to a nonstatistical hypothesis is concerned, Popper (1973:58–59; see also footnote 1) clarified,

In fact, nothing can be further removed from my aims. I do not think that degrees of verisimilitude, or a measure of truth content, or falsity content (or, say, degree of corroboration, or even of logical probability) *can ever be numerically determined*, except in certain limiting cases (such as 0 and 1).

Specified that way, the limiting cases are those near complete falsification or verification of a hypothesis — hardly a specification that can be met in the empirical world of experience.

How, then, is it possible to confound Fisher's likelihood function with Popper's notion of degree of corroboration? Popper (1976:Appendix IX) proved to his satisfaction "... that the identification [of his notion] of degree of corroboration or [Carnap's notion of] confirmation with [statistical] probability (and even with likelihood) is absurd on both formal and intuitive grounds: it leads to self-contradiction."

Why have $C(x, y)$ and $P(x, y)$ been confounded so persistently? ... perhaps under the influence of the mistaken view that science, unable to attain certainty, must aim at a kind of 'Ersatz' — at the highest attainable probability ... [But] *Science does not aim, primarily, at higher probabilities. It aims at a high information content, well backed by experience.* (Popper, 1976:Appendix IX)

Popper rejects the probabilification of science and the goal of maximal (statistical) probability, because these are part of the verificationist strategy. For Popper, the content of a theory increases with its (logical) improbability. In contrast, the degree of corroboration for a statistical theory is only of interest if the attempt is being made to refute it, that is, to reduce its probability to 0 or close to 0, because nothing is easier than to select statistical data in support of a statistical hypothesis (Popper 1976:Appendix IX).

Sophisticated Methodological Falsificationism and the Sociology of Science

Is it possible to get more sophisticated (Kluge, 1997), in order to salvage systematics from a nondeductivist interpretation? Is it possible to turn systematics into a research program, at the core of which the logically subordinated hierarchy is accepted by *fiat* or *a priori*? Could we then move ahead and, acknowledging the fact that unequivocal falsification is not possible in practice, shift the emphasis to corroboration? Finally, could we use the finite number of cladograms that obtain for any given number of terminals as the competing theories of a dynamic research program, which have to be evaluated relative to one another in terms of relative degrees of corroboration? Evidently, in doing so, the emphasis shifts from Popper to Lakatos (1974). In the context of this philosophical model, the basic problem of empiricism, that is, the acceptance of basic statements reporting on observations, remains the same — whether those observations are found to be in compliance with, or in contradiction to, the theory from which a prediction was derived (Lakatos, 1974:127). Popper solved this problem methodologically, essentially by declaring it a decision to be taken in the court of public opinion (or rather, in the court of the scientific establishment, so as not to get too close to Feyerabend's relativism), and Lakatos has offered nothing better in this regard. He merely shifted the emphasis from falsifying to corroborating instances, or from falsification to corroboration, thereby transforming Popper's asymmetry into a dichotomy, or rather symmetry (see footnote 22).

Popper always emphasized the conventional aspect of the decision that is taken when a basic statement is accepted. Lakatos (1974:12, 131) acknowledges that "sophisticated methodological falsificationism" learned from the conventionalists the "importance of decisions in methodology," and from empiricists it "inherited the determination to learn primarily from experience." This evidently must involve decisions on observational reports, or basic statements, because without such decisions there would be no way to decide whether a problem shift is progressive or degenerating (Lakatos, 1974:127). To this basic conventional element in science, Lakatos added another conventional element,[20] and that is the conventional decision not to attack the core of a research program, but only its protective belt. Whoever directs the *modus tollens* against the core of a research program excludes herself by that very fact from that research program (Chalmers, 1986). The unproblematic dogmatism that characterizes Popper's background knowledge thus becomes a conventional decision: As a member of a research program, I will not question its core, because if I do so, I have to bear the social consequences. This is a juncture where Popper did not want to go: the sociology of science. It is also the first juncture at which the language of science threatens to break down, as seems to be happening today between proponents of parsimony and proponents of likelihood methods.

According to Lakatos, a research program is designed to evaluate critically competing theories. This evaluation takes place in terms of degree of corroboration, which evidently leads back to Popper. Among the competing theories, the most highly corroborated should be the preferred one. The degree of corroboration depends on the severity of test, while the severity of test defines the explanatory power of a theory (Popper 1989:Addendum 2). The severity of test, in turn, depends on the relative empirical content, and the relative empirical content is larger, the more a theory transcends background knowledge. Indeed, Popper's notion of the relative content of a theory corresponds exactly to the criterion on the basis of which Lakatos (1974:117–121) evaluates competing theories in a progressive (rather than degenerative[21]) research program. The new knowledge claimed by a theory is what defines its empirical content, and the bolder the theory is, the greater the severity of test.

Lakatos needs a critical evaluation of theories in his system just as much as Popper, and just like Popper he did not allow *ad hoc* auxiliary hypotheses that are not independently testable to distort such critical evaluation of competing theories. Both Lakatos and Popper understood hypotheses in the sense of universal propositions and both employed the *modus tollens* form of argument. Whether the emphasis is on corroboration or falsification, this very distinction cannot be had outside Popper's solution to the logical problem of induction, for it only can provide that very asymmetry, and hence the basis on which to distinguish falsificationism from verificationism.[22]

---

[20] Lakatos (1974) argues that Popperian falsificationism, in whichever guise, subscribed to four kinds of conventional, methodological decisions. These are (1a) to make, by *fiat*, some singular statements unfalsifiable (p. 106) (not correct); (1b) to accept singular statements (correct); (2) to tentatively accept background knowledge as unproblematic (correct); (3) to provide rules under which probabilistic theories may be rejected (correct); and (4) given the *ceteris paribus* clause, to make a decision as to whether a falsifying instance can count as falsification of the specific theory under test (correct). As long as Lakatos (1974) separates corroboration from confirmation (p. 115), sophisticated methodological falsificationism cannot avoid decisions (1b, 2, 3) (p. 127). Lakatos' argument (p. 125) for redundancy of decision (4) in sophisticated falsificationism is contradicted by his remarks that the articulation of all background knowledge is a piecemeal process (p. 131), subject to conventionally drawn limits. The fifth decision discussed by Lakatos concerns the elimination of syntactically metaphysical statements, or series of statements thereof, from scientific discussion. In that regard, Lakatos (p. 126) specifies, "If an auxiliary theory ... produces novel facts (that is, it is 'independently testable') Cartesian mechanics should be regarded as good, scientific ... ." According to Popper, it would not be scientific, but it would be good, as it would generate science historically (see footnote 6). To protect the hard core of a research program from the *modus tollens* by *fiat* is an expanded version of decision 1a not inherent in Popper's philosophy of science.

[21] Note that accordingly, a research program starts to degenerate if it becomes unable to generate new facts in the sense of corroborated conjectures that transcended given background knowledge when they were first conjectured. This follows from the fact that once corroborated, knowledge claims are relegated to background knowledge. Sophisticated methodological falsificationism therefore does not provide a basis for the artificial reduction of background knowledge to the necessary and sufficient.

Sober (1988; see also Kitts, 1977) already provided conclusive arguments that parsimony analysis lacks the fundamental asymmetry of falsification versus verification that characterizes hypothetico-deductive systems *sensu* Popper. The lack of asymmetry of verification versus falsification in parsimony analysis makes it impossible to distinguish characters as falsifiers of alternative hypotheses of relationships, as opposed to characters supporting (or confirming; Sober, 1988) the most parsimonious hypothesis of relationships. There prevails a symmetry of verification versus falsification in parsimony analysis, a hallmark of all inductive systems according to Popper (1979:256). Farris' (1983) paper has become the classic defense of parsimony analysis in falsificationist terms, because it links parsimony to explanatory power, but the missing asymmetry is evident even in this paper, as sentences such as the following make clear (Farris, 1983:11): "In either case the decision is made by accepting the stronger body of evidence over the weaker, and *ad hoc* hypotheses of homoplasy are required to the extent that evidence must be dismissed in order to *defend* the conclusion." A Popperian falsificationist will not defend a conclusion by dismissing contrary evidence *ad hoc* or by appealing to a stronger body of evidence, especially if the strength of the body of evidence is measured in numbers of confirming instantiations (of evidence). Instead, a falsificationist will seek a greater severity of test by decreasing the logical playground of a hypothesis in order to increase its empirical content and hence render it more severely testable by forcing it to make more precise predictions. This, evidently, cannot be achieved in systematics.

From a Popperian perspective, the appeal to a stronger body of evidence in terms of numbers of characters is highly problematical, in as much as it is impossible to fully describe particulars in universal terms. As long as universal terms rather than proper names are employed, any attempt at a most exacting description of a particular will logically still be applicable to an infinite class of particulars. Conversely, whereas a particular can instantiate universals, those universals cannot exhaust the particularity of that particular. This means that every particular can be described by an infinity of universal terms. Again, this is a logical problem, which originates from the logical relations between universals and particulars. It is not a problem of empirical similarity or differences of things, but there are important consequences to these considerations: They reveal that any particular does not consist of characters, that it is we who bestow characters on particulars, and that we do so arbitrarily (Popper 1979:375–376).

If character delimitation is arbitrary, though, it becomes evident that there cannot be, in a logical sense, any such thing as character saturation, because every character — once conceptualized — can be subdivided, theoretically (but not in empirical practice; Körner, 1970) infinitely so (Vane-Wright [2001:592] deplores the lack of a sharp "stopping rule" for practical character subdivision.). On the basis of the principle of continuity, the subdivision of characters is only limited by limited semantic or mathematical skills. The character evidence can always be bent toward the preferred hypothesis of relationships by subdivision of those characters that were initially found in its support. Patterson (1982) praised the test of congruence because it rendered *a priori* character weighting obsolete: "Characters weigh themselves" was his slogan (1982). Today, from a Popperian perspective, one would have

---

[22] Lakatos' emphasis on corroboration can be viewed as turning Popper's asymmetry into a dichotomy, indeed symmetry, of falsification and corroboration. This follows from what Popper (1974:1005ff) identified as Lakatos' central thesis, which is that "exactly the most admired scientific theories simply fail to forbid any observable state of affairs" (Lakatos 1974:100). Lakatos proceeds to recount the story of an imaginary account of planetary misbehavior in support of his thesis (the story is a variation of the discovery of Neptune). Ironically, this story is built on the class of forbidden statements derived from Newtonian mechanics (why else would this be a case of planetary misbehavior?), and merely portrays a researcher who is unwilling to concede, or accept, the occurrence of a forbidden instance. Popper (1974:1005ff) took very serious issue with, and refuted, Lakatos' central thesis.

to conclude that characters do not test each other in parsimony analysis; they merely outweigh or outnumber each other.

Bonde (1974:563) early on recognized the absence of the asymmetry of falsification versus verification in parsimony analysis, and he concluded,

> The difference ... may seem merely a semantic one — the two methods must in principle lead to the same results — but philosophically it is quite important to recognize testability as more valuable (more 'scientific') than confirmation (the difference to some degree reflects that between natural sciences as Popper would have preferred them to be, and sciences as they are in the actual world ...).

This seems to imply that Popper's logic of scientific discovery is possible in theoretical sciences only, whereas the empirical world of the systematists is tied to nondeductivist procedures. Popper did, indeed, note that "from the point of view of a theory of cognition, theoretical sciences are far more interesting. The problem of induction also applies to them only (Popper, 1979:242).

The Basic Problem of Systematics

We appear to have hit rock bottom in the theoretical foundation of parsimony analysis in terms of Popper's falsificationism. We have arrived at the second juncture where the language of science threatens to break down, as is evident from stagnant debates in systematics (Rieppel and Kearney, 2001, 2002). However, is it possible to gain something from Bonde's insight, namely that it is important to maintain a falsificationist language, only because it is more valuable? In the absence of an asymmetry between verification and falsification, it would seem to be possible, at least semantically, to maintain a falsificationist language (that is symmetrical to verificationist language).

To maintain a falsificationist language can be more valuable only if it leads to the insight that the impossibility of testing (in its weakest sense of "critical; discussion") assumptions of homology versus *ad hoc* hypotheses of homoplasy independently constitutes a serious problem. This problem leads to the search for independent tests of character hypotheses — independent, that is, of the test of congruence. It also leads to a different understanding of Hennig's auxiliary principle than is implied in parsimony analysis: Assume homology in the absence of contrary evidence, but contrary evidence is to be derived from a critical discussion of character hypotheses in themselves, not merely from the reciprocal relationships among all characters. A falsificationist language leads to the basic problem of systematics, indeed, the basic problem of all empirical science in general (*Basisprobleme sensu* Popper, 1979), because characters will have to be seen as basic statements *sensu* Popper.

Basic statements are singular statements rooted in the world of experience that may falsify a given hypothesis. Recognizing, however, as Popper did, that there is no theory-free observation, there is no way to know for sure whether a basic statement indeed falsifies a theory or whether the basic statement might not be false in itself. This results in the requirement that basic statements are to be testable in themselves, that is, they are to be formulated and treated as hypotheses of a lower level of generality. Yet carried to its logical conclusion, this requirement results in an infinite regress, or "degress" *sensu* Popper (1976:§29), because any basic statement can only be justified by another basic statement (Popper, 1979:434; see also the problem of the transcendence inherent in any description discussed above).

Although unproblematic from a logical point of view, because no requirement nor any desire to prove any statement is involved, this degress still requires that in order to get on with the practical business of empirical science, the necessity will arise at some point to accept a basic statement, not because it is true, but as a matter of convention or, as Popper describes it, as a matter of a decision to be taken in the court of public opinion.[23] This may smack of dogmatism, but again, it is an unproblematic kind of dogmatism, since anybody can subject any accepted basic statement at any time to renewed critical discussion. However, since intersubjective agreement is to be achieved relative to the (provisional) acceptance of basic statements, the method that is used to arrive at these basic statements must be made explicit, and — according to Popper — basic statements must refer to repeatable instances. Disagreement about basic statements cannot be based on subjective conviction or belief, and if disagreement exists about the methods employed to obtain acceptable basic statements, then new methods will have to be devised and submitted to critical discussion.

Because character hypotheses are essentially descriptive statements, they cannot be applied to repeatable instances in the same sense that basic statements can in experimental sciences, which renders their critical discussion more difficult. Rieppel and Kearney (2002) suggested the classical criteria of homology, specifically topology and connectivity, as methodological tools in the critical discussion of character hypotheses *qua* basic statements, tools that also allow the rejection of character hypotheses. Wiley (1975) appealed to "morphological testing" in the context of a critical discussion of characters, allowing for predictions such as maintenance of topological correspondence or connectivity at ever finer levels of structural analysis, including structural analysis throughout ontogeny. Evidently, this approach is not foolproof, and because it also is threatened by an infinite regress, it makes it necessary for the court of public opinion to decide whether or not a certain character hypothesis should, here and now, be provisionally accepted. At least this approach provides the jury with some methodological rules upon which to base its deliberations, methodological rules that go beyond the mere counting of what theoretically constitutes infinitely divisible character delimitations. Maybe topology and connectivity, although historically eminently successful, are not the ultimate rules upon which to base the critical discussion of character hypotheses, but if alternative rules of character conceptualization are to be invoked, they must be explicitly formulated and subjected to critical discussion.

Stagnant debates in contemporary systematics will remain so if systematics does not solve its basic problem. Ever more and faster algorithms can be devised to generate hypotheses of relationships and ever more statistical measures can be used to place confidence limits on those hypotheses. These do not, however, solve the basic problem of systematics, which is the nature of character hypotheses, and the problem of their critical discussion.

Synthesis

Popper (1976) formulated his philosophy very much in contradistinction to the Vienna Circle and the logical positivism that it generated (Ayer, 1952). This is perhaps most strikingly illustrated with reference to logical probability. For Carnap, logical probability corresponded to degree of confirmation (Carnap, 1995; Körner, 1970); for Popper, logical probability is the inverse of the degree of corroboration. Where Carnap felt that hypotheses could be

---

[23] The appeal to the court of public opinion reflects back on Carnap's criterion for the truth of protocol statements, which needs to be established by "accredited observers, notably the scientists of our era" (Ayer, 1959:233). The court of public opinion establishes intersubjective agreement on the observational report, all the while assuring conformity of the use of language with the rules governing the meaning of words. Popper (1979:133–134; *Robinsoneinwand* [see discussion by Foster, 1992]) acknowledged that there is no logical reason why Robinson Crusoe should not be able to develop a science of physics comparable to our own, even if he spoke a different language, but he rejects the characterization of science by its results rather than by its method, and asserts that Robinson would sooner or later have gone down the wrong road in his scientific endeavor, adhering to a hypothesis that would have failed if exposed to intersubjective scrutiny.

confirmed more and more (Mayhall, 2002:11), Popper claimed that degree of confirmation provided a historical account only of the past performance of a theory, with no predictive value as to its future performance. Nevertheless, Popper's (1976) falsificationism shares a lot of common ground with the verificationist program of logical positivism (Ayer, 1952). Both philosophical systems operate — in practice — in a layered world of multiple competing theories coupled with background assumptions or initial condition; both use theories, or hypotheses, as bases of predictions; both acknowledge that ultimate, or finite, verification or falsification of any specific theory cannot be achieved in practice; and both agree that conventional rules or considerations must play a role in any decision to declare a specific theory as — relatively speaking — true or false. The most critical issue that sharply separates Popper's (1976) falsificationism from verificationism is how he deals with the logical problem of induction.

For logical positivists, the problem of induction, "as it is ordinarily conceived" (Ayer, 1952:50), is a "fictitious problem," for it offers only two alternatives. These are the deduction of an empirical proposition either from a purely formal (analytic) principle or from an empirical principle. In the former case, one is accused of committing the error of supposing that a proposition about an observable state of affairs can be deduced from a tautology. In the second case one sets out to prove what one had assumed in the first place (Ayer, 1952:49), but since the principle of the uniformity of nature only restates that the past is a reliable guide to the future, and since not even a theoretical solution to the problem of induction (as ordinarily conceived) seems possible, the problem of induction reveals itself as a pseudo-problem, because "all genuine problems are at least theoretically capable of being solved" (Ayer, 1952:50). What remains is the probabilification of science (i.e., treating scientific hypotheses as statistical hypotheses) and the symmetry of inductive systems in which the degree of verification (rather, confirmation) corresponds to the inverse of the degree of falsification (rather, disconfirmation) of a hypothesis (Mayhall, 2002; Sober, 1988).

Popper (1976) believed that he provided a theoretical (logical) solution to the problem of induction, but in order to do so he had to reformulate it in terms of universal propositions. This resulted in the recognition of the asymmetry of verification (which, on logical grounds, is never possible) versus falsification (which, on logical grounds, is possible) as Popper directed the *modus tollens* against induction. Popper's (1976) solution works in theory, in logical space, but not in practice — hence Popper's (1979:242) preference for theoretical science and his mention that the problem of induction applies to it only. This is because in practice, the negative existential statement that can logically be deduced from a universal proposition needs to be applied to, or identified with, an empirical base (Körner, 1959). However, this application, or identity, will always remain incomplete — hence the layered world of competing theories and the need for conventional elements in any decision to consider any specific theory to be false.

Parsimony analysis as it is practiced in cladistics reveals the symmetry of verification (confirmation *sensu* Sober, 1988) and falsification (disconfirmation *sensu* Sober, 1988) that is a hallmark of inductive systems (Popper 1979:256). The degree of support that characters lend to a specific hypothesis of relationship mirrors the degree of disconfirmation of alternative hypotheses of relationships. Parsimony analysis is thus inherently inductive and hence probabilistic (Vane-Wright, 2001; see also Rieppel, 1988:166).

This is well documented by the appeal to "total evidence" (Kluge, 1989), which is a methodological requirement invoked for cases in which inductive reasoning yields contradictory (or inconsistent) results (e.g., Hempel, 1965:63–64). The necessity for an appeal to total evidence highlights the inconsistent (i.e., contradictory) distribution of characters across taxa, whereby parsimony is a method that sorts out characters and terminals in terms of a logically subordinated hierarchy in a pattern that maximizes support for inclusion but minimizes conflict. By allowing conflict, parsimony analysis accounts for the messiness of the empirical world, in contrast to clique analysis (Estabrook, 1972; Estrabrook et al., 1977), which dismisses conflict with overlap. However, by allowing conflict with overlap, parsimony analysis also acknowledges that we are not able to pick out only those characters that carry phylogenetic information in all cases (if we could we would have the true tree of life). There can be no doubt, however, that parsimony analysis results in the most economical classification of organisms based on all data at hand.

The theory of evolution pictures a continuous genealogical nexus, within which species or taxa can be nothing but interconnected components or chunks (Hull, 1999a). The logically subordinated hierarchy will impose sharp boundaries on that continuum, yet this cannot fully succeed. The representation of patterns of phylogenetic relationships will therefore necessarily be blurred. One can view the two pictures, one of the continuity of change, the other of the discontinuity of hierarchical relationships, as complementary (Rieppel, 1988). If looked at from this perspective, the complementarity inherent in patterns versus process analysis should warn against too much of a realistic interpretation of our theories of the history of Earth relationships.

Acknowledgments

I am very grateful to David Williams and Peter Forey, who provided this paper with a home and offered important input. Additional invaluable help with this text, which is greatly acknowledged, came from Richard I. Vane-Wright, Maureen Kearney, Kevin Padian and Peter Wagner, as well as from my two sons, Michael and Lukas Rieppel. All deficiencies of this paper are, of course, my own responsibility.

References

Ayer, A.J., *Language, Truth, and Knowledge*, Dover, New York, 1952.

Ayer, A.J., Verification and experience, in *Logical Positivism*, Ayer, A.J., Ed., Free Press, Glencoe IL, 1959.

Ayer, A.J., Reply to F. Miró Quesada, in *The Philosophy of A.J. Ayer*, Hahn, L.E., Ed., Open Court, La Salle, IL, 1992.

Ball, I.R., Implication, conditionality and taxonomic statements, *Bijdr. Dierkd.*, 52, 186–190, 1982.

Bonde, N., Review of 'Interrelationships of Fishes,' *Syst. Zool.*, 23, 562–569, 1974.

Boyd, R., Confirmation, semantics, and the interpretation of scientific theories, in *The Philosophy of Science*, Boyd, R., Gasper, P., and Trout, J.D., Eds., MIT Press, Cambridge, MA, 1991, pp. 3–35.

Boyd, R., Homeostasis, species, and higher taxa, in *Species: New Interdisciplinary Essays*, Wilson, R.A., Ed., MIT Press, Cambridge, MA, 1999, pp. 141–185.

Brady, R.H., Parsimony, hierarchy, and biological implications, in *Advances in Cladistics*, Vol. 2, Platnick, N.I. and Funk, A., Eds., Columbia University Press, New York, 1983, pp. 49–60.

Brady, R.H., On the independence of systematics, *Cladistics*, 1, 113–126, 1985.

Carnap, R., *An Introduction to the Philosophy of Science*, Gardner, M., Ed., Dover, New York, 1995.

Cartmill, M., Hypothesis testing and phylogenetic reconstruction, *Z. Zool. Syst. Evol.*, 19, 73–96, 1981.

Chalmers, A.F., *Wege der Wissenschaft*, Springer, Berlin, 1986.

Darwin, C., *On the Origin of Species by Means of Natural Selection, or, The Preservation of Favoured Races in the Struggle for Life*, John Murray, London, 1859.

DeQueiroz, K. and Poe, S., Philosophy and phylogenetic inference: a comparison of likelihood and parsimony methods in the context of Karl Popper's writings on corroboration, *Syst. Biol.*, 50, 305–321, 2001.

De Waal, C., *On Peirce*, Wadsworth/Thomson, Belmont, CA, 2001.

Dupré, J., On the impossibility of a monistic account of species, in *Species: New Interdisciplinary Essays*, Wilson, R.A., Ed., MIT Press, Cambridge, MA, 1999, pp. 3–22.

Edwards, A.W.F., *Likelihood: Expanded Edition*, The Johns Hopkins University Press, Baltimore, 1992.

Estabrook, G.F., Cladistic methodology: a discussion of the theoretical basis for the induction of evolutionary history, *Annu. Rev. Ecol. Syst.*, 3, 427–456, 1972.

Estabrook, G.F., Strach, J.G., and Fiala, K.L., An application of compatibility analysis to the Blackiths' data on orthopteroid insects, *Syst. Zool.*, 26, 269–276, 1977.

Faith, D.P. and Trueman W.H., Towards an inclusive philosophy for phylogenetic inference, *Syst. Biol.*, 50, 331–350, 2001.

Farris, S.J., The logical basis of phylogenetic analysis, in *Advances in Cladistics*, Vol. 2, Platnick, N.I. and Funk, V.A., Eds., Columbia University Press, New York, 1983, pp. 7–36.

Farris, J.S., Kluge, A.G., and Carpenter, J.M., Popper and likelihood versus "Popper,*"* *Syst. Biol.*, 50, 438–444, 2001.

Foster, J., The construction of the physical world, in *The Philosophy of A.J. Ayer*, Hahn, L.E., Ed., Open Court, La Salle, IL, 1992, pp 179–197.

Foucault, M., *Die Ordnung der Dinge*, Suhrkamp, Frankfurt, 1971.

Gaffney, E.S., An introduction to the logic of phylogeny reconstruction, in *Phylogenetic Analysis and Paleontology*, Cracraft, J. and Eldredge, N., Eds., Columbia University Press, New York, 1979, pp. 79–111.

Ghiselin, M.T., A radical solution to the species problem, *Syst. Zool.*, 23, 536–554, 1974.

Gibson, R.F., Jr., *The Philosophy of W.V. Quine: An Expository Essay*, University Presses of Florida, Tampa, 1982.

Grant, P.R. and Grant B.R., Unpredictable evolution in a 30-year study of Darwin's finches, *Science*, 296, 707–714, 2002.

Gregg, J.R., *The Language of Taxonomy*, Columbia University Press, New York, 1954.

Hempel, C.G., *Aspects of Scientific Explanation and Other Essays in the Philosophy of Science*, Collier, Macmillan, London, 1965.

Hennig, W., *Phylogenetic Systematics*, University of Illinois Press, Urbana, 1966.

Hesse, H., *The Glass Bead Game*, Holt, Rinehart and Winston, Austin, Texas, (1946) 1970.

Hull, D.L., Are species individuals? *Syst. Zool.*, 25, 174–191, 1976.

Hull, D.L., On the plurality of species: questioning the party line, in *Species: New Interdisciplinary Essays*, Wilson, R.A., Ed., MIT Press, Cambridge, MA, 1999a, pp. 23–48.

Hull, D.L., The use and abuse of Sir Karl Popper, *Biol. Philos.*, 14, 481–504, 1999b.

Kitts, D.B., Karl Popper, verifiability, and systematic zoology, *Syst. Zool.*, 26, 185–194, 1977.

Kluge, A.G., A concern for evidence and a phylogenetic hypothesis of relationships among *Epicrates* (Boidae, Serpentes), *Syst. Zool.*, 38, 7–25, 1989.

Kluge, A.G., Sophisticated falsification and research cycles: consequences for differential character weighting in phylogenetic systematics, *Zool. Scripta*, 26, 349–360, 1997.

Kluge, A.G., Philosophical conjectures and their refutation, *Syst. Biol.*, 50, 322–330, 2001a.

Kluge, A.G., Parsimony with and without scientific justification, *Cladistics*, 17, 199–210, 2001b.

Kluge, A.G. and Farris, J.S., Quantitative phyletics and the evolution of anurans, *Syst. Zool.*, 18, 1–32, 1969.

Körner, S., *Conceptual Thinking: A Logical Enquiry*, Dover Books, New York, 1959.

Körner, S., *Erfahrung und Theorie*, Suhrkamp, Frankfurt, 1970.

Kuhn, T.S., Reflections on my critics, in *Criticism and the Growth of Knowledge*, Lakatos, I. and Musgrave, A., Eds., Cambridge University Press, Cambridge, 1974a.

Kuhn, T.S., Logic of discovery or psychology of research, in *Criticism and the Growth of Knowledge*, Lakatos, I. and Musgrave, A., Eds., Cambridge University Press, Cambridge, 1974b.

Kulenkampff, J., David Hume (1711–1776), in *Klassiker der Philosophie*, Vol. I, Höffe, O., Ed., C.H. Beck, Munich, 1981, 434–456.

Lakatos, I., Falsification and the methodology of scientific research programs, in *Criticism and the Growth of Knowledge*, Lakatos, I. and Musgrave A., Eds., Cambridge University Press, Cambridge, 1974.

Lee, M.S.Y., Divergent, evolution, hierarchy and cladistics, *Zool. Scripta*, 31, 217–219, 2002.

Lorenz, K., *Die Rückseite des Spiegels*, Piper, Munich, 1973.

Mallet, J., The speciation revolution, *J. Evol. Biol.*, 14, 887–888, 1973.

Mayhall, C.W., *On Carnap*, Wadsworth/Thomson, Belmont, CA, 2002.

Nagel, E., Principles of the theory of probability, in *Foundations of the Unity of Science*, Vol. I, Neurath, O., Carnap, R., and Morris, C., Eds., The University of Chicago Press, Chicago, 1971.

Mayr, E., *The Growth of Biological Thought*, Belknap Press of Harvard University Press, Cambridge, MA, 1982.

Panchen, A.L., *Classification, Evolution, and the Nature of Biology*, Cambridge University Press, Cambridge, 1992.

Patterson, C., Morphological characters and homology, in *Problems of Phylogenetic Reconstruction*, Joysey, K.A., Friday, A.E., Eds., Academic Press, London, 1982, pp. 21–74.

Pears, D.F., Ayer's views on meaning-rules, in *The Philosophy of A.J. Ayer*, Hahn, L.E., Ed., Open Court, La Salle, IL, 1992.

Platnick, N.I., Philosophy and the transformation of cladistics, *Syst. Zool.*, 28, 537–546, 1979.

Platnick, N.I. and Gaffney, E.S., Evolutionary biology: a Popperian perspective, *Syst. Zool.*, 27, 137–141, 1978.

Piaget, J., *The Child and Reality*, Frederick Muller, London, 1972.

Popper, K.R., *The Poverty of Historicism*, Routledge, Kegan Paul, London, 1961.

Popper, K.R., *Objective Knowledge: An Evolutionary Approach*, Clarendon Press, Oxford, 1973.

Popper, K.R., Replies to my critics, in *The Philosophy of Karl Popper*, Vol. 2, Schilpp, P.A., Ed., Open Court, La Salle, IL, 1974a, 961–1197.

Popper, K.R., Autobiography, in *The Philosophy of Karl Popper*, Vol. 1, Schilpp, P.A., Ed., Open Court, La Salle, IL, 1974b, 1–181.

Popper, K.R., *Logik der Forschung*, J.C.B. Mohr (Paul Siebeck), Tübingen, Germany, 1976.

Popper, K.R., Part I, in *The Self and its Brain: An Argument for Interactionism*, Popper, K.R. and Eccles, J.C., Springer International, Heidelberg, 1977, 3–223.

Popper, K.R., *Die beiden Grundprobleme der Erkenntnistheorie*, J.C.B. Mohr (Paul Siebeck), Tübingen, Germany, 1979.

Popper, K.R., Evolution, *New Sci.* 87 (21 August), 611, 1980.

Popper, K.R., *Realism and the Aim of Science*, Routledge, London, 1983.

Popper, K.R., *The Open Universe: An Argument for Indeterminism*, Routledge, London, 1988.

Popper, K.R., *Conjectures and Refutations*, Routledge, London, 1989.

Rieppel, O., Species are individuals: a review and critique of the argument, *Evol. Biol.*, 20, 283–317, 1986.

Rieppel, O., Pattern and process: the early classification of snakes, *Biol. J. Linn. Soc.*, 31, 405–420, 1987.

Rieppel, O., *Fundamentals of Comparative Biology*, Birkhäuser Verlag, Basel, Switzerland, 1988.

Rieppel, O. and Kearney, M., The origin of snakes: limits of a scientific debate, *Biologist*, 48, 110–114, 2001.

Rieppel, O. and Kearney, M., Similarity, *Biol. J. Linn. Soc.*, 75, 59–82, 2002.

Rosenberg, A., *The Structure of Biological Science*, Cambridge University Press, Cambridge, 1985.

Sanderson, M.J. and Donoghue, M.J., Patterns of variation in levels of homoplasy, *Evolution*, 43, 1781–1795, 1989.

Sanderson, M.J. and Donoghue, M.J., The relationship between homoplasy and confidence in a phylogenetic tree, in *Homoplasy: The Recurrence of Similarity in Evolution*, Sanderson, M.J. and Hufford, L., Eds., Academic Press, San Diego, CA, 1996, pp. 67–89.

Schrödinger, E., *What is Life? With Mind and Matter and Autobiographical Sketches.* Cambridge University Press, Cambridge, 1969.

Sober, E., *Reconstructing the Past: Parsimony, Evolution, and Inference,* MIT Press, Cambridge, MA, 1988.

Tscha Hung, Ayer and the Vienna Circle, *The Philosophy of A.J. Ayer,* Hahn, L.E., Ed., Open Court, La Salle, IL, 1992, 279–300.

Vane-Wright, R.I., Species concepts, in *Global Biodiversity,* Groombridge, B., Ed., Chapman & Hall, London, 1992.

Vane -Wright, R.I., Taxonomy, methods of, in *Encyclopedia of Biodiversity,* Levin, S., Ed., 5, 589–606. Academic Press, New York, 2001.

Weiner, J., *Frege,* Oxford University Press, Oxford, 1999.

Wiley, E.O., Karl R. Popper, systematics, and classification: a reply to Walter Bock and other evolutionary taxonomists, *Syst. Zool.,* 24, 233–243, 1975.

Wiley, E.O., *Phylogenetics: The Theory and Practice of Phylogenetic Systematics,* John Wiley & Sons, New York, 1981.

Wittgenstein, L., *Tractatus Logico-Philosophicus* (trans. Ogden, C.K.), Routledge, Kegan Paul, London, 1922.

Appendix I – An Exercise in the Logic of Phylogenetic Systematics

All quotes below are taken from Kluge (2001b). L and R refer to the left and right columns, respectively, on the pages indicated.

Explanation

p. 199R: "... explanation is achieved by deducing effect from cause, in light of an explaining law or general theory ..."

p. 200L: "Scientific explanation occurs when the elements in a hypothesis are connected to the causal mechanism(s) responsible for a phenomenon."

Comment: Two elements are apparent in these quotes: deduction (of effect from cause) and connection (of the elements of a hypothesis with causal mechanisms underlying observable phenomena). In light of what is to follow below, these two elements will be interpreted in terms of Hume's notion of causality (which presumably violates Kluge's [2001b] intent but addresses the quote of Hume discussed below): A constant conjunction of phenomena is suggestive of causality, that is, of a connection of cause with effect. According to Hume, however, causal relations cannot be obtained through observation; instead, the relation of causality obtains from the (past) experience of constant conjunction of events. Likewise, Hume did not consider the conjunction of the statements that describe two events to be analytical (in the sense that the first statement logically entails the second, or, in Popper's [1983:132] words, the *explicans logically* entails the *explicandum*). According to Hume, our trust in causal relations is habitually conditioned — a habit, moreover, that relies on the principle of uniformity of nature: Because the sun rose every day ever since there were humans on the planet Earth, we hold it for certain that the sun will rise again tomorrow, although the negation of that sentence is not self-contradictory (Kulenkampff, 1981).

In Popperian falsificationism, a scientific explanation consists of the reduction of the known (the observable state of affairs, singular statement) to the unknown (a natural law, universal proposition) (Popper, 1989:63). A causal explanation consists, in general, of a "set of statements one of which describes the observable state of affairs to be explained (the *explicandum*), while the others, the explanatory statements, form the 'explanation' in the narrower sense of the word, (the *explicans* of the *explicandum*)" (Popper, 1983:132). To causally explain some state of affairs therefore means to deduce a sentence that describes the state of affairs from laws and initial conditions (Popper, 1976:§12). In order for this to be possible, the law must consist of a universal statement or proposition, and the state of affairs deduced from it must be repeatable, given identical initial conditions. "If we speak of causality, we always speak of the regularity, of the lawfulness of states of affairs (events), not of a unique, non-repeatable coincidence of certain singular states of affairs (events)" (Popper, 1979:103). If causal explanation means to deduce some state of affair from a natural law, the *explicans* must logically entail the *explicandum* (Popper, 1983:132).

Justification

p. 200L: "Empirical scientists claim their research leads to increased knowledge, through explanation ... It seems equally fair to say, however, that few researchers actually ... *justify* their inferential methods and results ... Hume (1739) long ago recognized the importance of justification, arguing that knowledge must be distinguished from belief and concluding that knowledge is belief based on rational justification ..."

Comment: If the *explicans* does not logically entail the *explicandum*, that is, if the belief in causal relationships is habitual, based on the past experience of a constant conjunction of events (as was argued by Hume; see above), then the concept of causal relations requires positive justification if some (future) effect is to be inferred from past experience of a constant conjunction of events.

The asymmetry of falsification versus verification discovered by Popper in his solution to the logical problem of induction transcends Hume's skepticism. As was shown by Popper, a justification of any theory on grounds of positive reasons is impossible without taking recourse to "special knowledge" (Popper, 1983:19, 21, 54–55, 1989:29). Instead of justifying a theory by providing positive reasons, it is — in principle — possible to defend (rather than justify) the preference for one over other competing theories on the basis of critical reasons (Popper, 1983:21).

p. 200L: The "nature and importance of rational justification" of a scientific method is illustrated by an "example of enumerative induction."

Comment: Since enumerative induction is not hypothetico-deductivist, the example is irrelevant to a Popperian, or falsificationist, interpretation of cladistics. A scientific theory justifies a method only insofar as the method depends on the theory in the sense that the method must generate an observational report that is relevant to the theory. If in scientific explanation the *explicans* logically entails the *explicandum*, as was argued by Popper, it must be the theory that determines what observations are relevant, not the method (Körner, 1970).

p. 200R : "Descent with modification" (law or general theory) plus cladogram (cause) allow the explanation of synapomorphy (*qua* homology due to inheritance).

Comment: The cladogram as structure is *a priori* relative to the theory (phylogeny) it mediates, not a cause. The explanation of synapomorphy as homology lies in descent, with modification. Taken as a law or general theory (of biology), descent, with modification forbids little more than spontaneous generation. This biological law might be used to establish the factual nature of evolution, in the sense that all organisms descend from other organisms, with modification. However, having established that certainty does not allow the generation of (logically entailed) predictions about how evolution proceeds, neither in terms of pattern (how are organisms related to one another?), not in terms of process (what are the causes of descent, with modification?): The *explicans* (descent, with modification) does not logically entail the *explicandum* (state of affairs *qua* character distribution on a tree; event *qua* inheritance).

p. 203R: "This leaves unanswered what guiding theory is used to collect data ..."

Conclusion: With regard to Darwin's theory of evolution (Darwin, 1859) and phylogenetic parsimony as the method of choice for phylogeny reconstruction, it needs to be shown what exactly the theoretical components of descent, with modification are that mandate parsimony analysis and how the cladogram generated by parsimony analysis connects to decent, with modification. In a Popperian context, this connection would have to be a deductive one, that is, one rooted in logic.

Descent

p. 209L: "... cladists, whose primary concern is knowledge of the empirical world of which species are a part."

Comment: In Popperian falsificationism, knowledge of the empirical world requires a deductive link between a theory and a singular statement that can be related to an observable state of affairs, given initial conditions. The statement "Species are a part of the empirical world" requires a (ostensive) concept of species that can be applied to an empirical base (Körner, 1959). Since the attachment of an ostensive concept to an empirical base occurs here and now (*hic et nunc*), the only ostensive species concept is Mayr's nondimensional species concept (Mayr, 1982). The phylogenetic species concept is a nonostensive concept.

p. 201L: "... the 'descent' principle means that species evolve from other species ..."

p. 206R: "To the contrary, without a concept of species, there is no compelling reason to believe the state relations of different characters should conform to a single hierarchy."

Conclusion: A species concept is required beyond the general theory of descent, with modification to justify phylogenetic parsimony in search of the single historical hierarchy. The deductive connection between the cladogram generated by phylogenetic parsimony (character distribution among species) and the general theory of descent, with modification would require:

- A universal species concept, applicable to all organisms
- That would give rise to a universal law of speciation (where speciation is lawfully dichotomous)
- Which could then be used in the attempt to establish a deductive link (given specified initial conditions) between that universal explanatory theory of speciation (cause) and the grouping of species in a single hierarchy (effect)

These requirements remain unfulfilled.

Appendix II: The True Tree

All great scientific theories that were formulated as a universal proposition have sooner or later been falsified (in a technical sense): Kepler < Newton < Einstein is the classical example. So it seems that truth cannot be had in science, for what is true today cannot be false tomorrow. If a statement believed to be true yesterday is actually false, it must be admitted that an erroneous statement was made yesterday – hence it could not have been true yesterday (Popper, 1976:§84). Indeed, this notion of truth is governed by bivalent logic and cannot be had in empirical science (*Wahrheit*, Popper, 1979:xxv).

But there may be other approaches to truth in science. For example, Newton's mechanics is contradicted by Einstein's theory in certain special cases, but Newton's mechanics work perfectly well in the everyday world of experience. It therefore must be a theory that is highly corroborated. Einstein's theory is even more highly corroborated, as it successfully answers questions that Newton's theory cannot answer. Since both theories are highly corroborated, each must have a mathematical probability that exceeds 0.5. The combined theories therefore must have a mathematical probability that is greater than 1, which is impossible. This shows that degree of corroboration cannot be a matter of probability calculus (Popper, 1979:ix, note11). It also shows that Newton's theory, in spite of being (logically) falsified, need not to be rejected, because it still works in everyday life and even beyond. Euclidean geometry is technically falsified, but it still works in everyday life, in relatively small space.

So Popper developed the intuitive idea that whereas Newton's theory captures some truth, Einstein's theory captures all the truth that is also captured by Newton's and more. Intuitively, one might say that that, somehow, Einstein's theory comes closer to the truth than Newton's (corresponding to the realists' idea of progressive science). This, then, can be taken as the basis for an argument that the notion of truth can be a regulative principle in science. The idea is that the belief in truth motivates scientists to keep testing their theories or develop new theories, because this process serves an ever-closer approximation of scientific knowledge to truth. However, this is an understanding of truth as a regulative concept only, which in itself is metaphysical, and as Popper happily conceded, "... it is perfectly possible to argue in favor of the ... criterion of progress in science without ever speaking about the truth of its theories" (Popper, 1989:223).

The aim of science thus becomes the approximation of truth, even if it is perfectly understood that there can never be any empirical knowledge of truth, in the sense of absolute, objective truth (Popper, 1976:§85). Popper felt uneasy about this notion of truth, however, because although it is a notion commonly used in everyday language, it was not clearly understood what it could mean relative to science, and it could certainly not be linked to Popper's logic of scientific discovery. That changed when Popper met Tarski and read his papers (Popper, 1973:319–340, 1974a), for there he found a notion of truth rooted in logic, or semantics. This concept of truth is not one of subjective certainty (*Gewissheit*) but one of absolute or objective truth (*Wahrheit* according to Popper [1979:xxv]). "The dangerous confusion or muddle which has to be cleared up is that between truth in the realist sense — the 'objective' or 'absolute' truth — and truth in the subjectivist sense as that in which I (or we) 'believe'" (Popper, 1989:402).

According to Tarski, a statement made in an object language such as "*Der Mond ist aus grünem Käse gemacht*," is true, that is, it can be asserted as corresponding to fact if and only if it can be asserted in a metalanguage. "The statement that 'Der Mond ist aus grünem Käse gemacht' is true if and only if, the moon is made of green cheese" (Popper, 1973:45). Evidently, this holds true only if the object language fact-describing statement is faithfully represented by its metalinguistic translation. In essence, then, the metalanguage asserts the truth of the statement made in the object language under an intended interpretation or interpretation with respect to a pretence (Boyd, 1991; Weiner, 1999), and these may change. This conception of truth has also been characterized as semantic (e.g., Kuhn, 1974a:26), which is why it should be noted that Tarski's truth theorem states truth conditions only, not what is true. Because the object language sentence can be asserted if and only if the metalanguage sentence can be asserted, which in turn can only be asserted if and only if the meta-meta-language sentence can be asserted, there results an infinite hierarchy of languages in the attempt to assert truth in observational (empirical) space (i.e., in the absence of an extralinguistic criterion of truth).

Tarski's truth theorem provided Popper with a bridge to develop his concept of verisimilitude, an intuitive (Popper, 1973:52, note 20) measure for the approximation of a theory to truth: The stronger the theory, the greater its empirical content, the greater is its truth content or verisimilitude unless its falsity content is also greater (Popper, 1973:53). Combining Tarski's notion of truth with that of the logical content of a statement, Popper (1973:47) obtained verisimilitude as the intersection of the truth content of a theory with the system of all true statements of the language in which the theory is formulated (Popper, 1974a:1102). Verisimilitude cannot be an empirical concept, because observational reports cannot be rendered as true statements. Nothing could be further from Popper's intention than to use verisimilitude as a numerical measure of approximation to truth (Popper, 1974a:1101, 1973:335).

Absolute measurement of verisimilitude is impossible, because it would require knowledge of the absolute, objective truth (*Wahrheit*). Assume, though, a theory with a small logical probability and hence a high logical improbability (i.e., one that significantly transcends background knowledge, generating bold, new, but testable, predictions). Assume further that this theory is put to a severe test, and it survives this test in spite of its high logical improbability. From such a successful first survival of a severe test, one would have to conclude that this incident of corroboration cannot be just coincidental (as was assumed by logical positivism). Such a coincidence would seem too improbable given the high logical improbability of the theory. No doubt, the incident would generate the suspicion that there might be some underlying lawfulness involved, as might indeed be corroborated by further testing. Now, this underlying lawfulness must at least be approximated by the initial suspicion that some underlying lawfulness is involved.

This is the early thought experiment (Popper, 1979:151, note3) that Popper picked up again at a later stage and developed into his concept of verisimilitude. The history of the concept of verisimilitude thus reveals its proximity to the notion of degree of corroboration ("*sekundäre Hypothesenwahrscheinlichkeit*;" Popper, 1979:150, 151, note3). This first instance of survival already confers a relatively high degree of corroboration on that theory. However, verisimilitude is not meant to be measured in terms of degree of corroboration, but only to represent its methodological counterpart (Popper, 1974a:1011). Corroboration, after all, is only a historical account of the past performance of a theory, while verisimilitude *qua* approximation to the objective or absolute truth is timeless (Popper, 1989:402).

However, accepting the fact that in the context of Popper's analysis of the logic of scientific discovery, theories take the form of strictly universal statements; accepting that the empirical content of theories, the severity of test, their degree of corroboration, their explanatory power and ultimately their verisimilitude, all depend on the degree to which predictions generated from these theories transcend current background knowledge; accepting that falsification requires an asymmetry that can only be established relative to universal statements and that falsification takes the form of a *modus tollens* argument; and finally, accepting that only logic can result in true statements, that observational statements or statistical probabilities can never result in true statements, it must then also be accepted that the notions of truth (logical truth or *Wahrheit*) and verisimilitude are not applicable to phylogeny reconstruction.

The true tree cannot be had as a result of empirical research.

Chapter 5

Hennig's Phylogenetic Systematics Brought Up to Date

Johann-Wolfgang Wägele

Abstract

Some of the tools cladistics use today for phylogeny inference were not available in Willi Hennig's time. These tools can easily be incorporated into the Hennigian method, which may be called "phylogenetic cladistics" to distinguish it from "phenetic cladistics." Phylogenetic cladistics requires character analysis prior to tree construction to discern between characters of high and low probability of homology; it implies *a priori* weighting. Character analysis also allows the polarization of characters before tree construction. It is not necessary to refer to a fully resolved tree to establish a hypothesis of monophyly for a subgroup of terminal taxa. The cladistic step, the transformation of a data matrix into a topology, is only one of several steps necessary to substantiate a hypothesis of phylogenetic relationships. An important last step frequently missing in published analyses is to examine the plausibility of the evolutionary scenario implied by a given tree topology. The circularity of the phenetic congruence test for homologies is emphasized. It cannot replace the *a priori* evaluation of character quality. Hennig did not offer a theory for *a priori* character weighting. However, theoretical bases for the development of new objective methods of weighting does exist and are presented in this paper.

Introduction

Phylogenetic systematics is a complex science that requires the application of several methods and theoretical concepts. Willi Hennig (1913–1976), who laid an important part of the foundations of modern systematics, coined the basic terminology and many of the concepts used today, especially the construction of phylogenetic trees by the search for sister group relationships. His phylogenetic systematics (Hennig, 1950, 1966, 1982) contains a theoretical superstructure that can be adapted to modern computerized methods and to the analysis of both morphological and molecular data. The general acceptance of his terms and methods can be explained by the strictly logical argumentation he presented in his theoretical publications and also by the convincing examples he and his early followers worked out (Hennig, 1953, 1966, 1969; Ax, 1963; Brundin, 1968; Schlee, 1971; Andersen, 1978). In this contribution, a synthesis of Hennig's ideas with the more recent developments of cladistics is outlined, and the difference between phenetic cladistics and modern phylogenetic systematics (phylogenetic cladistics) is discussed.

What is Missing from Hennig's Original Methodological Repertoire?

The theory of phylogenetic systematics developed gradually and is still evolving (Wägele, 1996a). Hennig also had predecessors who prepared the ground: Ernst Haeckel, for example, drew phylogenetic trees (Haeckel, 1866) as a reaction to Darwin's theory of the evolution of species, Fritz Müller (1864) was already using outgroup comparison to find derived character states, Danielle Rosa (1918) developed the principles of cladistic argumentation (Hennig was possibly not aware of Rosa's work), and Konrad Lorenz (1941) independently developed the idea of the argumentation scheme used by Hennig (Figure 5.1). Hennig's concepts matured between the publication of *Grundzüge einer Theorie der Phylogenetischen Systematik* (1950) and *Phylogenetic Systematics* (1966; the original German text of the latter was not published until 1982). Richter and Meier (1994) described the development of Hennig's terminology and concepts. After the publication of *Phylogenetic Systematics* (1966) a lively debate began concerning the relevance and application of his ideas; at the

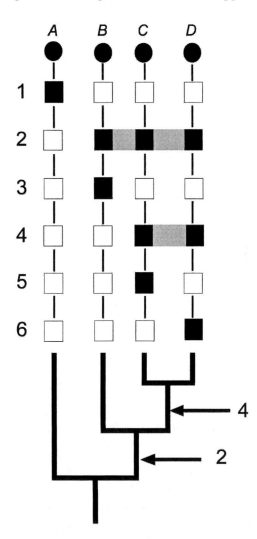

FIGURE 5.1   Hennig's argumentation scheme is simultaneously a dendrogram and a data matrix showing apomorphic and plesiomorphic character states. In this example, characters 1, 3, 5 and 6 are trivial (autapomorphies), while characters 2 and 4 are informative (synapomorphies); homoplasies are not shown.

same time, the use of numerical methods began. The debate, and the development of methods, is still going on, especially since the rise of molecular systematics.

Comparing Hennig's original method with modern applications of phylogenetic systematics, some steps of a complete phylogenetic analysis and some theoretical arguments have emerged that were neglected by Hennig, or were simply unknown to him. These are:

1. Numerical *a priori* character weighting and its theoretical justification
2. The search for the shortest trees in tree space and the use of the cladistic version of the principle of parsimony, as implemented in computer programs
3. The cladistic outgroup comparison
4. The construction of unrooted topologies
5. The relevance of distance- and maximum-likelihood methods for phylogeny inference

Character Weighting

In modern phylogenetic systematics the different value (utility) of characters as bearers of phylogenetic information is expressed by weighting (more on this below). Hennig was aware that the choice of putative synapomorphic characters was based on the idea that they were probably more often interpreted as homologies rather than analogies. For this purpose criteria are needed to discern between useful and less useful similarities. To explain this, Hennig cited Woodger: "… we choose a pairing which will bring the maximum number of parts into identity correspondence with respect to certain of their morphological properties" (Hennig, 1966:94, 1982:97; Hennig found Woodger's statement in Cain and Harrison, 1958). Hennig considered the best evidence for homology to be the complexity of similar morphological patterns. Of Remane's criteria for identification of homologies (Remane, 1952; also discussed, e.g., in Wiley, 1981), Hennig preferred the criterion of complexity, that of the specific quality of characters (Schlee, 1971). This means that characters that can easily be discerned owing to their specific patterns of details have a higher probability of being correctly identified as homologies than do simple patterns. Hennig stressed the importance of complexity, in the sense of the sum of its details, as early as 1950, though his statement is not very clear:

> Jede Eigenschaft der Holomorphe, jede Übereinstimmung und jeder Unterschied zwischen den Organismen wiegt also in der phylogenetischen Systematik nicht absolut, sondern sie gewinnen ihr Gewicht, mit dem sie als Zeugen für den Grad der phylogenetischen Verwandtschaft auftreten, nur durch ihre Stellung im Gesamtgefüge der die Holomorphe des Organismus ausmachenden Einzeleigenschaften (Hennig, 1950:185).[1]

Hennig and Schlee (1978; published shortly after Hennig's death) stated more clearly that characters rich in fine structure are better because the probability of parallelisms and chance similarities is lower. However, even though he mentioned the weight of characters, he did not propose any numerical weighting simply because he did not use any cladistic optimality criterion, even though the importance of weighting had already been discussed (e.g., Kluge and Farris, 1969).

[1] "Each property of the holomorph, each correspondence and each difference between organisms thus does not weigh absolutely in phylogenetic systematics, but they gain their weight with which they appear as witnesses for the degree of phylogenetic relationship only through their position in the whole composition of single properties of the organism's holomorph." [A holomorph is the sum of all characters occurring during ontogeny and adult life] (my translation).

The Criterion of Parsimony

Hennig did not emphasize the importance of the principle of parsimony. Its utility as an optimality criterion for tree search calculation was introduced and discussed by other authors (see Camin and Sokal, 1965; Kluge and Farris, 1969; Farris et al., 1970; Fitch, 1971; Andersen, 1978; Panchen, 1982; more details in Forey et al., 1992; Swofford et al., 1996) and was originally an element of numerical and phenetic taxonomy. The phylogenetic computer algorithm of Kluge and Farris (1969) was based on the work of Wagner (1961), not on Hennig, even though Wagner's method of grouping organisms was similar to Hennig's. Kluge and Farris (1969) did not cite Hennig, but in Farris et al. (1970) Hennig's terminology and his statement on the relevance of the number of apomorphies for selection among competing hypotheses of monophyly is incorporated into their justification.

Hennig, however, used the principle of parsimony in other steps of his analysis. To construct trees, Hennig selected the hypotheses of monophyly supported by the largest number of putative synapomorphies. However, he did not incorporate a methodical search in tree space for alternative, equally parsimonious topologies. To compare two competing hypotheses of monophyly, the construction of a complete topology for all terminal taxa is not necessary. The more evidence (number of good putative apomorphies) found in favor of a putative monophylum, the better the hypothesis:

> Je mehr sicher als apomorph zu deutende Merkmale (nicht: Merkmale über-
> haupt!) bei einer Anzahl verschiedener Arten vorhanden sind, desto besser ist die
> Annahme begründet, daß diese Arten zusammen eine monophyletische Gruppe
> bilden (Hennig, 1982:121).[2]

This assumption is based on the relationship between the number of similarities in two patterns and the probability that a common cause exists (see below). It is more parsimonious to assume that a specific series of events (mutations) occurred only once than to assume chance similarity and thus a higher number of parallel events.

A second, equally important, application of the principle of parsimony is the analysis of the plausibility of dendrograms (see below). Hennig stressed that all available data, morphological, ecological, physiological and geographical, should fit the same scenario. He named this the "principle of mutual (or reciprocal) illumination," which also relies on the principle of parsimony. Some authors (e.g., Wiley, 1981; Patterson, 1982; De Pinna, 1991) point out that mutual (reciprocal) illumination is the same as the cladistic congruence test; this is a restricted interpretation. In the sense of Hennig's work, the fit of all available data, including those that have not been used for tree construction (ecology, physiology, geographical distribution) must be considered in an explanation of the phylogeny and evolution of any group of organisms. This is an aspect that has to be included in the synthesis of available methods of phylogenetic systematics.

The cladistic parsimony principle, as understood by most users of computer programs, is utilized only in the search for the shortest tree. I have frequently heard cladists say that it should be impossible to discover an optimal tree topology by hand when many species are being considered, because there are millions of alternative topologies to be considered (e.g., about two million unrooted topologies for only 10 terminal taxa). As there exist more than 900,000 known species of insects and the number of alternative topologies that

---

[2] "The more characters that are safely identified as being apomorphic (not: characters in general!) are present in a number of different species, the better founded is the assumption that these species form together a mono-phyletic group" (my translation). A similar translation appears in Farris et al. (1970: 174). This translation, and that in Farris et al., differs slightly (but significantly) from that appearing in Hennig (1966: 121) but not from the 1999 reprint of that book.

theoretically have to be considered are beyond the reach of today's computers, it is nothing short of a miracle that Hennig was able to propose a phylogeny for all the insects (Hennig, 1969 [Hennig, 1981, English translation]). Nevertheless, that phylogeny is still accepted today by most entomologists (e.g., the divergence between Entognatha and Ectognatha) and is used in textbooks (e.g., Brusca and Brusca, 1990).

Hennig's phylogeny is possible because a search of the complete tree space is not necessary whenever valuable apomorphies that support monophyla are discovered. As soon as a few monophyla are well corroborated, the number of alternative topologies that have to be compared is drastically reduced (Schlee, 1971). If in a data set with 10 species, a monophylum of 4 species is discovered owing to the presence of apomorphies of high probability of homology, 6 terminal taxa remain. This reduces tree space by about 1999 million alternative unrooted topologies that can be safely ignored. If, for example, there are good arguments in favor of the monophyly of the Pterygota (e.g., presence of wings of a specific complex structure), tree space is drastically reduced when the focus of interest is the basal divergences of the insect tree, because all combinations of winged insects with apterygotes (wingless insects) can be ignored, along with the possible topologies within the Pterygota. Thus, to explain the basal dichotomies in insect evolution and to show that the Apterygota are paraphyletic, only 6 terminal taxa are needed instead of 900,000 (the 6 taxa are Pterygota, Zygentoma, Archaeognatha, Diplura, Protura and Collembola). Of course, a prerequisite is that there are good arguments in favor of the monophyly of each of these terminal taxa. The typical cladistic search in tree space is highly inefficient from the point of view of investment of effort, because most partitions of a given species list (putative monophyla) that are being considered have absolutely no support in a given data matrix and can be safely ignored.

Outgroup Comparison

There is a great difference between Hennig's outgroup comparison and the numerical cladistic outgroup comparison (Figure 5.2). Typical cladistic outgroup comparison consists of the selection of one or more outgroup taxa to root a tree that was constructed from unpolarized characters (e.g., Meacham, 1984). Afterwards, the evolution of characters is mapped on the resulting dendrogram. This is *a posteriori* polarization of characters. Hennig, however, used *a priori* information to distinguish between plesiomorphic and apomorphic character states and thus always constructed rooted trees. For this reason he discussed polarization of characters but did not propose any strategies for rooting topologies. Today, analysis of molecular sequence data is a case where outgroup comparison for *a priori* character analyses has no tradition (but see Fuellen and Wägele, 2000). However, when using molecular data, the same traps occur as in comparative morphology when rooting and polarization is done *a posteriori* (the "plesiomorphy trap": Wägele, 1996b, 2000).

Therefore, in a Hennigian analysis, wherever possible, polarization of character states is part of the character analyses that precedes tree construction, and the information thus gained is used to find sister groups (If additional phylogenetic information is available prior to tree construction, why not use it [Donoghue and Maddison, 1986]?). An outgroup, *sensu* Hennig, consists of virtually all organisms that do not belong to a monophylum. The presence of characters described as putative apomorphies of ingroup taxa are examined, in principle in all known organisms, and the analysis is not restricted to just a few outgroup taxa represented in a data matrix. Hennig (1966) furthermore proposed other criteria, rather than just the mere search of character states in outgroups (geological character precedence in fossils, chorological character precedence, ontogenetic character precedence and correlation of transformation series). An outgroup, then, is best represented by a hypothetical taxon with all-zero characters (zero representing the plesiomorphic state). This contrasts with the

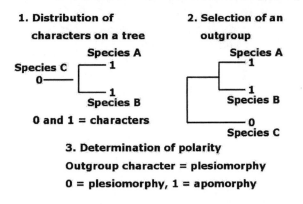

Cladistic determination of character polarity

1. Distribution of characters on a tree

0 and 1 = characters

2. Selection of an outgroup

3. Determination of polarity

Outgroup character = plesiomorphy

0 = plesiomorphy, 1 = apomorphy

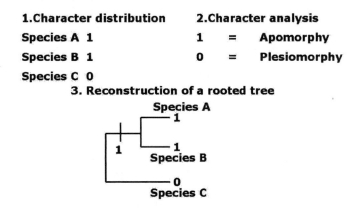

Hennigian determination of character polarity

1.Character distribution

Species A 1

Species B 1

Species C 0

2.Character analysis

1 = Apomorphy

0 = Plesiomorphy

3. Reconstruction of a rooted tree

FIGURE 5.2   Differences between the cladistic outgroup comparison and Hennig's outgroup comparison (see text for further details; modified after Wägele, 2000).

numerical cladistic outgroup designation, which implies that a plesiomorphic character state is not found by character analysis, but it is assumed that most character states of the selected outgroups are plesiomorphic (which often is not the case). This may lead to mistakes (Wägele, 1994).

Unrooted Topologies

Hennig did not try to construct tree-like diagrams with unordered or unpolarized characters. Unrooted topologies do not show the evolution of characters, but they may contain partitions of taxa that reflect real shared properties of organisms and therefore are informative.

Distance Methods

Hennig was very critical of clustering procedures, as mere morphological similarity is not a reliable criterion for relationships. He doubted, for example, that morphometric data contained useful phylogenetic information and pointed out that the rate of evolution differs in different organisms (Hennig, 1982:91). In this context he cited Bigelow:

> Unequal rates of evolution will produce cases in which overall basic similarity does not correspond with recency of common ancestry. Slow evolution in two phyletic lines will tend to produce an appearance of recent common ancestry when actual common ancestry is remote in time. (Bigelow, 1958)

This statement is still valid in the epoch of molecular systematics. A Hennigian principle is that similarities are only relevant when they are classified into homologies and analogies, and among the homologies, plesiomorphies and apomorphies must be discerned. A major problem for distance methods is that trivial characters (autapomorphies of terminal taxa) strongly influence pair-wise distances (Wägele, 1996c). Sober (1988) comments that there exist "two pitfalls for pheneticism": the phenomenon of homoplasy (in the sense of analogy or convergence) and the occurrence of autapomorphies.

Phylogenetic Cladistics: The Synthesis between Hennig's Method and Numerical Cladistics

In Germany, the development of phylogenetic systematics continued after Hennig's death with two important textbooks written by Peter Ax (1984, 1987 [English translation of 1984], 1988). Ax explained in a clear and precise way the core of Hennig's method and included a description of *a priori* outgroup comparison, paragraphs on the importance of the estimation of the probability that characters are homologies and the use of the principle of parsimony to find the shortest tree. In contrast to Hennig's later textbook (1966, 1982), Ax's work contains more concise instructions to learn methods for practical research and includes some of the more recent theoretical discussions of Anglo-American scientists. However, numerical cladistic methods were not included, though younger taxonomists had already started to use computer programs. I will not examine the historical evolution and integration of methods in detail but will present my personal view of what a modern version of Hennig's method should be. I prefer to name this collection of analyses "phylogenetic cladistics," to stress the difference from pure cladism (pattern cladistics or phenetic cladistics).

With the development of computer software, new methods became available that were unknown to Willi Hennig. The most important aspects are:

1. Implementation of a more differentiated weighting of characters.
2. Automatic construction of topologies and a method that will, in principle, consider all the alternative topologies (for the time being, if there are no more than 20 taxa).
3. Algorithms that implement a cladistic application of the principle of parsimony allowing the use of exact optimality criteria and (in principle) the exact search for the shortest trees
4. The distribution of character states can be compared over many different and large topologies, the number of state changes and homoplasies can be easily found and characters can be mapped easily onto different topologies.
5. Different data sets and topologies can be compared numerically.

There is no contradiction between Hennig's ideas and the use of computers. The opposite is true: the implementation of the cladistic method by computer is welcomed. Yet this does not mean that earlier analyses carried out without computers are worthless. Hennig's classification of insects, for example, is still accepted in its essential structure (Hennig, 1969, 1981).

It is very important to note that the methods implemented in computer programs replace only one of several steps of a phylogenetic analysis. The cladistic step is a purely deductive procedure that leads from a data matrix to one or more selected dendrograms. Phylogenetic cladistics, however, requires several other steps that are equally important and frequently

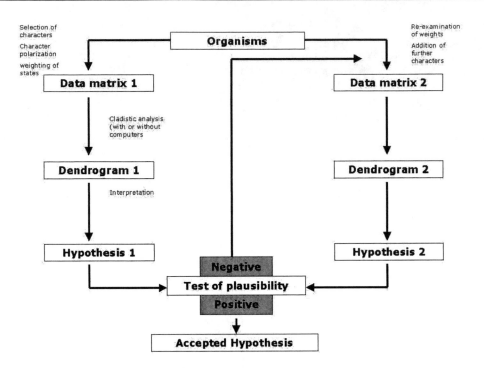

FIGURE 5.3   Diagram with the steps of a phylogenetic cladistic analysis. Note that the cladistic step is only one of several and that, in principle, computers are useful but not always absolutely necessary tools (translated from Wägele, 2000).

neglected. Figure 5.3 is a summary of those steps. The sequence of a complete analysis would be the following:

1. Search for similarities that occur among the organisms of interest. Whenever higher taxa are used as terminals, only characters of ground patterns (states of the last common ancestor) that were reconstructed with the methods or arguments of phylogenetic cladistics in a previous analysis are used.
2. Perform a character analysis for the selected similarities. Discover characters of high probability of homology and weigh these higher than those characters of low probability. Characters suspected to evolve easily more than once should be excluded whenever possible or given low weight.
3. Compare the states of selected characters with all known organisms, especially with those that are closely related to the group of interest, the ingroup. Define the states occurring only in species or subgroups of the ingroup as apomorphic states. Use an all-zero ancestor as the outgroup whenever all character states can be polarized *a priori* or code characters as polarized whenever polarity seems to be obvious.
4. Describe explicitly all arguments used in the character analysis (steps 2 and 3) and enter the results in a data matrix.
5. Hypotheses that characters are apomorphies are used to support hypotheses of monophyly; putative synapomorphies support sister-group relationships.
6. Test the compatibility of hypotheses of monophyly: They should form a hierarchy that fits a single tree. Incompatibilities falsify hypotheses of monophyly. However, of two incompatible monophyla, the one supported by distinctly better evidence (more apomorphies with higher probability of homology) can be accepted. Thus, the final tree with the largest number of well-supported monophyla should be the most parsimonious one.

Steps 5 and 6 constitute the cladistic step. This part of the analysis may be undertaken with a cladistic computer program.

Many publications conclude by presenting "the best" cladogram. However, this is not a complete phylogenetic analysis. Further steps are needed:

7. Check with data not used for tree construction if the result is plausible. For example, compare with other data sets. The corresponding trees should have the same topology as the first one. Above all, describe the fit of further information (ecology, physiology and biogeography) to the evolutionary scenario implied by a given tree.
8. Explain contradictions that occur in step 7 and reexamine all previous steps to discover possible sources of error. Discuss the latter.

There are no fundamental differences to the analyses used by Hennig. The theoretical superstructure is the same. However, many options offered by software packages are certainly new. Morphologists familiar with Hennig's method may wish to use the modern implementation of phylogenetic cladistics. They need to appreciate the method of presentation by data matrices and familiarize themselves with the workings of the various computer programs. Those scientists unfamiliar with Hennig's method, however, must appreciate that the mere use of modern software does not guarantee a good and complete phylogenetic analysis. To avoid the pitfalls of phenetic cladistics, attention should be paid to steps 1 to 4 and 7 to 8.

Phenetic Cladistics: Elegant Analyses with Many Sources of Errors

First I need to explain phenetic cladistics. Recently, Scott-Ram (1990), citing Sneath and Sokal (1973), defined phenetics as a classification based on a set of phenotypic characteristics derived from the objects or organisms under study, resulting in a hierarchical system that reflects those similarities without consideration of their evolutionary relationships. However, cladists, constructing data sets from similar character states, will usually assume that their result is a phylogenetic tree, depicting certain aspects of evolution. Nevertheless, as long as similarities are not classified according to their estimated phylogenetic information content, the analysis will remain a phenetic one.

Methodologies differ in their susceptibility to error. Phenetic cladistics is clearly the poorer alternative to phylogenetic cladistics. It is a mistake to believe that by using phylogenetic software (e.g., PAUP; Swofford, 1998) one acquires all the methods necessary for a complete phylogenetic analysis. Phenetic cladistics, as defined here, is a reduced set of procedures, in the worst case composed only of steps 1, 5 and 6 above, the latter two steps being represented by a parsimony algorithm. The result of step 1 would be a data matrix (characters vs. taxa), which is processed via the program and completed when one or more of the most parsimonious cladograms are found. The major differences of the complete procedure is (1) the absence of a thorough character analysis, which includes a quality control for the evidence prior to tree construction, and (2) the lack of a test of the congruence between the cladogram and further available data to check its plausibility, which can falsify a proposed phylogeny or raise new questions. Note that the character analysis of phylogenetic systematics is not the control of the fit of characters to a topology (e.g., as in Andersen, 1978), but an *a priori* examination of character quality (i.e., probability of homology; see below).

One historical fact that contributed to the reduction of phylogenetics to the cladistic step was the dogma that different qualities of hypotheses of homology can only be discovered after tree construction. Patterson's statement that "the ... most decisive test of homology is by congruence with other homologies" (1988) is widely accepted by many cladists. Congruence in this sense is the fit of characters to the shortest tree. This appears to free the

species:	D	C	B	A
characters	1	1-2	1-2-3	1-2-3-4
homologies of A + B:			1-2-3	1-2-3
synapomorphies of A + B:			3	3
autapomorphy of A:				4

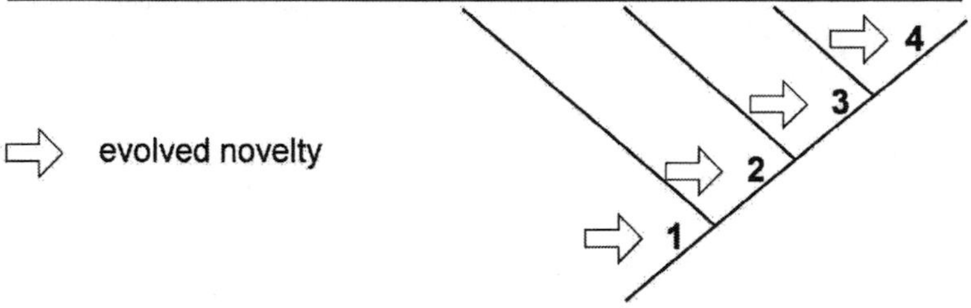

⇨ evolved novelty

FIGURE 5.4   In Hennigian terminology, *homology* and *synapomorphy* are not the same. In this example, characters 1 to 3 are homologies of the species A and B. However, only character 3 is a synapomorphy of the monophylum (A, B).

investigator from checking character quality with an *a priori* analysis; parsimony programs only identify which character changes occur once on the most parsimonious tree and therefore are apomorphic homologies. For Nelson (1994), "as seen by cladists, a particular homology is a function, or derivative, of a particular tree." The same attitude is expressed, for example, in the statement of Goloboff (1993) who thought that *a priori* weighting is not necessary. Andersen (1978:247) wrote, "The purpose of a cladistic analysis is to reconstruct the branching sequence (cladogenesis) in the phylogeny of a group of organisms ... ." The result of this philosophy is that in the publications cited above, and many textbooks on cladistics, the quality of the data entered into a data matrix and how this quality can be estimated is never addressed; chapters on the quality of morphological characters are rare in the current literature. (The situation in molecular systematics is somewhat different, because there is a discussion of the information content of alignments, for example, Charleston et al., 1994; Graybeal, 1994; Lento et al., 1995; Lyons-Weiler et al., 1996; Philippe and Laurent, 1998; Wägele and Rödding, 1998; Wägele et al., 1999).

Some confusion exists because certain authors equate homology with synapomorphy (Patterson, 1982), a usage that certainly (and rightly so) would have been rejected by Hennig. Both terms are needed because it is not possible to state without exact knowledge of a phylogeny that a specific character found in several organisms has been inherited from a common ancestor. Furthermore, the terms *apomorphy* and *plesiomorphy* both are used for characters inherited from a common ancestor and both terms are used for special distributions of homologies in phylogenetic trees (Figure 5.4). Also confusing is the fact that a homology can be called *topographical identity* (positional homology in alignments) or *character state identity* (nucleotide identity in alignments; see Brower and Schawaroch, 1996). These are two different patterns, also named *frame homology* and *detail homology* by Riedl (1975). An apomorphy is a novel detail that is identified within a frame (a gene, an organ, an appendage). Therefore, to compile a data matrix the probability of homology has to be discussed for each character in at least two steps (Figure 5.5): For the two-state character "lower molar with anterior shearing edge present or absent" the frame has to be homologous (e.g., the molar in the lower jaw in two skulls) and the novelty has to be within this frame (the shearing edge). Both hypotheses have to be discussed in character analysis, the topographic identity (i.e., frame) having priority over the character state identity (i.e., detail) (Brower and Schawaroch, 1996).

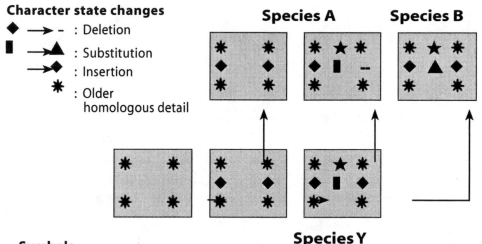

Character state changes

◆ ⟶ − : Deletion

■ ⟶ ▲ : Substitution

⟶ ◆ : Insertion

✳ : Older homologous detail

Symbols

▲ : Autapomorphy of B

✳◆ : Plesiomorphy of Y, A, B, also symplesiomorphy of A and B

★ : Synapomorphy of A and B

▢ : Frame homology of A and B

FIGURE 5.5   To compare characters, the "frames" and "details" need distinguishing, with homology assessed for both types of characters. Frames can be, for example, appendages, genes or sequence positions; corresponding details would be, for example, single bones, introns or nucleotides (after Wägele, 2000).

Phenetics means, in this context, the clustering of organisms owing to similarities. In principle, Hennig's methods are designed to classify similarities into those that are informative and those that are not, discerning between homologies and analogies, using only homologous apomorphies as evidence of monophyly, a major difference from phenetic cluster analyses. Hennig and others pointed out that clustering algorithms can be used to calculate dendrograms. However, autapomorphies of terminal taxa will increase the distance of real neighbors so that the true sister groups will not be discovered in many cases. As discussed by Scotland (1992), the mere absence of characters (a plesiomorphic state or secondary loss) might support clusters that are not monophyletic. No distinction is made between plesiomorphies and apomorphies.

If similarities are entered into a data matrix without a previous analysis of character quality, there is always the possibility that characters are chance similarities, convergences or parallelisms. A data matrix may consist of pure noise if characters like the number of hairs or pigment cells are used. If characters are selected without consideration of the probability that they are homologies, then parsimony analysis is phenetic because only the number of similar entries determines the structure of the topology, and not the probability that similar patterns have a common cause. If, however, a cladistic analysis is preceded by a character analysis based on a theory of the phylogenetic information content of characters (see below), then there is a synthesis of the Hennigian method with modern cladistics. Figure 5.6 shows some major differences between phenetic and phylogenetic cladistics.

As character analyses are usually missing in phenetic cladistics, several sources of errors may exist (see also Wägele, 1994):

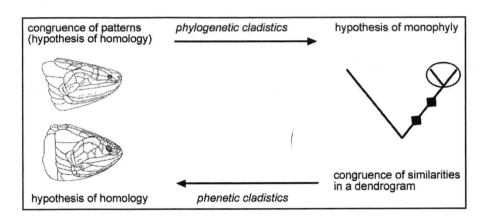

FIGURE 5.6   A major difference between phylogenetic cladistics and phenetic cladistics is that the quality of the empirical data is evaluated before the cladistic step in the Hennigian approach.

1. *Some characters may not be based on real properties of organisms but are the result of erroneous observations.* An example is Zrzavý et al.'s (1998) coding of the character "metameric coelomic cavities" as absent in arthropods, and as the same state for taxa of the Cycloneuralia (part of the former nemathelminths). Examination of arthropod embryos shows that arthropods do have a metameric coelom, although it is secondarily modified during ontogeny (a great difference from the Cycloneuralia).

2. *Characters coded for higher taxa are not ground pattern features but autapomorphies of subordinated groups.* An example is in Brusca and Wilson (1991), where a styliform endopod of pleopod 1 is considered to be an apomorphy of the Oniscidea, the terrestrial Isopoda (woodlice). However, this is only a character of the more derived groups, while the Ligiidae still have a broad endopod similar to that of outgroup isopods (Erhard, 1997).

3. *Plesiomorphies of the ingroup are erroneously coded as apomorphies.* For example, Schram (1991) coded a character named "orthogonal nervous system with anterior nerve ring and several longitudinal cords" as an apomorphy of the Nemathelminthes and the Chaetognatha, the plesiomorphic condition being a "poorly polarized nerve net." Since orthogonal nervous systems occur outside the Nemathelminthes (e.g., in Mollusca, Articulata and Platyhelminthes), this is either a plesiomorphy or a weak character prone to convergence.

4. *Characters are used that may contain more noise than phylogenetic signal (characters of low probability of homology) and are not weighted to express their value.* For example, reduction of epidermal microvilli, reductions of motile cilia and reductions of endodermal cilia are considered to be synapomorphies of the Ecdysozoa (Zrzavý et al., 1998). Simple reductions of this type are not very specific characters and occur frequently in unrelated organisms.

5. If the chosen outgroup taxon is highly derived, many of its character states are autapomorphic and not plesiomophic, and plesiomorphies of the ingroup are not recognized as such. For example, if amphipod crustaceans are used as an outgroup to study the evolution of isopod crustaceans, the laterally flattened body may appear to be plesiomorphic within the Isopoda (present in the Phreatoicidea and in the Amphipoda), and the broad, leaf-like pleopods 4 and 5 may be coded as an apomorphy of the Isopoda (absent in Amphipoda). However, both features of the Amphipoda are not the ancestral state but autapomorphies of amphipods. For a correct a priori character analysis more than one outgroup taxon has to be considered. In principle, all organisms not belonging to the ingroup constitute the outgroup studied to discover the most probable plesiomorphic state.

Many cladistic publications are problematic, as the data are not fully understandable, either because character states are not explained, sometimes not even listed, or the data matrix is not published or is hardly readable (e.g., in Briggs and Fortey, 1989; Wills et al., 1995). In addition, the most relevant apomorphies should be mapped onto the accepted cladogram.

The Necessity of *a priori* Weighting: Why is Character Quality the Same as Probability of Homology?

I will not discuss all previous definitions of *homology* or how to identify homologies, as there exists a large body of literature on the subject (e.g., Remane, 1952; Patterson, 1982, 1988; de Queiroz, 1985; Wagner, 1989, 1994; De Pinna, 1991; Rieppel, 1992, 1994; Haszprunar, 1992; Mindell, 1991; Nelson, 1994; Hall, 1994; Panchen, 1994), and my own views have been published elsewhere (Wägele, 1996b, 2000). Similarly, this is not the place to review all the literature on character weighting (e.g., Kluge and Farris, 1969; Farris, 1969; Neff, 1986; Sober, 1986; Wheeler, 1986; Sharkey, 1989; Fitch and Ye, 1991; Goloboff, 1993; Minelli, 1993; DeSalle and Brower, 1997; Wenzel and Siddall, 1999). The following is intended to present a theoretical basis for *a priori* weighting.

"The principle of parsimony says, that the most probable cladogram is that containing the least number of parallelisms or convergences ..." (Andersen, 1978). This simple statement contains an important principle of cladistics: The shortest cladogram is the best one and has something to do with probability. Probability of what? Different probabilities might be considered:

1. The aim is to find the cladogram for which the probability that it reflects the true history of the organisms in question is highest. The cladistic method implies that this probability is optimal in the shortest topology estimated from a given data set, tree length being measured as number of character state changes.
2. If each character state change on the branches of a topology is counted once and with the same weight, then the probability that a change occurs should logically be the same for all characters. Only this assumption justifies equal consideration of each character.
3. In the rooted true tree, a character state change that occurs on a branch indicates that the corresponding novelty is an apomorphic homology of all distal taxa connected to this branch.
4. Therefore, the probability that a character change occurred on a branch is the same as the probability that a shared novel character (or an apomorphic state) is a homology. Therefore, if each character state change is treated equally, then the probability of homology of each novelty should be the same.

The conclusion from this reasoning is that cladistic application of the principle of parsimony implies an important first principle, namely that for equally weighted characters the probability of homology for each putative novelty should be the same for all characters. Only this justifies counting each change in the same way. The innate existence of this assumption in cladistics, and its methodical consequences, has been ignored in the past by many cladists.

A rule of logic is that first principles cannot be tested by deductions that themselves rely on the same first principles. An important issue is, therefore, the question of whether the probability that a character is a homology can be estimated using a cladistic analysis, that is, if it is possible to identify homologies *a posteriori* (Patterson's test of congruence). The answer is complex (Wägele, 2000).

If a first principle of cladistics is that characters should have more or less the same probability of homology, then this cannot logically be tested by a cladistic analysis. Such an analysis would lead to a circular argument. For example, Schram (1991), after a cladistic analysis of the Metazoa, obtained a sister group relationship between Tracheata and Ony-chophora. The only character supporting this group was coded as "whole-limb mandible present." The typical phenetic argument would be based on circularity:

1. The data matrix contains a character called "whole-limb mandible," present only in Tracheata and Onychophora. This implies the assumption of the sameness of the whole-limb mandible.
2. Owing to the absence of conflicting evidence in the data matrix (not in nature) the most parsimonious tree contains the monophylum {Trachata, Onychophora}.
3. Congruence suggests that this character is a homology because it fits the most parsimonious tree.

This is clearly a circular argument. This particular tree was only obtained from coding the whole-limb mandible as a similarity of Onychophora and Tracheata. Other authors are convinced that the whole-limb mandible does not exist in the Tracheata, that the tracheate mandible evolved in a different segment than that of the Onychophora, that it is not a whole-limb appendage and that the Tracheata belong to the monophylum Mandibulata, characterized by specific mouthparts (e.g., Lauterbach, 1980; Wägele, 1993; Scholtz et al., 1998; Ax, 1999). In other words, it is decided *a priori* that the mouthparts of onychophorans and tracheatans are not homologous, and therefore analyses yield other topologies.

My conclusion is that the *a posteriori* identification and weighting of homologies leads to circular arguments. In the case of Schram's analyses, the mandibles of Onychophora and Tracheata are coded as if they were "the same thing," that is, the assumption of homology is contained in the data matrix and therefore it appears in the tree because there was no conflicting evidence. The same problem is well known in molecular systematics. The topology of dendrograms depends on the alignment (i.e., hypotheses of homology of nucleotides), and the construction of an alignment is usually considered to be independent of tree construction. A closer look at this example reveals that positional homology in alignments is the same as frame character homology in a morphological data matrix (the single column, if the rows contain the species names).

The *a posteriori* identification of homologies makes sense when a tree is already supported by good characters (i.e., characters of high probability of homology) and the topology is then used to describe the evolution of weak characters (those that did not contribute to the structure of the topology or had little influence owing to their low weight). Weak characters may be reductions of eyes in deep sea animals, for example. If all members of a well-corroborated monophylum are blind, then it is probable that the last common ancestor lacked eyes and that blindness is a homology (i.e., *a posteriori* determination of homology for a weak character).

The congruence test may fail for plesiomorphies: Since plesiomorphies may be present in distantly related taxa and can be absent in organisms that branch off in between these in a phylogenetic tree, the character may appear as a homoplasy in an optimal tree even though it is a homology. For example, cladistic congruence does not help to decide if the similar composition of the mandible seen in some malacostracan crustaceans (especially in peracarids) and for example, in larvae of Ephemeroptera is a plesiomorphic homology or a convergence. The Ephemorpteran mandible may be an atavism and hence a homology.

The congruence test tells us nothing about the quality of a single hypothesis of homology. If there are conflicts in the form of homoplasies, then the source of errors has to be examined (errors might have occurred in the *a priori* character analyses, in weighting or in algorithms). It might be that the correct topology has been found and that the homoplasies do indicate false hypotheses of homology, but it might also be that the topology is wrong and the numerous putative synapomorphies are in reality weak characters and the few homoplasies are the real homologies. The test does not reveal which kind of mistake occurred. Finally, homology statements should be possible for serial organs or repetitive genes that occur in a single individual and for genes transferred horizontally to organisms of entirely different phylogenetic origin. The test of congruence is not applicable here (De Pinna, 1991; Haszprunar, 1998); other criteria are needed.

Therefore, the congruence test does not have the power that Patterson (1982), and other cladists, have claimed for it and cannot replace *a priori* character analysis. Kluge and Farris (1969), for example, included in their analysis of anuran evolution long paragraphs with *a priori* character analyses where the selection and quality of characters is discussed.

A Theoretical Basis for *a priori* Character Weighting

> If valid means of character weighting can be found, they will tend to improve our chances of inferring the correct phylogeny. (Kluge and Farris, 1969)

> The principle of parsimony degenerates to numerical taxonomy, if characters are not weighted … . (Remane, 1983)

> … parsimony requires that the characters already should be weighted. (Sober, 1986)

> It is clear that parsimony works best with 'good' data … . (Mishler, 1994)

Hennig did weigh characters, but in a simple way, by excluding some from the analysis (weight 0) and including others (weight 1). In the meantime, several methods for differentiated character weighting have been proposed. I do not intend to discuss all the criteria but will concentrate on some basic theoretical considerations.

Weighting should reflect an estimation of the probability that characters used for phylogenetic cladistics are homologies. For example, if there is good evidence to postulate that feathers evolved once and it seems to be highly improbable that such a structure appears by chance a second time in an outgroup, the character "feathers present" may receive the necessary weight to allow all bearers of this character in a cladistic analysis to be members of the same monophylum. How, though, can we avoid unfounded *ad hoc* weighting? In practice there are unsolved problems that need further research. Which weights should be selected for which probability of homology? Can the latter be quantified? An absolute estimation of this probability may be impossible, but at least there are good grounds for the ranking of characters beginning with "very trustworthy" characters (such as the feathers of birds) and ending with "useless" characters (such as the number of hairs in mammals). In the following I will briefly discuss why it is, in principle, possible to make this distinction.

Replacing the word *character* with the word *pattern*, one can discern between simple patterns, composed of a few details, and more complex patterns. The principle used by many systematists, including Hennig, is the assumption that the similarity of many details in two complex patterns evolved with higher probability only once than that the similarity evolved in two simple patterns, which might have evolved more than once by chance alone. This assumption has a probabilistic basis (Wägele, 2000). If we assume a simplified model where two words (representing two patterns) are formed by chance from a pool of letters

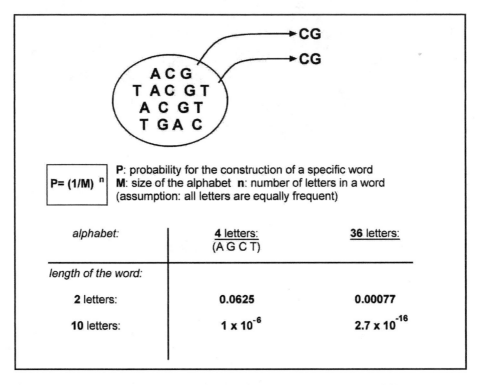

$$P = (1/M)^n$$

P: probability for the construction of a specific word
M: size of the alphabet n: number of letters in a word
(assumption: all letters are equally frequent)

alphabet:	4 letters: (A G C T)	36 letters:
length of the word:		
2 letters:	0.0625	0.00077
10 letters:	1×10^{-6}	2.7×10^{-16}

FIGURE 5.7   The probability that two patterns (in this case the two letters "CG") are constructed by chance depends in this simplified model only on the size of the alphabet and the length of the word.

from an alphabet with M letters (all letters occurring with the same frequency) and without preference for single letters, the probability P that two words are identical is $P = (1/M)^n$, n being the length of the word (Figure 5.7). The relevance of this law is evident in molecular genetics when scientists search for primer sequences that will allow the amplification of homologous genes in different organisms: The longer the primer is, the higher the probability that homologous sequences are amplified (as long as the corresponding primer-binding areas are conserved).

Since DNA is based on a small alphabet, short words that are not homologous occur frequently (see, e.g., target sequences of restriction enzymes). The same law has been used in comparative morphology, even though the theoretical reasons have not been explained explicitly. Complexity of characters is the most significant criterion of homology. Of course, the probability that complex organs evolve in a particular way depends on parameters that are not known to us, namely all factors influencing their evolution (including developmental constraints, population size, ecological factors, irregular changes of climate and other vital parameters). However, although the correct parameters for an accurate model of character evolution are unknown, we may assume single point mutations occur more frequently than a series of very specific mutations. Novelties based on mutations that probably occur frequently are with low probability homologies, and complex patterns are more likely to have evolved only once than several times convergently.

In practice, the complexity of patterns can be measured by the number of discernible details. This is easy in molecular data, and the approach is used in spectral analysis, where monophyla with a higher number of apomorphic positions are considered to have better support (e.g., Lento et al., 1995; Wägele and Rödding, 1998; Wägele, 2000). Morphologists have the difficulty that the letters of their alphabet may consist of very different details (position of nerves in a brain, of arteries in a circulatory system, of organelles in a cell, etc.). The alphabet consists of the number of different details used by the scientist to describe a character. The advantage, in comparison with molecular data, is that the alphabet is potentially very large (which increases the probability of homology). The disadvantage is that the description of these details often requires tedious work that cannot be automated (e.g., preparation and comparison of histological sections).

A discussion of the complexity of characters used in real data sets has recently been prepared to examine the Ecdysozoa hypothesis (Wägele et al., 1999; Wägele and Misof, 2001). To illustrate the principle, the following artificial example might be useful.

In Table 5.1, a fictitious scientist gives the character "insect head" a higher weight than the character "complex insect eye," with the lowest weight assigned to the character "pigment granule." The characters can be given different weights according to the number of identical details discerned by the scientist (the real number of components remains unknown). The more details that are identified, the higher the weight. This is justified by the assumption that homologies can be corroborated with higher certainty when their fine structure is known, while superficially examined characters might not be similar in detail. In the table, many other details (such as the structure of the mouthparts) can be added to increase the weight of the characters. The following ranking of characters is given in Table 5.1.

If two patterns are postulated to be homologous owing to the presence of identical details, one has to verify that the details really are the same. Therefore, it is often better to use patterns smaller than the "insect head," where the structure is easier to grasp and the homologies are easier to discuss. For the cladistic analysis the effect is the same whether a character is weighed 10 times due to 10 identical details or if 10 details are coded separately and each assigned a weight of 1. In practice, there exists a large hierarchy of encaptic details,

TABLE 5.1   The use of described homologous details as a measure of complexity

Character	Complexity	Details Discerned
Insect head	13	1. Lateral complex eyes
		2. With crystal cone in ommatidia
		3. One pair of antennae
		4. Three frontal eyes
		5. Clypeus and labrum inserting frontally
		6. Three pairs of mouthparts inserting ventrally
		7. Endoskeletal elements forming a tentorium
		8. Caudal foramen occipitale
		9. Postoccipital suture
		10. Occipital suture
		11. Mouth placed between first pair of mouthparts
		12. Labial glands
		13. Chemosensory sensilla trichodea on antenna.
Insect ommatidium	5	1. Cuticular cornea
		2. Number of corneagene cells
		3. Number of crystal cone cells
		4. Number and position of pigment cells
		5. Number and shape of retinular cells
Pigment granules	3	1. Average size
		2. Position within cell
		3. Chemical composition of contents

for example, organs are composed of tissues and cells, and cells are composed of organelles and molecules. However, morphological details are not the only components worthy of consideration. There are also shapes, connections and developmental patterns, that is, all variations for which we suspect that specific genetic information is necessary. Shapes and lateral contacts are, for example, important in identifying the bones in a skull.

The probability that a pattern evolves twice by chance depends not only on the number of elements of the pattern but also on the process that forms the pattern. In the alphabet model above, things are simplified if a constant process is assumed and no bias exists in the selection of letters. Some criteria for weighting are derived from assumptions about rate differences of character evolution: highly variable characters, for example, get lower weight, because similarities can evolve easily by chance several times. Whenever some evidence (not derived from an inferred topology) for the parameters of processes exists, this can be used for *a priori* weighting. However, in most cases it is not possible to use knowledge about the singular historical process of evolution (rate of genetic drift, mutation rate, effect of selection, effect of variations of population size on a particular population during a defined period of history).

Using the pattern analysis described herein we weigh the amount of information (in the sense of the number of identities that suggest homology) that has been uncovered from the structure of individuals. The more information is available, the higher is the probability that homologies are identified correctly. The amount of information depends on the structures compared (Are they composed of many discernible details?) and on the activity of the researcher (Is he familiar with the objects? Did he examine the objects carefully? Neff [1986:122] asked, "How much do we think we know about this character?"). In principle, the identified details are used in the same way as the letters in the word of length n constructed from the alphabet of the size M in the model explained above to estimate the probability that the identity of two patterns is the product of chance alone.

Even though the argumentation remains fuzzy in comparative morphology, because the real size of the alphabet (i.e., the possible alternatives for details in a pattern) and the rate variations are not known, the weighting of complexity is a useful approximation with a sound theoretical foundation. Adopting this concept we can reject Patterson's statement (1982:58) that "hypotheses of homology are conjectures whose source is immaterial to their status."

Discussion

Farris (1986) asked, "but what, then, is 'Hennigian'?" He suggested that crucial principles are the monophyly of taxa and grouping according to synapomorphy. This certainly distinguishes phylogenetic from nonphylogenetic approaches. However, Hennig's repertoire is more extensive and includes character analysis to discern between different classes of similarities and to dismiss useless and misleading data. Since, in principle, a data matrix containing characters for different minerals can be analyzed with PAUP to obtain a dendrogram, the application of cladistic techniques alone does not make an analysis phylogenetic.

Many authors equate Hennig's phylogenetic systematics with cladistics (e.g., Mishler, 1994). Strict boundaries between numerical cladistics and Hennigian schools of thought do not exist, but there are two extremes: there are those who do not use computers and construct a single tree by hand and those who use PAUP but restrict their analyses to the steps of phenetic cladistics. There certainly exist important differences when Hennig's repertoire is compared with that of Patterson, for instance. To bring Hennig's phylogenetic systematics up to date requires its integration into numerical cladistic methods. Several publications have attempted this (see, e.g., many of the contributions in the *Zoological Journal of the Linnean Society*). However, some numerical concepts remain missing from phylogenetic systematics (e.g., the consideration of alternative tree topologies), while some of Hennig's concepts remain missing from phenetic cladistics (e.g., *a priori* character analyses). In addition, some ideas need further research and have yet to mature (e.g., how to estimate data quality *a priori*).

The discussion between cladists and their critics is not new. Panchen reviewed the issues in 1982 (Panchen, 1982). He attacked the restriction to just parsimony, by the optimization of synapomorphies of cladograms, and pointed out that "in the real world parallel evolution is common." What he did not offer was a theory that allows the classification of characters into those that are predisposed to parallelism or chance similarity and those that are not, and therefore he thought that "cladists may have to admit the persistent intuitive element in classification." Today, intuition (in the sense of unfounded *ad hoc* assumptions) can be avoided in phylogenetic cladistics using objective criteria of character weighting.

Scott-Ram (1990) discussed the differences between transformed cladistics (*sensu* Platnick, 1979; Nelson, 1979; Nelson and Platnick, 1981) and phylogenetic cladistics. According to Scott-Ram (p. 135), in phylogenetic cladistics a cladogram shows "abstract hypotheses of relationships which do not correspond exactly to the ancestral-descendant pattern of the unknown segment of the evolutionary tree ... ," while in transformed cladistics a cladogram "only depicts branching patterns" that are summaries of character patterns. Such a distinction does not exist in the concept of phylogenetic cladistics outlined above. Of course, a cladogram always represents certain aspects of the patterns present in a data matrix, but at the same time the relationships summarized in a cladogram may depict real ancestor-descendant relationships and real divergence events that occurred during the phylogeny of organisms. If the data are excellent, then so are the hypotheses of relationships derived from the best cladogram.

A significant point made here is that without a thorough *a priori* character analysis, cladistic research is more phenetic than phylogenetic. The importance of character analysis has been stressed by other critics of cladistics. Bryant (1989), for example, developing the thoughts of Neff (1986), recognizes two levels of a phylogenetic analysis where a hypothetico-deductive argumentation is necessary: (1) character analyses plus the cladistic *step* (outlined above) (level one), and (2) the macro-evolutionary analysis (level two). This is equivalent to Hennig's views on data analysis. The necessity of both levels is recognized in this paper (Figure 5.8).

The first inductive step is the collection of single observations from nature, which are summarized in a data matrix. This already contains hypotheses, namely those of the homology of similarities. How these hypotheses can be corroborated has been briefly outlined above. Each hypothesis of apomorphy (the homology of putative novelties) yields a testable prediction: all bearers of the same set of homologous novelties should appear in the same monophylum after tree construction. The test is carried out by construction of a tree from this data matrix. The transformation of the matrix into a dendrogram is a purely deductive step that can be automated because it is composed of strictly logical operations and is therefore the best tool for testing the compatibility of hypotheses of monophyly. Homoplasies indicate that either the hypotheses of homology are equivocal or that the dendrogram does not reflect the true phylogeny. Note again that the deductive test only finds compatible hypotheses of monophyly derived from hypotheses of homology — it says nothing about the quality of single characters.

The second level proposed by Bryant (1989) is what I name the "test of plausibility," which is equivalent to Bryant's macroevolutionary analysis. A dendrogram implies new hypotheses: hypotheses about the evolution of organs (e.g., from leg to wing), modes of nutrition (e.g., from carnivory to parasitism) and reproduction and hypotheses about organism migration and the causes of vicariant distributions. Knowledge concerning the biology, physiology, etc. of organisms should be compatible with the phylogeny implied by a den-

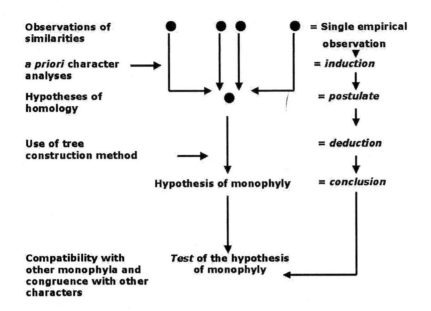

Observations of similarities   = Single empirical observation

a priori character analyses →   = *induction*

Hypotheses of homology   = *postulate*

Use of tree construction method →   = *deduction*

Hypothesis of monophyly   = *conclusion*

Compatibility with other monophyla and congruence with other characters   *Test* of the hypothesis of monophyly

FIGURE 5.8   Phylogenetic cladistics is a hypothetico-deductive method; analyses restricted to the construction of a dendrogram from a data matrix are not.

drogram. Only if all available information fits the evolutionary scenario derived from a dendrogram can one reduce the search for possible mistakes or misinterpretations.

For example, the fossil Crustacean *Rehbachiella* was thought to belong to the branchiopods on the basis of morphological characters (Walossek, 1993). Additional information on the mode of life of branchiopods shows that all extant species are highly specialized small crustaceans, of which most of the less derived organisms are adapted to live in ephemeral epicontinental habitats (such as ponds that are dry in summer). They all have the ability to produce resting eggs resistant to different degrees of desiccation and sunlight. The evolutionary scenario derived from these data is the colonization of, and adaptation to, ephemeral epicontinental waters by the last common ancestors of all extant Branchiopoda. The fossil *Rehbachiella*, however, has been found in sediments of marine origin. Thus, the differences between the evolutionary scenario and the empirical data inspire a search for possible errors. Either *Rehbachiella* is not a branchiopod (the phylogenetic hypothesis has to be revised) or the assumption that the more primitive branchiopods evolved in ephemeral freshwater habitats and that adaptations such as resting eggs are homologies is wrong (a new scenario has to be developed).

De Pinna (1991) distinguishes two steps in the identification of homologies. The first step is the generation of a hypothesis, which he names "primary homology"; the second step is the legitimation of the hypothesis, which may become a secondary homology. Legitimation is a test of congruence in the sense of Patterson (1982), that is, a cladistic analysis. Consequently, what De Pinna (and other cladists) calls "character analysis" is just the deductive cladistic step of phylogenetic cladistics (steps 5 and 6 above) and not an analysis of data quality before the data matrix is compiled. I agree however, that "homology itself is assumed prior to the analysis, based on similarities ..." and that "... the discovery of cladistic analysis in fact is non-homology" (De Pinna 1991:375). However, one should be more careful and say that the result of a cladistic analysis is the discovery of noncompatible hypotheses of homology. This incompatibility should encourage the reexamination of the quality of the data. What is rare in the cladistic literature is discussion of concepts of data quality.

Wiley (1975) also advocates phylogenetic cladistics. For Wiley, the initial suspicion that two patterns may be homologous is tested in two phases. In the first phase, potential falsifiers are usually morphological similarities or dissimilarities. Absence of identity in a larger number of details would motivate the sorting out of weak characters. This certainly is an *a priori* examination of data quality. Wiley calls the search for further identities in two patterns a "test." The reasoning is that the first hypothesis allows predictions (there should be more identical details in structures considered to be homologous) that can be tested to corroborate or weaken the hypothesis (e.g., when the structures are only similar superficially, but not in detail). Wiley's first phase is part of the *a priori* character analysis (steps 1 and 2 above) of phylogenetic cladistics. Wiley's second phase of testing is the cladistic step, where other hypotheses based on different empirical data are the falsifiers (Haszprunar [1998] also recommends a similar two-step procedure).

The logical relationship between probability of homology and weighting has been implicitly recognized. The simplest form is the omission of characters that are considered to be prone to convergence. According to Bryant (1989), for example, a hypothesis of homology must be corroborated by the discovery of identities in shape, constituent tissues, topographic relationships, details of ontogenetic development, etc., that is, with the help of all discernible details of a pattern. "A hypothesis of homology between the avian and chiropteran wing can be rejected during character analysis because, despite superficial similarities, the profound differences in the detailed structure of these wings reduce the feasibility that they could have evolved from the same wing like structure" (Bryant, 1989:220–221). This example also demonstrates that the primary justification for any hypothesis of homology is not congruence but the empirical observation of patterns existing in nature.

Weighting of characters according to their fit on a tree (Farris, 1969; Goloboff, 1993: "There is no necessity to estimate the weights prior to the analysis.") could be considered a typical procedure for phenetic cladistics, because the quality of similarities is not estimated from empirical observation of their organic structure in individuals but from the numerical effect of character states coded in a data matrix. *A posteriori* or successive weighting is a circular procedure (Neff, 1986) because the effects of patterns contained in a data matrix are used to weigh single characters: The result of successive weighting (namely that a character is with high probability a homology) is already contained in the first assumptions of the cladistic analysis (only homologous character states can be used to support a monophylum). For the same reason, cladistic statistics (e.g., consistency index, retention index) are not true criteria for the measurement of data quality, as held by Nelson (1994).

However, the idea of successive weighting is derived from an empirical observation that can not be neglected. As stated by Kluge and Farris (1969), "... we would expect that characters with high variability within OTUs would be more prone to convergence than would other characters ... ." Phylogenetic systematics needs to consider this observation but avoid any circular argumentation. To achieve this, variable characters should get low weight without reference to a topology. This can be (and has been) done with variable positions in alignments. If a partition of terminal taxa is supported by a high number of conserved (binary) positions, other positions that have more than two states or that show deviations from the consensus character state of the ingroup (more variable or noisy positions) get lower weight (e.g., Füllen and Wägele, 2000). Note that partitions (splits) can be compared without reference to any tree topology. Down-weighting of third codon positions is based on the same argument. Morphologists have been using the same approach by selecting characters of low variability for their data matrices (Sober, 1986).

The integration of numerical cladistic methodology with the concepts and procedures proposed by Hennig is common practice and started with some of the first computerized contributions (e.g., Kluge and Farris, 1969). The result will be a complex collection of methods that deserves the name *phylogenetic cladistics*. It is clear that the framework erected by Hennig will survive in the epoch of modern life sciences.

Acknowledgments

I thank Dr. David Williams for the invitation to participate in the third biannual conference of the Systematics Association in London (September 2001) and for correcting the manuscript. Dr. Barbara Holland was so kind to search for mistakes in the first version of this contribution.

References

Andersen, N.M., Some principles and methods of cladistic analysis with notes on the uses of cladistics in classification and biogeography, *Z. Zool. Syst. Evol.forsch.*, 16, 242–255, 1978.

Ax, P., Relationships and phylogeny of the Turbellaria, in *The Lower Metazoa*, Dougherty, E. and Brown, Z.N., Eds., University of California Press, Berkeley, 1963, pp. 191–224.

Ax, P., *Das Phylogenetische System*. G. Fischer, Stuttgart, Germany, 1984.

Ax, P., *The Phylogenetic System: The Systematization of Organisms on the Basis of Their Phylogenesis* (trans. Jefferies, R.P.S.), Wiley, Chichester, UK, 1987.

Ax, P., *Systematik in der Biologie*, G. Fischer, Stuttgart, Germany, 1988.

Ax, P., *Das System der Metazoa II. Ein Lehrbuch der Phylogenetischen Systematik*, G. Fischer, Stuttgart, Germany, 1999.

Bigelow, R.S., Classification and phylogeny, *Syst. Zool.*, 5, 145–146, 1958.

Briggs D.E.G. and Fortey, R.A., The early radiation and relationships of the major arthropod groups, *Science*, 246, 241–243, 1989.

Brower, A.V.Z. and Schawaroch, V., Three steps of homology assessment, *Cladistics*, 12, 265–272, 1996.

Brundin, L., Application of phylogenetic principles in systematics and evolutionary theory, in *Current Problems of Lower Vertebrate Phylogeny*, Nobel Symposium, Orvig, T., Ed., Stockholm: Wiksell and Almquist, Interscience, 1968, pp. 473–495.

Brusca, R.C. and Brusca, G.J., *Invertebrates*, Sinauer, Sunderland, MA, 1990.

Brusca, R.C. and Wilson, G.D.F., A phylogenetic analysis of the Isopoda with some classificatory recommendations, *Mem. Qld. Mus.*, 31, 143–204, 1991.

Bryant, H.N., An evaluation of cladistic and character analyses as hypothetico-deductive procedures, and the consequences for character weighting, *Syst. Zool.*, 38, 214–227, 1989.

Cain, J.H. and Harrison, G.A., An analysis of the taxonomist's judgement of affinity, *Proc. Zool. Soc. London*, 131, 85–98, 1958.

Camin, J.H. and Sokal, R.R., A method for deducing branching sequences in phylogeny, *Evolution*, 19, 311–326, 1965.

Charleston, M.A., Hendy, M.D., and Penny, D., The effects of sequence length, tree topology, and number of taxa on the performance of phylogenetic methods, *J. Comput. Biol.*, 1, 133–151, 1994.

De Pinna, M.C.C., Concepts and tests of homology in the cladistic paradigm, *Cladistics*, 7, 367–394, 1991.

De Queiroz, K., The ontogenetic method for determining character polarity and its relevance to phylogenetic systematics, *Syst. Zool.*, 34, 280–299, 1985.

DeSalle, R. and Brower, A.V.Z., Process partitions, congruence, and the independence of characters: inferring relationships among closely related Hawaiian *Drosophila* from multiple gene regions, *Syst. Biol.*, 46, 751–764, 1997.

Donoghue, M.J. and Maddison, W.P., Polarity assessment in phylogenetic systematics: a response to Meacham, *Taxon*, 35, 534–545, 1986.

Erhard, F., Das pleonale Skelett-Muskel-System von *Titanethes albus* (Synocheta) und weiterer Taxa der Oniscidea (Isopoda), mit Schlußfolgerungen zur Phylogenie der Landasseln, *Stuttg. Beitr. Na.kd,* A550, 1–70, 1997.

Farris, J.S., A successive approximations approach to character weighting, *Syst. Zool.,* 18, 374–385, 1969.

Farris, J.S., On the boundaries of phylogenetic systematics, *Cladistics,* 2, 14–27, 1986.

Farris, J.S., Kluge, A.G., and Eckardt, M.J., A numerical approach to phylogenetic systematics, *Syst. Zool.,* 19, 172–189, 1970.

Fitch, W.M., Towards defining the course of evolution: minimum change for a specified tree topology, *Syst. Zool.,* 20, 406–416, 1971.

Fitch, W.M. and Ye, J., Weighted parsimony: does it work?, *Phylogenetic Analysis of DNA Sequences,* Miyamoto, M.M. and Cracraft, J., Eds., Oxford University Press, New York, 1991, pp. 147–154.

Forey, P.L., Humphries, C.J., Kitching, I.J., Scotland, R.W., Siebert, D.J. and Williams, D.M., *Cladistics: A Practical Course in Systematics.* Oxford Science Publications, Clarendon Press, Oxford, 1992.

Fuellen, G. and Wägele, J.W., Phylogeny inference by minimum conflict, in *Classification and Information Processing at the Turn of the Millenium,* Gaul, W. and Decker, R., Eds., Proceedings of the 23rd Annual Conference of the Gesellschaft für Klassifikation, University of Bielefeld, Germany, 2000, pp. 377–84.

Goloboff, P.A., Estimating character weights during tree search, *Cladistics,* 9, 83–91, 1993.

Graybeal, A., Evaluating the phylogenetic utility of genes: a search for genes informative about deep divergences among vertebrates, *Syst. Biol.,* 43, 174–193, 1994.

Haeckel, E., *Generelle Morphologie der Organismen: Allgemeine Grundzüge der organischen Formen-Wissenschaft, mechanisch begründet durch die von Charles Darwin reformierte Descendenz-Theorie,* Georg Reimer, Berlin, 1866.

Hall, B.K., Homology and embryonic development, *Evol. Biol.,* 28, 1–37, 1994.

Haszprunar, G., The types of homology and their significance for evolutionary biology and phylogenetics, *J. Evol. Biol.,* 5, 13–24, 1992.

Haszprunar, G., Parsimony analysis as a specific kind of homology estimation and the implications for character weighting, *Mol. Phylogenet. Evol.,* 9, 333–339, 1998.

Hennig, W., *Grundzüge einer Theorie der phylogenetischen Systematik,* Deutscher Zentralverlag, Berlin, 1950.

Hennig, W., Kritische Bemerkungen zum phylogenetischen System der Insekten, *Beitr. Entomol.,* 3, 1–85, 1953.

Hennig, W., *Phylogenetic Systematics,* University of Illinois Press, Urbana, 1966.

Hennig, W., *Die Stammesgeschichte der Insekten,* Waldemar Kramer, Frankfurt, 1969.

Hennig, W., *Insect Phylogeny,* (trans. Pont, A.C., with revisionary notes by Schlee, D.), John Wiley, Chichester, UK, 1981.

Hennig, W., *Phylogenetische Systematik,* Paul Parey, Berlin, 1982.

Hennig, W. and Schlee, D., Abriß der phylogenetischen Systematik, *Stuttg. Beitr. Natkd.,* Ser. A, 319, 1–11, 1978.

Kitching, I. J., Forey, P. L., Humphries, C. J., and Williams, D.M., *Cladistics: The Theory and Practice of Parsimony Analysis,* 2nd ed., Systematics Association, Publication No. 10, Oxford University Press, Oxford, 1998.

Kluge, A.G. and Farris, J.S., Quantitative phyletics and the evolution of anurans, *Syst. Zool.,* 18, 1–32, 1969.

Lauterbach, K.E., Schlüsselereignisse in der Evolution des Grundplans der Mandibulata (Arthropoda), *Abh. Nat.wiss. Ver. Hamb.,* NF 23, 105–161, 1980.

Lento, G.M., Hickson, R.E., Chambers, G.K., and Penny, D., Use of spectral analysis to test hypotheses on the origin of pinnipeds, *Mol. Biol. Evol.,* 12, 28–52, 1995.

Lorenz, K., Vergleichende Bewegungsstudien an Anatinen, *J. Ornithol.,* (spec. vol.) 3, 194–293, 1941.

Lyons-Weiler, J., Hoelzer, G.A., and Tausch, R.J., Relative apparent synapomorphy analysis (RASA). I. The statistical measurement of phylogenetic signal, *Mol. Biol. Evol.*, 13, 749–757, 1996.

Meacham, C.A., The role of hypothesized direction of characters in the estimation of evolutionary history, *Taxon*, 33, 26–38, 1984.

Mindell, D.P., Aligning DNA sequences: homology and phylogenetic weighting, in *Phylogenetic Analysis of DNA Sequences*, Miyamoto, M.M. and Cracraft, J., Eds., Oxford University Press, London, 1991, pp. 73–89.

Minelli, A., *Biological Systematics. The State of the Art*, Chapman and Hall, London, 1993.

Mishler, B.D., Cladistic analysis of molecular and morphological data, *Am. J. Phys. Anthropol.*, 94, 143–156, 1994.

Müller, F., *Für Darwin*, Wilhelm Engelmann, Leipzig, Germany, 1864.

Neff, N.A., A rational basis for *a priori* character weighting, *Syst. Zool.*, 35, 110–123, 1986.

Nelson, G.J., Cladistic analysis and synthesis: principles and definitions with a historical note on Adanson's *Famille des Plantes* (1763–1764), *Syst. Zool.*, 28, 1–21, 1979.

Nelson, G.J., Homology and systematics, in *Homology: The Hierarchical Basis of Comparative Biology*, Hall, B.K., Ed., Academic Press, San Diego, CA, 1994, pp. 101–49.

Nelson, G.J. and Platnick, N.I., *Systematics and Biogeography*, Columbia University Press, New York, 1981.

Nielsen, C., *Animal Evolution: Interrelationships of the Living Phyla*, Oxford University Press, Oxford, 2001.

Panchen, A.L., The use of parsimony in testing phylogenetic hypotheses, *Zool. J. Linn. Soc.*, 74, 305–328, 1982.

Panchen, A.L., Richard Owen and the concept of homology, in *Homology: The Hierarchical Basis of Comparative Biology*, Hall, B.K., Ed., Academic Press, San Diego, CA, 1994, pp. 21–62.

Patterson, C., Morphological characters and homology, in *Problems of Phylogenetic Reconstruction*, Joysey, K.A. and Friday, A.E., Eds., Academic Press, London, 1982, pp. 21–74.

Patterson, C., Homology in classical and molecular biology, *Mol. Biol. Evol.*, 5, 603–625, 1988.

Philippe, H. and Laurent, J., How good are deep phylogenetic trees?, *Curr. Opin. Genet. Dev.*, 8, 616–623, 1998.

Platnick, N.I., Philosophy and the transformation of cladistics, *Syst. Zool.*, 28, 537–546, 1979.

Remane, A., *Die Grundlagen des natürlichen Systems, der vergleichenden Anatomie und der Phylogenetik*, Akademische Verlagsgesellschaft, Geest and Portig, Leipzig, Germany, 1952.

Remane, J., The concept of homology in phylogenetic research: its meaning and possible applications, *Palaeontol. Z.*, 57, 267–269, 1983.

Richter, S. and Meier, R., The development of phylogenetic concepts in Hennig's early theoretical publications (1947-1966), *Syst. Biol.*, 43, 212–221, 1994.

Riedl, R., *Die Ordnung des Lebendigen*, Paul Parey, Hamburg, Germany, 1974.

Rieppel, O., Homology and logical fallacy, *J. Evol. Biol.*, 5, 701–715, 1992.

Rieppel, O., Homology, topology, and typology: the history of modern debates, in *Homology: The Hierarchical Basis of Comparative Biology*, Hall, B.K., Ed., Academic Press, San Diego, CA, 1994, pp. 63–100.

Rosa, D., *Ologenesi: Nuova teoria dell'evoluzione e della distribuzione geografica dei viventi*, Bemporad e Figlio, Florence, Italy, 1918.

Schlee, D., Die Rekonstruktion der Phylogenese mit Hennig's Prinzip, *Aufs. Reden Senckenb. Nat.forsch. Ges.*, 20, 1–62, 1971.

Scholtz, G., Mittmann, B., and Gerberding, M., The pattern of distal-less expression in the mouthparts of crustaceans, myriapods and insects: new evidence for a gnathobasic mandible and the common origin of Mandibulata, *Int. J. Dev. Biol.*, 42, 801–810, 1998.

Schram, F.R., Cladistic analysis of metazoan phyla and the placement of fossil problematica, in *The Early Evolution of Metazoa*, Simonetta, A.M. and Morris, S.C., Eds., Cambridge University Press, Cambridge, 1991, pp. 35–46.

Schram, F.R. and Hof, C.H.J., Fossils and the interrelationships of major crustacean groups, in *Arthropod Fossils and Phylogeny*, Edgecombe, G.D., Ed., Columbia University Press, New York, 1998, pp. 233–302.

Scotland, R.W., Cladistic theory, in *Cladistics: A Practical Course in Systematics*, Forey, P.L., Humphries, C.J., Kitching, I.J., Scotland, R.W., Siebert, D.J., and Williams, D.M., Clarendon Press, Oxford, 1992, pp. 3–13.

Scott-Ram, N.R., *Transformed Cladistics, Taxonomy and Evolution*, Cambridge University Press, Cambridge, 1990.

Sharkey, M.J., A hypothesis-independent method of character weighting for cladistic analysis, *Cladistics*, 5, 63–86, 1989.

Sneath, P.H.A. and Sokal, R.R., *Numerical Taxonomy*, W.H. Freeman, San Francisco, 1973.

Sober, E., Parsimony and character weighting, *Cladistics*, 2, 28–42, 1986.

Sober, E., *Reconstructing the Past: Parsimony, Evolution, and Inference*, Massachusetts Institute of Technology, Cambridge, MA, 1988.

Swofford, D.L., *PAUP*. Phylogenetic Analysis Using Parsimony (* and Other Methods), Version 4.0**, Sinauer Associates, Sunderland, MA, 1998.

Swofford, D.L., Olsen, G.J., Waddell, P.J., and Hillis, D.M., Phylogenetic inference, in *Molecular Systematics*, Hillis, D.M., Moritz, C., and Mable, B.K., Eds., Sinauer Associates, Sunderland, MA, 1996, pp. 407–514.

Wägele, J.W., Rejection of the "Uniramia" hypothesis and implications of the Mandibulata concept, *Zool. Jahrb., Abt. Syst.*, 120, 253–288, 1993.

Wägele, J.W., Review of methodological problems of 'computer cladistics' exemplified with a case study on isopod phylogeny (Crustacea: Isopoda), *Z. Zool. Syst. Evol.forsch.*, 32, 81–107, 1994.

Wägele, J.W., The theory and methodology of phylogenetic systematics is still evolving: a reply to Wilson, *Vie Milieu*, 46, 183–184, 1996a.

Wägele, J.W., First principles of phylogenetic systematics, a basis for numerical methods used for morphological and molecular characters, *Vie Milieu*, 46, 125–138, 1996b.

Wägele, J.W., Identification of apomorphies and the role of ground patterns in molecular systematics, *J. Zool. Syst. Evol. Res.*, 34, 31–39, 1996c.

Wägele, J.W., *Grundlagen der Phylogenetischen Systematik*, Verlag Dr. F. Pfeil, Munich, 2000.

Wägele, J.W. Erikson, T., Lockhart, P., and Misof, B., The Ecdysozoa: artifact or monophylum?, *J. Zool. Syst. Evol. Res.*, 37, 211–223, 1999.

Wägele, J.W. and Misof, B., On quality of evidence in phylogeny reconstruction: a reply to Zrzavý's defence of the "Ecdysozoa" hypothesis, *J. Zool. Syst. Evol. Res.*, 39, 165–176, 2001.

Wägele, J.W. and Rödding, F., *A priori* estimation of phylogenetic information conserved in aligned sequences, *Mol. Phylogenet. Evol.*, 9, 358–365, 1998.

Wagner, G.P., The biological homology concept, *Annu. Rev. Ecol. Syst.*, 20, 51–69, 1989.

Wagner, G.P., Homology and the mechanism of development, in *Homology: The Hierarchical Basis of Comparative Biology*, Academic Press, San Diego, CA, 1994, pp. 273–299.

Wagner, W.H., Problems in the classification of ferns, *Recent Adv. Bot.*, 1, 841–844, 1961.

Walossek, D., The Upper Cambrian *Rehbachiella* and the phylogeny of Branchiopoda and Crustacea, *Fossils Strata*, 32, 1–202, 1993.

Wenzel, J.W. and Siddall, M.E., Noise, *Cladistics*, 15, 51–64, 1999.

Wheeler, Q.D., Character weighting and cladistic analysis, *Syst. Zool.*, 35, 102–109, 1986.

Wiley, E.O., Karl R. Popper, systematics, and classification: a reply to Walter Bock and other evolutionary taxonomists, *Syst. Zool.*, 24, 233–243, 1975.

Wiley, E.O., *Phylogenetics: The Theory and Practice of Phylogenetic Systematics*, J. Wiley and Sons, New York, 1981.

Wills, M.A., Briggs, D.E.G., Fortey, R.A., and Wilkinson, M., The significance of fossils in understanding arthropod evolution, *Verh. Dtsch. Zool. Ges.*, 88, 203–215, 1995.

Zrzavý, J., Mihulka, S., Kepka, P., Bezdek, A., and Tietz, D., Phylogeny of the Metazoa based on morphological and 18S ribosomal DNA evidence, *Cladistics*, 14, 249–285, 1998.

Chapter 6

Cladistics: Its Arrested Development

G. Nelson

Citoyens, la Revolution est fixée aux principes qui l'ont commencée. Elle est finie. NB 15 Dec 1799

Abstract

Paleontology of the past is revived in molecular systematics of the present, in its search for ancestors and centers of origin. The revival ignores, or retreats from, the cladistic reform of paleontology of the 1970s, with historical roots in the work of Louis Dollo (1857–1931) and fossil lungfishes. The subsequent development of cladistics has been arrested, too, by computer implementations of character optimization and the ideology of total evidence, which reflects a phenetic rather than cladistic objective: the overall similarity of synapomorphy.

Introduction

Cladistic systematics has appeared in at least two recent historical arenas: numerical taxonomy and paleontology. Within each area of contention, cladistics achieved a reform — but to a degree only. Its success has been evident for a generation.

Numerical Taxonomy

In numerical taxonomy, measures of distance (overall similarity) were abandoned in favor of measures of evolutionary parsimony — or compatibility, or maximum likelihood.

Numerical taxonomy in its cladistic phase became animated through "the real progress in systematic theory over the past two decades ... the discoveries that phenetics is false; that parsimony [or compatibility, or likelihood] is essential; that exact solutions to parsimony problems [or compatibility problems, or likelihood problems] are possible" (Farris and Platnick, 1989:308). Animation flowed from the belief that (Farris, 1968:9) "perhaps the most impressive theoretical advantage of the parsimony criterion is that it is certain to give a correct tree, provided the data consist of a sufficient array of non-convergent characters," and from similar beliefs, with similar provisos, that compatibility, or maximum likelihood, might alternatively — even certainly — give a correct tree.

Paleontology

In paleontology, the search for ancestors was abandoned in favor of the search for the sister group, that is, the nearest relative, living or fossil.

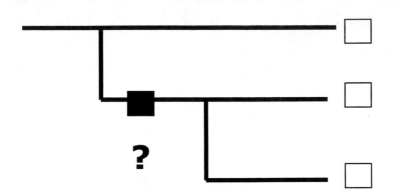

FIGURE 6.1   Three organisms, species, or taxa, and the possibility (?) that two are related more closely to each other than either is to the third.

Paleontology had been animated by the possibility that fossils have unique and overriding importance for the science of evolution — of phylogeny, the history of life. This inspiring view of fossils predominated in the early part of the twentieth century. It was exemplified by, among many others, William Diller Matthew (1871–1930), of the American Museum of Natural History. In 1927 he "founded the only separate department of paleontology in an American University" (the University of California at Berkeley) (Colbert, 1989:143). He saw fossils, when abundant enough, as a direct revelation of the "true genetic sequences," the continuous series of ancestors and descendants, not only of organisms but also of taxa (Matthew, 1925:288), reflecting the "law of direct genetic succession" (Osborn, 1911:583) — "*the profound hereditary phyletic movements* which can not be observed by the zoologist at all" (Osborn, 1915:224). For Matthew, fossils — the fossil record — had provided the facts of evolution, without which a true science of phylogeny (the history of life in time and in geographic space) was nonexistent and impossible.

Matthew's vision lives today, not only for fossils but also for DNA — more true genetic sequences and even truer sequences of ancestors and descendants, truer even than the fossil ones, except for fossil DNAs, of course. Revived within this living record, as in Matthew's fossil record, is the revelation of the real history of organisms in time and space — phylogeography (Avise, 2000).

Cladistics

Cladistics emphasizes that relationships, both of Recent and fossil life, are problematic in the same way, and that evidence, not revelation, is the relevant concern.

There are two general and persistent questions about evidence and its meaning, which are the basis of cladistics (Figure 6.1):

Question 1: In the simplest terms of three organisms, species, or taxa, what is the evidence that two are related more closely to each other than to the third? Never mind if they are known from fossil or living material, alone or in combination.

Question 2: If two of them are related more closely, what does this mean about the organisms, species, or taxa: about their evolution, classification, even their nomenclature and usefulness and interest to humans?

These are questions, not answers, because the spirit of cladistics is enquiry, not ideology, propaganda, sloganizing, or the husbandry of sacred cows and clams. At least, that is what the spirit was, once upon a time, when it achieved general relevance lasting to the present.

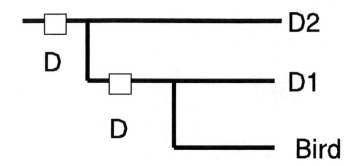

FIGURE 6.2   Two dinosaurs (D1-2) and a modern bird, showing the bird related more closely to one dinosaur (D1, *Dromaeosaurus*) than to the other (D2, *Allosaurus*).

When the questions are given up in favor of particular answers to them, however tentatively or forcefully embraced the answers might be, so given up to that extent, too, is the spirit of enquiry. The arenas become vacated. They have been rather vacant for a while now. If he were alive, Mark Twain might remark on the state of the cows and clams — how peacefully and quietly in small and isolated groups they lie about the premises.

Two Dinosaurs and the Bird

In research, evidence comes first. In news items and publicity generated for the media, meaning comes first.

I begin with meaning, an example taken from a website: Are birds really dinosaurs? Yes, says the average paleontologist, according to the news from DinoBuzz from U.C. Berkeley (February 2002: ucmp.berkeley.edu/; Padian and Horner, 2002). Accordingly, birds are dinosaurs because birds have evolved from dinosaurs: "The evidence is overwhelmingly in favor of birds being the descendants of a maniraptoran dinosaur, probably something similar (but not identical) to a small dromaeosaur. What is this evidence?"

Reduced to a minimum (Figure 6.2), the evidence is that a modern bird relates more closely to one dinosaur (D1, e.g., *Dromaeosaurus*) than to another (D2, e.g., *Allosaurus*). Guided by these relationships, things at the tips of branches — names and natures in this case — drop to nodes. Biological data, too, drop to nodes. Nodes are reified as ancestors. Excuse me if I read a definition. "To reify: To treat an abstraction as a concrete material object."

Out of the mists of time the ancestors arise as dinosaurs (D in Figure 6.2). They might have feathers, but we are told that they definitely are nonbirds. They have jaws with teeth, not toothless jaws with a rhamphotheca (a horny bill).

From this viewpoint, the search for ancestors is realized in imagination, where Jurassic Park comes alive. The evidence (the pathway to the park) is paraphyly, or if you prefer, symplesiomorphy. Deceased in 1930 (Abel, 1931a), Matthew would have thrilled to hear the news (from his own department, too) and to learn where birds have their geographic center of origin, where their ancestors must once have lived — their own little Garden of Eden, hypothesis though it be — whence birds have dispersed over the surface of the earth.

Well, are birds really dinosaurs? There are several other questions that have the same general import, for example:

Are fishes really invertebrates?
Are humans really apes?
Are animals really plants?
Are angiosperms really gymnosperms?
Are Greeks really barbarians?

And so on. The list is endless. Examine each case. Find that fishes nest within, or among, invertebrates; humans, within apes; animals, within plants; angiosperms, within gymnosperms; and Greeks, alas, within barbarians.

From one viewpoint, call it transformational or even evolutionary, none of this is problematic in the least. It is simple and mindless truism about evolution and its evidence, which like religion, to keep one's friends — or fellow clams — one simply cannot do without: "Birds arose from nonbirds and humans from nonhumans. No person who pretends to any understanding of the natural world can deny these facts any more than he or she can deny that the world is round, rotates on its axis, and revolves around the sun" (Lewontin, 1981).

Accordingly, fishes evolved from invertebrates, humans from apes, even birds from dinosaurs. To suggest otherwise is tiresome nonsense, safely ignored, and best kept away from children and impressionable youth. In a Donald Duck cartoon strip of February 17 1986, one reads (Nielsen, 1987:Fig. 14):

"Uncle Donald, did we evolve from apes?"
"Of course not! We evolved from pterodactyls."

Beyond the teachings of the church within our culture, what else is there, after all, to fill the void of general understanding that accompanies each new generation?

From a second viewpoint, call it relational or even cladistic if you would provoke the fundamentalists (Kluge and Wolf, 1993), evidence might relate birds to some dinosaurs more closely than to others, but this does not mean that birds are dinosaurs, or that birds have evolved from them. It means that of them, there are two kinds at least, maybe more. For them it implies two kinds of biology — two kinds of jaws with teeth, one related most closely to jaws of birds. It means a discovery about dinosaurs, not a discovery about birds: "In the 150 years since Richard Owen named Dinosauria for a few bones of *Iguanodon*, *Megalosaurus* and *Hylaeosaurus*, the result of the efforts of paleontologists has been to discover what dinosaurs are, not what birds are" (Patterson, 1993).

What is the discovery? Dinosaurs are not monophyletic but paraphyletic; they evidently are not one group, but two or more, with similar relationships to modern birds, as if they were merely diverse groups of primitive birds. For Patterson (1993; also Ax, 1989:41), "Perhaps surprisingly, dinosaurs turn out to be kinds of bird, a theory of relationships that is most clearly and economically expressed by including them within Aves."

What does this mean for phylogeny? Life's history is more complex than previously was evident. More groups, not fewer, have evolved along the way to modern birds.

What does this mean for dropping biological data, even names, to nodes, which are then reified as ancestors? The process generates symplesiomorphy; that is its purpose. To imagine dinosaurs as ancestral birds is inexact, misleading, and ultimately mistaken, but if not dinosaurs, what were they, these common ancestors of dinosaurs and birds? Pterodactyls?

Uncle Donald may believe something that we do not — that pterosaurs, sometimes considered the sister group of dinosaurs and birds (Dingus and Rowe, 1998:160), are paraphyletic. Or he may regard as ancestral an extinct sister group, whether it be paraphyletic or not. That it be extinct and related is enough for a judgement that accepts, as evidence of ancestry, an interpretation of that kind.

Paleontology of the Parts

My intent here is only to show that nodes, when reified as ancestors (symplesiomorphy), do not, and perhaps cannot, work perfectly as mechanisms of human understanding, even when relationships are unproblematic, when there is no conflicting evidence, of whatever sort, about them. It need be no surprise that reified nodes do not work perfectly or even exactly. Why, after all, should they be expected to do so?

To associate biological data with nodes has been a concern of cladistics from its beginning, even if its beginning be located in Aristotle. Today, it is commonplace to read that relationships are tested by homologies through congruence, as optimized on the best-fitting tree — that scheme of relationships that implies the fewest evolutionary origins, or transformations, of the characters (one of the exact solutions to a parsimony problem, or to a compatibility problem, or to one of maximum likelihood).

From the relational viewpoint, however, this litany of so-called cladism reads like Matthew's paleontology, the search for the true sequence of ancestors and descendants, not of whole organisms but of their parts. It might be true that the search for ancestral parts seemed, for a time, to work better, and that the better working explains the success of the cladistic reform of paleontology. It might be true, too, that cladistics itself developed as a merely abstract paleontology of the parts — one in which the outlook of paleontology was subsumed at a less prominent, even subliminal, level. If so, another round of reform may prove necessary to a future generation. For a few of us aging radicals, such has already proven necessary.

Spezialisationskreuzungen or Its Chevauchement

Here the problem stems from an idea basic to the litany: that an organism is like an automobile, a mix of parts, some of ancient design, some recent. For this character combination — a prerequisite for recognizing monophyletic sister groups — Willi Hennig (1949:138, 1950:142, 1966:194; also 1936:557) borrowed from vertebrate paleontology the term *specialization-crossings*, a translation of *Gesetz der Spezialisationskreuzungen* from Abel (1911:245, 1912:637; also 1907a:190, 1909:245, 1910:49, 1913:122, 1917:124, 1920:21, 1921:148, 1924:21, 1929:259), which is a translation of *chevauchement des specialisations* from Dollo (1895:88, 1900:397, 1912:132; and correspondence mentioned by Gould [1970:192], Plate [1912:925, 1933:1015], Kälin [1936:654, 1941:15], Schindewolf [1937:207, 1942:117, 1950:Fig. 251, "cross-specializations" in 1993:Fig. 3.106], Kuhn [1942:77], Gross [1943:244], Remane [1952:283], Zimmermann [1954:57, 1960a:316, 1962:132, 1963:543, 1967:101], Günther [1956:46], Osche [1966:829], Rieppel [1979:538], Richter and Meier [1994:215], Schmitt [2001:329]) and *evolution entre-croisée* from Nierstrasz (1936:Fig., p. 674, a graphic scheme similar to the character matrix of numerical taxonomy, reproduced in Hennig [1950:Fig. 31], reinterpreted in Meyer-Abich [1943:14, Fig. 6, also 1964:56, Fig. 4, 1970:84], who commented on the history of this concept from Goethe onward — as a matrix and as an alternative scheme approaching the advancement index of Sporne (1954) and the ground-plan divergence of Wagner (1980:186, 190), who claimed his invention of the "1940s and 1950s" was "one of the oldest, if not the oldest, formalized methods for making cladograms." Takhtajan (1954:42; "evolutionary age difference of characteristics") commented on the history of this concept from Lamarck onward (see also Mayr 1972:88; Schindewolf, 1955, 1957:49, 1969:78; Osche, 1966:876; Gould, 1970:192; Mayr, 1982:613; Wagenitz, 1996:174). Related terms were De Beer's "mosaic of characters" (1954a:48), "mosaic evolution" (1954b:163, 1969:134) and Watson's rule (1954b:170, 1969:148). Schindewolf also commented on the history of this concept from Mendel onward: *Merkmalsmosaik* from Gross (1956:32; see also Hennig, 1984:137) and *mosaikartige Verteilung* and *mosaikartige Merkmalsverteilung* from Königsmann (1975:107, 110).

Lungfishes : From Dollo to White to Miles' Stones

Dollo's *chevauchement* (overlap of characters) derived from his study of fossil and living lungfishes and their varied combinations of characters, such that no known taxon is apparently ancestral to any other. Hypothetical ancestors are the only ones in Dollo's summary cladogram (1895:89; Brien, 1951:95, figure revised in Abel, 1929:259; comment also in Abel 1911:245, 1912:637, 1917:129), which Simpson, for reasons not evident, (1953:220, cf. 1944:195) "rightly considered a triumph of the first half century of phylogenetic paleontology" (Abel 1907b:74, 1931b:328; Westoll, 1949:121).

In a muted reprise of Dollo's theme 70 years later, fossil lungfishes were featured in a presidential address to the Linnean Society of London, delivered by Errol White after a long career as Keeper of Paleontology at the British Museum (Natural History). With no reference to Dollo, and apparently with no inspiration from him, White nevertheless lamented (1966:8),

> And this brings me to the real point of this address ... . In my experience with fossil fishes, while one can see the general drift of evolution readily enough, when it comes to pin-pointing the linkages, whether it be at generic level or at that of a higher group, the links are invariably either missing altogether or faulty, that is to say always one or more characters are out of phase ... .

In the same year, the cladistic reform of paleontology began in reaction to Lars Brundin's monograph on austral midges (Brundin, 1966) and the English translation of Hennig (1966).

Inspired by Hennig's earlier papers in German, Brundin's monograph included an influential 50-page introduction, with a summary of Hennig's ideas and original critiques of vertebrate paleontology and biogeography (Wanntorp, 1993; Fittkau, 1995). By invitation in June, 1967, in Stockholm's Kungliga Vetenskapsakademien, the entomologist Brundin addressed the Nobel Symposium on the Current Problems of Lower Vertebrate Phylogeny — mainly of fossil fishes — on the topic of phylogenetic principles (Brundin, 1968). "The heart of Brundin's paper was one message: 'phylogenetics is the search for the sister group.' That message eventually got through ..." (Patterson, 1989:472).

Soon after came the 1972 Symposium on Interrelationships of Fishes, organized for the Linnean Society of London by Humphry Greenwood, Roger Miles, and Colin Patterson, all of the British Museum (Natural History). All three had attended the Nobel Symposium of 1967. According to Patterson (1995:12),

> our hidden agenda was cladistics, to get as many groups of fishes as possible worked over in the new cladistic framework. The symposium volume came out in 1973 [Greenwood et al.]. We didn't manage to raise a complete cast of cladists but I think this was the first multi-author volume, anywhere in biology, in which the overall message is cladistics. It has a certain historical significance.

Lungfishes and their varied combinations of characters again were featured in White's department, this time in the hands of his young successor working on Paleozoic fishes in the British Museum. In 1973 the relationships of lungfishes were featured in the symposium on the *évolution des vertébrés* (Miles, 1975), one in the series of *colloques internationaux* sponsored by the Centre National de la Recherche Scientifique (CNRS), this one organized in Paris by Jean-Pierre Lehman, Professor at the Muséum National d'Histoire Naturelle. He, too, had attended the Nobel and Linnean symposia.

The CNRS symposium was later seen as the beginning of "the cladistic revolution among French paleontologists" (Goujet, 2000:79). The published proceedings, concerning all major groups of vertebrates, were notable also because of the spirited discussion of Niels Bonde

(Geologisk Museum, Copenhagen), another paleoichthyologist who had attended the Nobel and Linnean symposia. He predicted that (Bonde, 1975:312; cf. Jefferies, 2001) "in some twenty years Hennig's methods and his 'phylogenetic system' will be the dominant theory — perhaps including some mathematical approximations. The rest of numerical taxonomy will still be in use for special purposes ... ."

Finally, lungfishes became the subject of the first modern cladistic monograph of a vertebrate group with an extensive fossil record "involving only hypothetical common ancestors" (Miles, 1977:302), where the search for ancestors was abandoned in favor of the search for the sister group (reflecting the message that eventually got through). Ten years on, this publication was described as "the only comprehensive cladistic analysis of the Dipnoi" (Marshall, 1987:151). These milestones, and other related events, were the basis for the chronology of the cladistic revolution given by Patterson (1989) with hindsight allowed by a second Nobel Symposium and Brundin as one of its organizers, 20 years after the first in 1967. For Patterson (1997:4; also Bonde, 1974, 2000; Janvier, 1986, 1996:317; Tassy, 1986), the revolution "began in the late 1960s, accelerated in the 1970s, and was virtually complete by the eighties" (Nelson, 2000:20). With ironic understatement, Brundin, whose critique (1966:11–64) began the revolution, agreed (1988:366): "Little by little some paleontologists have perceived that Hennig's principles of phylogenetic systematics meant a revolution to their science."

Heterobathmie

For the same idea of character combination but in a general form suitable for sorting the characters, Hennig later switched to the term *heterobathmy* — different steps on a stair (1965:107, 1968:263, 1974:279, 1975:244, 1982:189, 1984:137). This term (from the Greek *bathmos,* a step or stair) he borrowed from the Armenian botanist Armen Takhtajan, from a book translated from the Russian, published in Germany in 1959 (Takhtajan, 1959:11, 1991:227, 1997:5; Zimmermann, 1962:132, 1963:543, 1965:11, 1967:101; Zimmermann and Schneider, 1967:343; Königsmann, 1975:107; Ax, 1984:119, 1987:109; Meeuse, 1982:2, 1984:11, 1986:186; Meyen, 1987:365; Sudhaus and Rehfeld, 1992:109; Darlu and Tassy, 1993:33; Wagenitz, 1996:174; Forterre and Philippe, 1999:875; Classen-Bockhoff, 2001:1165; Dupuis, 1986:230 refers to Darwin's similar idea; with reference to Hennig, 1969:20 and 1981:6, Kristensen and Nielsen [1979:74] named *Heterobathmia*, a genus of primitive moths).

Among English speakers, neither of the two borrowed terms ever caught on (cf., Hill and Crane, 1982:297 for an explanation of the term [heterobathmy, Hill, 1986:124] and for comment on related ideas of Arber and Parkin [1907:35] and of Davis and Heywood [1963:34]; Schoch, 1986:25, 335 for "heterobathmy of characters" [a concept "fundamental to cladistic analysis"]; Schuh, 2000:69 for "heterobathmy of synapomorphy").

The mix — the heterobathmy, unproblematic in itself — requires sorting and arranging the parts in proper order up and down the steps, according to their origins. Afterwards, the parts, as evidence of relationship, may then be properly understood as the marks of synapomorphy, in contrast to symplesiomorphy and convergence: each synapomorphy at the proper taxonomic level, each upon the proper step of, and thereby associated with, a more or less inclusive taxon. One lesson often overlooked is that *taxon* and *synapomorphy* are different names for the same thing: the same step of the same stair — the same relationship.

Merkmalsphylogenie

For the sorting and arranging, Hennig borrowed yet another term from botany: "character phylogeny," the study of homologies or transformation series (*Merkmalsphyletik,* or *Semo-*

phyletik, and *Merkmalsphylogenie* and *Organphylogenie* from Zimmermann, 1930:427, 1931:981, 984, 1001, 1934:382, 1943:26, 1948:70, 1949:vii, 1952:456, 1953:4, 1954:59, 1957:164, 1959:480, 1960a:321, 1960b:18, 1962:132, 1966:160, 1967:103; Lam, 1935:99, 1936:180; Remane, 1952:283; Günther, 1956:49; Osche, 1966:827; Königsmann, 1975:109; Dupuis, 1979:38; Donoghue and Kadereit, 1992). Wagenitz (1996:234) gives earlier usages: *heterobathmischen Merkmals-Analyse* from Günther (1971:76), *semophylesis* from Lam (1948:111, 1959:42; Meeuse, 1966:12), "semiphylogeny" from Zimmermann (1969:819).

For Zimmermann (1957:164), "Im Grunde ist alle exact Phylogenetik Merkmalsphyl-ogenetik" ("the only true phylogenetic analysis is that of characters" — the true paleontology of the parts). But, from Hennig (1982:96, 1966:93), "Mit 'Transformation' ist dabei natürlich an den realhistorischen Evolutionprozess zu denken und nicht an die Möglichkeit, die Merkmale im Sinne der idealistischen Morphologie formal auseinander herzuleiten." ("'Transformation' naturally refers to real historical processes of evolution, and not to the possibility of formally deriving characters from one another in the sense of idealistic morphology.")

Optimization as Idealistic Morphology

"die Merkmale ... formal auseinander herzuleiten" ("formally deriving characters from one another. ") What do you suppose he meant? Another kind of sorting and arranging? These, too, are functions of optimization, more paleontology of the parts, a numerical recipe whereby the characters of organisms and taxa are fitted to a tree at various nodes, which are themselves reified as ancestors, in part or in whole, some older and some younger, so that, at least in one's imagination, some evolve into others, and vice versa, depending on the particular tree used for the optimization.

Yes, optimization does formally derive characters from one another, in an idealistic morphology of all formal transformations imaginable, and their reversals, in the hope that some might intelligibly reflect, and thereby approximate, if not reveal, the real historical processes of evolution — whatever those processes might be. For Hennig, "... wir die phylogenetische Transformation eines Merkmales aber nicht direkt beobachten können ..." ("... we can never directly observe the phylogenetic transformation of a character").

Geography: Where Progression Would Rule

Other data dropped, or optimized, to nodes are geographic, and their revelation — what is reified — is the ancestral place, or evolutionary center of origin. Figure 6.3 specifies the minimal evidence, twice the minimal, and thrice the minimal to resolve a center of origin in Asia, not in North America or elsewhere, for, let us say, birds — where a real Jurassic Park might have been located.

Underlying the optimization is the principle formalized by Willi Hennig as the progression rule. He might have called it area phylogeny, phylogeny of geography, or even phylo-genetic geography (Takhtajan, 1969:143), but he did not. Like Matthew's true sequences, Hennig's rule is now reborn — as if it were for the first time — within phylogeography.

The formalization is original with Hennig, but the logical principle is not. It is found, for example, in the work of George Simpson, the chief disciple of Matthew, who referred to "a variety of earlier relatives" in the belief that "such evidence is more reliable than any evidence involving only earlier appearance" in the fossil record (Simpson, 1947:629).

Was Simpson an early progression ruler and phylogeographer? An early optimizer? And if Simpson was, then Matthew, with many others, is not far behind (Nelson and Ladiges, 2001).

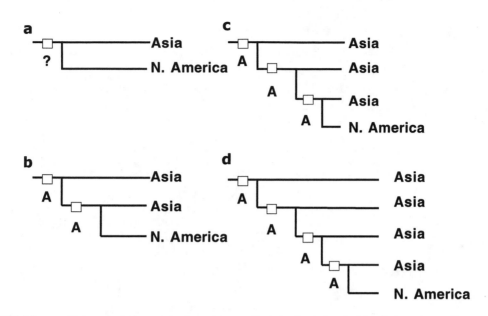

FIGURE 6.3   a, Two organisms, species, or taxa, one in Asia, the other in North America, and a possible center of origin (?). b–d, with additional relatives to in Asia, an Asian center of origin (A) as optimized according to Hennig's progression rule.

More on Optimization

I now address the numerical aspect of cladistics. There is a problem only when evidence conflicts — when there is homoplasy. But first, I am not a numericist. I never understood why, until just recently when I read the following (Blyth, 1960:32): "When a person has eight reasons for doing something, we know he is insincere. Even two is suspicious."

Algorithmic optimization, as implemented in current computer programs, fails to resolve nodes when it should and resolves nodes when it should not. Either that statement is true or optimization works perfectly; nothing more need be said about it, and we can all retire, peacefully and quietly among the cows and clams.

So what can be said?

Matrix 1 (Figure 6.4) is one of a series of matrices, purely hypothetical, that contain conflicting characters and evidence of relationship, in this case between taxa B, C, and D, which form a group that excludes taxon A. The outgroup is OG.

Matrix 1

	1	2	3
OG	0	0	0
A	0	0	0
B	1	1	0
C	1	0	1
D	0	1	1

FIGURE 6.4   Hypothetical Matrix 1, for four taxa (A to D), an outgroup (OG), and three conflicting characters (after Nelson, 1996:1, Matrix 1).

	1	2	3	1	2	3	1	2	3	1	2	3	
Matrix 2 (= Expanded Matrix 1)													
OG	0	0	0	0	0	0	0	0	0	0	0	0	etc.
A	0	0	0	0	0	0	0	0	0	0	0	0	etc.
B	1	1	0	1	1	0	1	1	0	1	1	0	etc.
C	1	0	1	1	0	1	1	0	1	1	0	1	etc.
D	0	1	1	0	1	1	0	1	1	0	1	1	etc.

FIGURE 6.5   Expanded Matrix 1, with 4 x 3 = 12 conflicting characters (after Nelson, 1996:2, Expanding Matrix 1).

Matrix 3

	1	2	3	4	5	6	7	8	9	10
OG	0	0	0	0	0	0	0	0	0	0
A	0	0	0	0	0	0	0	0	0	0
B	1	1	1	0	0	0	1	1	1	0
C	1	0	0	1	1	0	1	1	0	1
D	0	1	0	1	0	1	1	0	1	1
E	0	0	1	0	1	1	0	1	1	1

FIGURE 6.6   Hypothetical Matrix 3 for 5 taxa (A to E), an outgroup (OG), and 10 conflicting characters (after Nelson, 1996:3, Matrix 3).

Consider character 1, based on a feature present in taxa B and C and absent in taxa A and D. Interpret the feature as evidence of relationship, that B and C are in a group and A and D are not. Then read the matrix from this point of view — the character as relationship: For character 1, the 1s show the relationship, or grouping, of BC.

In other words, the 1s represent not just the feature, but its significance, the relationship. Similarly, character 2 shows B and D related. Character 3 shows C and D related. The characters conflict, but agree in excluding taxon A from whatever group is evidenced. Programs see matrices of this type as uninformative. The programs find no synapomorphy for the group BCD.

Expanded matrices (Figure 6.5) give the same result. So do many other matrices in this series (Nelson, 1996). For Matrix 3 (Figure 6.6), the programs find no synapomorphy for the group BCDE.

For the group B-E, is there evidence of relationship or not? Jussieu, writing over two hundred years ago on the natural method and the buttercup family, Ranunculaceae, said, (1777:217, translated, also Stevens, 1994:277)

Parmi ces caractères, les uns sont constans, les autres peuvent quelquefois varier: chacun d'eux pris séparément se retrouve dans une ou plusieurs autres familles; mais leur assemblage ne s'observe que dans celle des renoncules: c'est cet assemblage qui en constitue le caractère essentiel & invariable.

Among these characters, some are constant, others vary: each character, taken separately, is found in one or many other families; but their combination is found only in the family of the buttercups; for that family, it is this combination that constitutes the synapomorphy [the essential and invariable character].

Matrix 4

	1	2	3
OG	0	0	0
A	1	0	0
B	1	1	0
C	0	1	1
D	0	0	1

FIGURE 6.7   Hypothetical Matrix 4 for four taxa (A to D), an outgroup (OG), and three conflicting characters.

	Matrix 5				**Matrix 6**				
	1	2	3	4	1	2	3	4	5
OG	0	0	0	0	0	0	0	0	0
A	1	0	0	0	1	0	0	0	0
B	1	1	0	0	1	1	0	0	0
C	1	1	1	0	1	1	1	0	0
D	0	1	1	1	1	1	1	1	0
E	0	0	1	1	0	1	1	1	1
F	0	0	0	1	0	0	1	1	1
G	—	—	—	—	0	0	0	1	1
H	—	—	—	—	0	0	0	0	1

FIGURE 6.8   Hypothetical Matrices 5 to 6 for six (A to F) and eight (A to H) taxa, outgroup, and four and five conflicting characters, respectively.

Is there evidence of relationship in character combination? If there is, then optimization does not always find it. Optimization, in principle, ignores it, slogans about "total evidence" notwithstanding (Kluge, 1998).

Matrix 4 (Figure 6.7) is one of a series of matrices, again purely hypothetical, that contain conflicting characters with, perhaps, no evidence of relationship. Nevertheless, the programs see matrices of this series as fully informative. For each matrix, the programs return a single resolved tree. Matrices 5 and 6 (Figure 6.8) are two more of this series, which can be extended indefinitely.

There is reason to view these matrices as uninformative, but I do not argue the point here. These few matrices show that the programs fail to find nodes where they should and they find nodes where they should not, so the theory of optimization, implemented in the programs, is not perfect (Platnick et al., 1997). If these matrices are not the best examples, then never mind. Someday, better ones will be found.

Finally, I mention a simple example of character conflict, two characters that overlap in Matrix 7 (Figure 6.9).

For these four taxa, there are 15 possible trees. Six trees best fit the matrix. In Figure 6.9d is one of the six trees, and on it is the optimization of Matrix 7. The optimization is equivalent to the new matrix, which shows what happens to character 2 (Figure 6.9b).

The optimization requires that character 2 be changed — reinterpreted — with the amount of change a measure of the fit of the character to the tree. There are two new characters where there was only one before, so for this character the fit is two rather than

	a) Matrix 7		b) New 7		c) Newer 7	
	1	2	2a	2b	2a	2b
OG	0	0	0	0	0	0
A	1	0	1	1	0	0
B	1	1	1	0	0	1
C	0	1	1	0	1	0
D	0	1	1	0	1	0

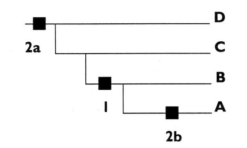

FIGURE 6.9   (a) Hypothetical Matrix 7 for four taxa (A to D), an outgroup (OG), and two conflicting characters; (b and c). Matrices (new 7, newer 7) representing alternative optimizations for character 2. (d). One of six best-fitting trees with new optimization.

one. The other five best-fitting trees also require change of the matrix, some trees for character number 1, some trees for character 2. The newer matrix shows what happens to character 2 when it is optimized on a different shortest tree (Figure 6.9c). As before, there are two newer characters.

There are two new characters in one case, two newer characters in the other. Two equals two. Therefore the amount of change is the same. Here we confront the beast in its lair: real progress and its exact solution to a parsimony problem. I regret that Mark Twain is not with us to share the experience. I wonder what he would say in response? Faced with hocus-pocus (Lopez, 2002), would he remain silent?

We can understand the tree that best fits complex data — a matrix larger than Matrix 7 — as the tree that requires the least change of the data when optimized. I hesitate to call the change distortion, but why not? That is what it is: The best-fitting tree least distorts the data, when optimized, when fitted to the tree.

Why distort the data at all? I know of only one answer that has been given so far: The distorted data are the True ones, the others not.

Biggest Is Best Is Truest Is Biggest Is ...

I have one further observation about the litany of so-called cladism, namely the mantra about homologies being tested through congruence — through optimization on the best fitting tree. Let us say this one, fitting a larger, a very large, matrix, of which Matrix 7 is just a part.

Today we read about primary homologies (Matrix 7) and secondary homologies, achieved by testing through congruence (the new matrices). The new are the ones that have been tested, confirmed, and legitimized, so to speak, and with the necessary corrections, too, as if the new ones were the real homologies and the old were not.

How do we test the secondary homologies? The real ones? The new ones? According to one viewpoint, sloganized as total evidence, the only way is to add more data to the matrix and for that bigger matrix to find the best fitting tree and if that tree is different, to optimize the data on it and examine the new distortion to see if there are newer and truer homologies.

In this way, bigger, if it changes at all, becomes newer and truer — and a great, and ever greater, source of numerical obfuscation, or, rather, numerical consolation.

Recall phenetics and its hypothesis of the matches asymptote. Recall also, the wishful thought (Sokal and Sneath, 1963:114) that, beyond a certain number, adding more characters to the matrix does not change the result — the best fitting tree (the phenogram): "You stop getting different answers or different resolutions when you add new data" (McLennan and Brooks, 2001:26).

With total evidence, need no such asymptote be expected? If it were we might witness the reemergence, even the vindication, of phenetics as the overall similarity of synapomorphy.

Interestingly, the matrix of total evidence soon outgrows the reach of exact solution — that element of real progress now generally abandoned. As it grows, the matrix of total evidence tends toward the biggest, the matrix with no real homologies at all, with all its characters homoplastic, when all of them change by optimization on the best fitting tree of an inexact solution, when the least amount of that better fit distorts each character, at least a little.

Imagine such better-fitting trees succeeding one another, as the biggest matrix grows bigger still, with each succeeding best fit determined by shifting pyramids of homoplastic characters, first distorted one way, then another. The vision is the stuff of nightmares — the death agony of systematics, not under a sky that of a sudden falls, but in the slow creep of miasmic entropy named total evidence.

Is it reassuring to learn that, from this viewpoint, homoplasy is good because it increases resolution (Källersjö et al., 1999)? Or that it is certain, or uncertain as the case may be, to give a correct tree, or even a mostly incorrect one? The era of the biggest matrix is at hand. That is an asymptote of a sort, and by that I mean nothing complimentary.

Mark Twain, if only he were here, might say in this case, then, that when the distortion is complete, it must be also perfect, and the cows and clams, all fat and happy, every one, and systematics as "phylogeny reconstruction" at last "respectable science" (Donoghue, 1994:405), as if the revolution were indeed over, having achieved "absolutely enormous," even "tremendous," progress (Donoghue, 2001:757, 2002:6) — having fulfilled the promise of the principles that began it.

The Buddha and the Bonaparte

Optimization as a procedure, or method, is intelligible, and can be learned and used, like a recipe, and its use encouraged, even enforced, by benevolence or decree of one or another Buddha or Bonaparte, self-appointed or incarnate. However, what about optimization as an element of theory to explain how and why systematics works and has worked as a human enterprise since Aristotle? That is another matter at a different level. There, intelligibility is not enough, and recipes are only grist for the mill. Credibility — the reasons for and against it — is important, too.

Cladistics is about relationships, about how the past relates to the present (as it truly does), about the past and present of earth and life, of the human endeavor toward their discovery — the eternal quest toward their understanding (Adams, 1969). Nothing less suffices: "The great illusion is that we are enlightened" (Blyth, 1960:22).

Acknowledgments

For information and comment I am grateful to Mary Andriani, Malte Ebach, Peter Forey, Daniel Goujet, Lance Grande, Wolfgang Hennig, Marianne Horak, Christopher Humphries, Philippe Janvier, Richard Jefferies, Jürgen Kellermann, Niels Kristensen, Norman Platnick, Olivier Rieppel, Michael Schmitt, Dennis Stevenson, John Wilkins, David Williams, and Mary Winsor.

References

Abel, O., Die Morphologie der Hüftbeinrudiment der Cetaceen, *Denkschr. Kais. Akad. Wiss., Math.-Nat.wiss. Kl.*, 81, 139–195, 1907a.

Abel, O., Die Aufgaben und Ziele der Paläozoologie, *Verh. Zool.-Bot. Ges. Wien*, 57, 67–78, 1907b.

Abel, O., Erster Diskussionsabend über einzelne phylogenetische Probleme am 18, November 1908. Diskussionsthema: Was verstehen wir unter monophyletischer und polyphyletischer Abstammung?, *Verh. Kais.-K. Zool.-Bot. Ges. Wien*, 59, 243–249, 1909.

Abel, O., Kritische Untersuchungen über die paläogenen Rhinocerotiden Europas, *Abh. Kais.-K. Geol. Reichanst.*, 20, Heft 3, 1–52, 1910.

Abel, O., Die Bedeutung der fossilen Wirbeltieren für die Abstammungslehre, in *Die Abstammungslehre: Zwölf gemeinverständliche Vorträge über die Deszendenztheorie im Licht der neueren Forschung, gehalten im Winter-Semester 1910/11 im Münchner Verein für Naturkunde*, Gustav Fischer, Jena, Germany, 1911, pp. 198–250.

Abel, O., *Grundzüge der Palaeobiologie der Wirbeltiere*, E. Schweizerbart, Stuttgart, Germany, 1912.

Abel, O., Neuere Weg phylogenetischer Forschung, *Verh. Ges. Dtsch. Nat.forsch. Ärzte, Leipz.*, 85, 116–124.

Abel, O., *Allgemeine Paläontologie*, G.J. Göschen, Berlin, 1917. (Reprint 1921, Walter de Gruyter, Berlin, Germany.)

Abel, O., *Lehrbuch der Paläozoologie*, Gustav Fischer, Jena, Germany, 1920.

Abel, O., Die Methoden der paläobiologischen Forschung, in *Handbuch der Biologischen Arbeitsmethoden*, Vol. 10, Abderhalden, E., Ed., Urban and Schwartzenberg, Berlin, Germany, 1921, pp. 129–312.

Abel, O., *Lehrbuch der Paläozoologie: Zweite, erweiterte Auflage*, Gustav Fischer, Jena, Germany, 1924.

Abel, O., *Paläobiologie und Stammesgeschichte*, Gustav Fischer, Jena, Germany, 1929 (Reprint 1980, Arno, New York).

Abel, O., William Diller Matthew 1871–1930: Ein Gedankblatt [and list of "Paläontologische Veröffentlichungen"], *Palaeobiologica*, 4, 1–24, 1931a.

Abel, O., Louis Dollo 7. Dezember 1857 — 19. April 1931. Ein Rückblick und Abschied [and list of "Veröffentlichungen"], *Palaeobiologica*, 4, 321–344, 1931b.

Adams, A.B., *Eternal Quest: The Story of the Great Naturalists*, G.P. Putnam's Sons, New York, 1969.

Arber, E.A.N. and Parkin, J., On the origin of the angiosperms, *J. Linn. Soc. Bot.*, 38, 29–80, 1907.

Avise, J.C., *Phylogeography: The History and Formation of Species*, Harvard University Press, Cambridge, 2000.

Ax, P., *Das Phylogenetische System (Systematisierung der Lebenden Natur Aufgrund Ihrer Phylogenese)*, Gustav Fischer, Stuttgart, Germany, 1984.

Ax, P., *The Phylogenetic System: The Systematization of Organisms on the Basis of their Phylogenesis*, John Wiley, Chichester, U.K., 1987.

Ax, P., The integration of fossils in the phylogenetic system of organisms, in *Phylogeny and the Classification of Fossil and Recent Organisms: Proceedings of a Symposium Organized by the Deutsche Forschungsgemeinschaft, Abhandlungen des Naturwissenschaftlichen Vereins in Hamburg*, NF, Vol. 28, Schmidt-Kittler, N. and Willmann, R., Eds., Hamburg, Germany: Paul Parey, 1989, pp. 27–43.

Blyth, R.H., *Zen and Zen Classics*. Vol. 1. From the Upanishads to Huineng, Hokuseido Press, Tokyo, 1960.

Bonde, N., Interrelationships of fishes [review of Greenwood et al. 1973], *Syst. Zool.*, 23, 562–569, 1974.

Bonde, N., Origin of the "higher groups": viewpoints of phylogenetic systematics, in *Problèmes Actuels de Paléontologie (Evolution des Vertébrés)*, Vol. 1, Colloques Internationaux du Centre National de la Recherche Scientifique, No. 218. CNRS, Paris, 1975, pp. 293–324.

Bonde, N., Colin Patterson: The greatest fish palaeobiologist of the 20th century, in *Colin Patterson (1933-1998): A Celebration of his Life*, Forey, P.L., Gardiner, B.G., and Humphries, C.J., Eds., The Linnean, Special Issue No. 2, The Linnean Society of London, London, 2000, pp. 33–38.

Brien, P., Notice sur Louis Dollo Membre de l'Académie, Né à Lille le 7 décembre 1857, mort à Bruxelles le 19 avril 1931, *Annu. Acad. R. Belg.*, 1951, 69–138, 1951.

Brundin, L., Transantarctic relationships and their significance, as evidenced by chironomid midges with a monograph of the subfamilies Podonominae and Aphroteniinae and the austral Heptagyiae, *K. Sven. Vetenskapsakad. Handl., Fjärde Ser.*, 11, 1–472, 1966.

Brundin, L., Application of phylogenetic principles in systematics and evolutionary theory, in *Current Problems of Lower Vertebrate Phylogeny: Proceedings of the Fourth Nobel Symposium held in June 1967 at the Swedish Museum of Natural History (Naturhistoriska riksmuseet) in Stockholm*, Ørvig, T., Ed., Almqvist and Wiksell, Stockholm, 1968, pp. 473–495.

Brundin, L., Phylogenetic biogeography, in *Analytical Biogeography: An Integrated Approach to the Study of Animal and Plant Distributions*, Myers, A.A. and Giller, P.S., Eds., Chapman and Hall, London, 1988, pp. 343–369.

Classen-Bockhoff, R., Plant morphology: the historic concepts of Wilhelm Troll, Walter Zimmermann and Agnes Arber, *Ann. Bot.*, 88, 1153–1172, 2001.

Colbert, E.H., *Digging into the Past: An Autobiography*, Dembner Books, New York, 1989.

Darlu, P. and Tassy, P., *La Reconstruction Phylogénétique: Concepts et Méthodes*, Collection Biologie Théorique No. 7. Masson, Paris, France, 1993.

Davis, P.H. and Heywood, V.H., *Principles of Angiosperm Taxonomy*, Oliver & Boyd, Edinburgh, 1963.

De Beer, G., *Archaeopteryx Lithographica: A Study Based upon the British Museum Specimen*, British Museum (Natural History), London, 1954a.

De Beer, G., Archaeopteryx and evolution, *Adv. Sci.*, 11, 160–170, 1954b (revised version in De Beer 1969).

De Beer, G., *Streams of Culture*, Lippincott, Philadelphia, 1969.

de Jussieu, A.L., Examen de la famille des renoncules, *Histoire de l'Académie Royale des Sciences, Année M. DCCLXXIII, Avec les Mémoires de Mathématique & de Physique, pour la Même Année*, L'Imprimerie Royale, Paris, 1777, pp. 214–240.

Dingus, L. and Rowe, T., *The Mistaken Extinction: Dinosaur Evolution and the Origin of Birds*, W.H. Freeman, New York, 1998.

Dollo, L., Sur la phylogénie des dipneustes, *Bull. Soc. Belg. Géol. Paléontol. d'Hydrol.*, 9, 97–128, 1895 (reprint in Gould 1980).

Dollo, L., Macrurus lecointei, poisson abyssal nouveau, recueilli par l'expédition antarctique belge, *Bull. Classe Sci., Acad. R. Belg.*, 6, 383–401, 1900 (reprint in Gould 1980).

Dollo, L., Les céphalopodes adaptés à la vie nectique secondaire et à la vie benthique tertiaire, *Zool. Jahrb.* Suppl., 15, 105–140, 1912 (reprint in Gould 1980).

Donoghue, M.J., Progress and prospects in reconstructing plant phylogeny, *Ann. Mo. Bot. Gard.*, 81, 405–418, 1994.

Donoghue, M.J., A wish list for systematic biology, *Syst. Biol.*, 50, 755–757, 2001.

Donoghue, M.J., Plant phylogeny and the greening of earth, in *Assembling the Tree of Life: Science, Relevance, and Challenges,* May 30–June 1, 2002, Abstracts, American Museum of Natural History, New York, 2002, pp. 6–7.

Donoghue, M.J. and Kadereit, J.W., Walter Zimmermann and the growth of phylogenetic theory, *Syst. Biol.*, 41, 74–85, 1992.

Dupuis, C., Permanence et actualité de la systématique: la "systématique phylogénétique" de W. Hennig (historique, discussion, choix de références), *Cah. Nat.*, 34, 1–69, 1979.

Dupuis, C., Darwin et les taxinomies d'aujourd'hui, in *L'Ordre et la Diversité du Vivant: Quel Statut Scientifique pour les Classifications Biologiques?*, Tassy, P., Ed., Fondation Diderot/Arthème Fayard, Paris, France, 1986, pp. 215–40.

Farris, J.S., The Evolutionary Relationships Between the Species of the Killifish Genera *Fundulus* and *Profundulus (Teleostei: Cyprinodontidae).* Ph.D. thesis, University of Michigan, Ann Arbor, 1968.

Farris, J.S. and Platnick, N.I., Lord of the flies: the systematist as study animal, *Cladistics*, 5, 295–310, 1989.

Fittkau, E.J., In memory of Lars Brundin 30 May 1907–18 November 1993, in *Chironomids: from Genes to Ecosystems*, Cranston, P., Ed., CSIRO, Melbourne, 1995, pp. 3–9.

Forterre, P. and Philippe, H., Where is the root of the universal tree of life?, *BioEssays*, 21, 871–879, 1999.

Goujet, D., The French connection, in *Colin Patterson (1933–1998): A Celebration of his Life*, Forey, P.L., Gardiner, B.G., and Humphries, C.J., Eds., The Linnean, Special Issue No. 2, Linnean Society of London, London, 2000, pp. 79–81.

Gould, S.J., Dollo on Dollo's law: irreversibility and the status of evolutionary laws, *J. Hist. Biol.*, 3, 189–212, 1970 (reprint in Gould 1980).

Gould, S.J., Ed., *Louis Dollo's Papers on Paleontology and Evolution*, Arno, New York, 1980.

Greenwood, P.H., Miles, R.S., and Patterson, C., Eds., Interrelationships of Fishes. Supplement No. 1, *Zool. J. Linn. Soc.*, Vol. 53, Linnean Society of London/Academic Press, London, 1973.

Gross, W., Paläontologische Hypothesen zur Faktorenfrage der Deszendenzlehre: Über die Typen- und Phasenlehren von Schindewolf und Beurlen, *Naturwissenschaften*, 31, Jahrgang, Heft 21/22, 237–45, 1943.

Gross, W., Über die "Watson'sche Regel," *Paläontol. Z.*, 30, 30–40, 1956.

Günther, K., Systematik und Stammesgeschichte der Tiere 1939–1953, *Fortschr. Zool.* N.F., 10, 33–278, 1956.

Günther, K., Abschliessende Zusammenfassung der Vorträge und Diskussionen, in Methoden der Phylogenetik: Symposion vom 12. Bis 13. Februar 1970 in I. Zoologischen Institut der Universität Erlangen-Nürnberg, Vol. 4, Siewing, R., Ed., Erlangen Forschungen, Reihe B: Naturwissenschaften, 1971, pp. 76–88.

Hennig, W., Über einige Gesetzmassigkeiten der geographischen Variation in der Reptiliengattung Draco L.: "parallele" und "konvergente" Rassenbildung, *Biol. Zent.bl.*, 56, 549–559, 1936.

Hennig, W., Zur Klärung einiger Begriffe der phylogenetischen Systematik, *Forsch. Fortsch.*, 25, 136–138, 1949.

Hennig, W., *Grundzüge einer Theorie der Phylogenetischen Systematik*, Deutscher Zentralverlag, Berlin, Germany, 1950 (Reprint 1980, Otto Koeltz, Koenigstein, Germany).

Hennig, W., Phylogenetic systematics, *Annu. Rev. Entomol.*, 10, 97–116, 1965 (reprints in Sober 1984, 1994).

Hennig, W., *Phylogenetic Systematics*, University of Illinois Press, Urbana, 1966 (reprints: 1979, 1999).

Hennig, W., *Elementos de una Sistemática Filogenética*, Editorial Universitaria, Buenos Aires, 1968.

Hennig, W., *Die Stammesgeschichte der Insekten*, Waldemar Kramer, Frankfurt am Main, Germany, 1969.

Hennig, W., Kritische Bemerkungen zur Frage "cladistic analysis or cladistic classification?" *Z. Zool. Syst. Evol.forsch.*, 12, 279–294, 1974.

Hennig, W., "Cladistic analysis or cladistic classification?": A reply to Ernst Mayr, *Syst. Zool.*, 24, 244–256, 1975.

Hennig, W., *Insect Phylogeny*, John Wiley, Chichester, 1981.

Hennig, W., *Phylogenetische Systematik*, Pareys Studientexte 34, Paul Parey, Berlin, Germany, 1982.

Hennig, W., *Aufgaben und Probleme Stammesgeschichlicher Forschung*, Pareys Studientexte 35, Paul Parey, Berlin, Germany, 1984.

Hill, C.R., The epidermis/cuticle and *in situ* spores and pollen in fossil plant taxonomy, in *Systematic and Taxonomic Approaches in Palaeobotany*, Spicer, R.A. and Thomas, B.A., Eds., Systematics Association Special Volume 31, Oxford University Press, Oxford, 1986, pp. 123–136.

Hill, C.R. and Crane, P.R., Evolutionary cladistics and the origin of angiosperms, in *Problems of Phylogenetic Reconstruction*, Joysey, K.A. and Friday, A.E., Eds., Systematics Association Special Volume 21, Academic Press, London, 1982, pp. 269–361.

Janvier, P., L'impact du cladisme sur la recherche dans les sciences de la view et de la terre, in *L'Ordre et la Diversité du Vivant: Quel Statut Scientifique pour les Classifications Biologiques?*, Tassy, P., Ed., Fondation Diderot/Arthème Fayard, Paris, 1986, pp. 99–120.

Janvier, P., *Early Vertebrates*, Oxford Monographs on Geology and Geophysics, No. 33, Oxford University Press, Oxford, 1996.

Jefferies, D., Phylogenetic systematics, *Paleontol. Newsl.*, 46, 60–62, 2001.

Kälin, J., Über einige Grundbegriffe in der vergleichenden Anatomie und ihre Bedeutung für die Erforschung der Baupläne im Tierreich, *Comptes Rendus, XIIe Congr. Int. Zool. Lisb.* 1935, II, 649–664, 1936.

Kälin, J., Ganzheitliche Morphologie und Homologie, *Mitt. Nat.forsch. Ges. Freibg.* (Schweiz), 3, Heft 1, 1–36, 1941.

Källersjö, M., Albert, V.A., and Farris, J.S., Homoplasy *increases* phylogenetic structure, *Cladistics*, 15, 91–93, 1999.

Kluge, A.G., Total evidence or taxonomic congruence: cladistics or consensus classification, *Cladistics*, 14, 151–158, 1998.

Kluge, A.G. and Wolf, A.J., Cladistics: what's in a word?, *Cladistics*, 9, 183–199, 1993.

Königsmann, E., Termini der phylogenetischen Systematik, *Biol. Rundsch.*, 13, 99–115, 1975.

Kristensen, N.P. and Nielsen, E.S., A new subfamily of micropterigid moths from South America: a contribution of the morphology and phylogeny of the Micropterigidae, with a generic catalogue of the family (Lepidoptera: Zeugloptera), *Steenstrupia*, 5, 69–147, 1979.

Kuhn, O., Typologische Betrachtungsweise und Paläontologie, *Acta Biotheoret. Ser. A*, 6, 55–96, 1942.

Lam, H.J., Phylogeny of single features, as illustrated by a remarkable new sapotaceous tree from British Malaya (Madhuca Ridleyi, n. sp.), *Gard. Bull. Straits Settl.*, 9, 98–112, 1935.

Lam, H.J., Phylogenetic symbols, past and present (being an apology for genealogical trees), Acta Biotheoret. Ser. A, 2, 153–194, 1936.

Lam, H.J., Classification and the new morphology, *Acta Biotheoret. Ser. A*, 8, 107–154, 1948.

Lam, H.J., Taxonomy: general principles and angiosperms, in *Vistas in Botany: A Volume in Honour of the Bicentenary of the Royal Botanic Gardens, Kew*, Turrill, W.B., Ed., International Series of Monographs on Pure and Applied Biology, Division Botany, Vol 2. Pergamon Press, London, 1959, pp. 3–75.

Lewontin, R., Evolution/creation debate: a time for truth, *BioScience*, 31(8), 559, 1981.

Lopez, D.J., Snaring the Fowler: Mark Twain debunks phrenology, *Sceptical Enq.*, 26 (no. 1, Jan–Feb, 33–36, 2002.

Marshall, C.R., Lungfish: phylogeny and parsimony, in *The Biology and Evolution of Lungfishes, Based on Proceedings of a Symposium Held During the American Society of Zoologists Meeting in Denver, Colorado, December 27, 1984*, Bemis, W.E., Burggren, W.W., and Kemp, N.E., Eds., J. Morphol. Supplement 1 (1987). Alan R. Liss, New York, 1987, pp. 151–162.

Matthew, W.D., Recent progress and trends in vertebrate paleontology, in *Annual Report of the Board of Regents of the Smithsonian Institution for the Year Ending June 30 1923* (Publication 2758), 1925, pp. 273–289.

Mayr, E., Lamarck revisited, *J. Hist. Biol.*, 5, 55–94, 1972.

Mayr, E., *The Growth of Biological Thought: Diversity, Evolution, and Inheritance*, Harvard University Press, Cambridge, 1982.

McLennan, D.A. and Brooks, D.R., Phylogenetic systematics: five steps to enlightenment, in *Fossils, Phylogeny, and Form: An Analytical Approach*, Adrain, J.M., Edgecombe, G.D., and Lieberman, B.S., Eds, Kluwer Academic/Plenum, New York, 2001, pp. 7–28.

Meeuse, A.D.J., *Fundamentals of Phytomorphology*, Ronald Press, New York, 1966.

Meeuse, A.D.J., *Origin of the Flowering Plants: Theories and Factual Evidence*. Caput-College 1981/1983, published by the author, Amsterdam, 1982.

Meeuse, A.D.J., Homology as an empirism, *J. Plant Anat. Morphol.*, 1, 9–24, 1984.

Meeuse, A.D.J., *Anatomy of Morphology*, E.J. Brill, Leiden, The Netherlands, 1986.

Meyen, S.V., *Fundamentals of Palaeobotany*, Chapman and Hall, London, 1987.

Meyer-Abich, A., Beiträge zur Theorie der Evolution der Organismen I. Das typologische Grundgesetz und seine Folgerungen für Phylogenie und Entwicklungsphysiologie, *Acta Biotheoret. Ser. A*, 7, 1–80, 1943.

Meyer-Abich, A., *The Historico-Philosophical Background of the Modern Evolution-Biology: Nine Lectures Delivered During October and November of 1960 at the Department of Zoology of the University of Texas in Austin, Texas USA*. Acta Biotheoretica, Supplementum Secundum, additum Actorum Biotheoreticorum Volumini XIII, E.J. Brill, Leiden, The Netherlands, 1964.

Meyer-Abich, A., *Die Vollendung der Morphologie Goethes durch Alexander von Humboldt: Ein Beitrag zur Naturwissenschaft der Goethezeit*, Vandenhoeck & Ruprecht, Göttingen, Germany, 1970.

Miles, R.S., The relationships of the Dipnoi, in *Problèmes Actuels de Paléontologie (Evolution des Vertébrés)*, vol. 1, Colloques Internationaux du Centre National de la Recherche Scientifique, No. 218, CNRS, Paris, 1975, pp. 133–157.

Miles, R.S., Dipnoan (lungfish) skulls and the relationships of the group: a study based on new species from the Devonian of Australia, *Zool. J. Linn. Soc.*, 61, 1–328, 1977.

Nelson, G., *Nullius in Verba*, published by the author, New York, 1996 (Reprinted in J. Comp. Biol., 1, 141–152, 1996).

Nelson, G., Ancient perspectives and influence in the theoretical systematics of a bold fisherman, in *Colin Patterson (1933-1998): A Celebration of His Life*, Forey, P.L., Gardiner, B.G., and Humphries, C.J., Eds., Linnean, Special Issue No. 2., The Linnean Society of London, London, 2000, pp 9–23.

Nelson, G. and Ladiges, P.Y., Gondwana, vicariance biogeography and the New York School revisited, *Aust. J. Bot.*, 49, 389–409, 2001.

Nielsen, J.L., Religion, evolution og abeprocesser, in *Udviklingsideens Historie: Fra Darwins Synthese til Nutidens Krise: Naturens Historie Fortaellere*, Bonde, N. and Hoffmeyer, J., Eds., Vol. 2, G.E.C. Gads Forlag, Copenhagen, 1987, pp. 469–495.

Nierstrasz, H.F., L'evolution entre-croisée chez les crustacés, *Mém. Inst. R. Sci. Nat. Belg.*, Ser. 2, 3, 667–677, 1936.

Osborn, H.F., Paleontology, in *The Encyclopaedia Britannica: A Dictionary of Arts, Sciences, Literature and General Information*, 11th ed., Vol. XX, Encyclopaedia Britannica, New York, 1911, pp. 579–591.

Osborn, H.F., Origin of single characters as observed in fossil and living animals and plants, *Am. Nat.*, 49, 193–239, 1915.

Osche, G., Grundzüge der allgemeinen Phylogenetik, in *Handbuch der Biologie, Band III/2, Allgemeine Biologie*, Gessner, F., Ed., Akademische Verlagsgesellschaft Athenaion, Frankfurt, 1966, pp. 817–906.

Padian, K. and Horner, J.R., Typology versus transformation in the origin of birds, *Trends Ecol. Evol.*, 17, 120–124, 2002.

Patterson, C., Phylogenetic relationships of major groups: conclusions and prospects, in The Hierarchy of Life: Molecules and Morphology in *Phylogenetic Analysis. Proceedings from Nobel Symposium 70 held at Alfred Nobel's Björkborn, Karlskoga, Sweden, August 29–September 2, 1988*, Fernholm, B., Bremer, K., and Jörnvall, H., Eds., Excerpta Medica Series, Elsevier, Amsterdam, 1989, pp. 471–488.

Patterson, C., Naming names, *Nature*, 366, 518, 1993.

Patterson, C., Adventures in the Fish Trade, unpublished ms. (annual address to the Systematics Association, London, December 6, 1995.)

Patterson, C., Molecules and Morphology, Ten Years On, unpublished ms. (address to the meeting on Molecules and Morphology in Systematics, Paris, France, March 24–28, 1997.)

Plate, L., Deszendenztheorie, in *Handwörterbuch der Naturwissenschaften, Zweiter Band, Blatt-Ehrenberg*, Teichmann, E., Ed., Gustav Fischer, Jena, Germany, 1912, pp. 897–951.

Plate, L., Deszendenztheorie, *Handwörterbuch der Naturwissenschaften, Zweiter Auflage, Zweiter Band, Blütenpflanzen — Dutrochet*, Gustav Fischer, Jena, 1933, pp. 989–1042.

Platnick, N.I., Humphries, C.J., Nelson, G. and Williams, D., Is Farris optimization perfect?, *Cladistics*, 12, 243–252, 1997.

Remane, A., *Die Grundlagen des Naturlichen Systems, der Vergleichenden Anatomie und der Phylogenetik*, Geest & Portig K.-G, Leipzig, Germany, 1952.

Richter, S. and Meier, R., The development of phylogenetic concepts in Hennig's early theoretical publications (1947–1966), *Syst. Biol.*, 43, 212–221, 1994.

Rieppel, O., The classification of primitive snakes and the testability of phylogenetic theories, *Biol. Zent.bl.*, 98, 537–552, 1979.

Schindewolf, O.H., Beobachtungen und Gedanken zur Deszendenzlehre, *Acta Biotheoret. Ser. A*, 3, 195–212, 1937.

Schindewolf, O.H., Entwicklung im Lichte der Paläontologie, *Biologie*, 11 Jahrg. Heft 5/6, 113–125, 1942.

Schindewolf, O.H., *Grundfragen der Paläontologie: Geologische Zeitmessung, Organische Stammesenwicklung, Biologische Systematik*, E. Schweizerbart, Stuttgart, Germany, 1950.

Schindewolf, O.H., Review of De Beer 1954b, *Zent.bl. Geol. Paläontol., Teil II, Hist. Geol. Paläontol.*, Jahrg. 1954, 358, 1955.

Schindewolf, O.H., Über Mosaikentwicklung, *Neues Jahrb. Geol. Paläontol., Mon.Heft, Jahrg.*, 1957, 49–52, 1957.

Schindewolf, O.H., Über den "Typus" in morphologischer und phylogentischer Biologie, *Abh. Math.-Nat.wiss. Kl., Akad. Wiss. Lit. Mainz, Jahrg.*, 1969, nr 4, 55–131, 1969.

Schindewolf, O.H., *Basic Questions in Paleontology: Geologic Time, Organic Evolution, Biological Systematics*, University of Chicago Press, Chicago, 1993.

Schmitt, M., Willi Hennig (1913–1976), in *Darwin & Co.: Eine Geschichte der Biologie in Portraits II*, Jahn, I. and Schmitt, M., Eds., C.H. Beck, Munich, 2001, pp. 316–343, 541–546.

Schoch, R.M., *Phylogeny Reconstruction in Paleontology*, Van Nostrand Reinhold, New York, 1986.

Schuh, R.T., *Biological Systematics: Principles and Applications*, Cornell University Press, Ithaca, NY, 2000.

Simpson, G.G., *Tempo and Mode in Evolution*, Columbia University Press, New York, 1944 (reprint with a new introduction, 1984).

Simpson, G.G., Holarctic mammalian faunas and continental relationships during the Cenozoic, *Bull. Geol. Soc. Am.*, 58, 613–688, 1947.

Simpson, G.G., *The Major Features of Evolution*, Columbia University Press, New York, 1953.

Sober, E., Ed., *Conceptual Issues in Evolutionary Biology: An Anthology*, MIT Press, Cambridge, MA, 1984.

Sober, E., Ed., *Conceptual Issues in Evolutionary Biology*, 2nd ed., MIT Press, Cambridge, MA, 1994.

Sokal, R.R. and Sneath, P.H.A., *Principles of Numerical Taxonomy*, W.H. Freeman, San Francisco, 1963.

Sporne, K.R., Statistics and the evolution of dicotyledons, *Evolution*, 8, 55–64, 1954.

Stevens, P.F., *The Development of Biological Systematics: Antoine-Laurent de Jussieu, Nature, and the Natural System*, Columbia University Press, New York, 1994.

Sudhaus, W. and Rehfeld, K., *Einführung in die Phylogenetik und Systematik*, Gustav Fischer, Stuttgart, Germany, 1992.

Takhtajan, A., *Essays on the Evolutionary Morphology of Plants*, Leningrad University, Leningrad, 1954 (translation of publication of American Institute of Biological Sciences, Washington, D.C.).

Takhtajan, A., *Die Evolution der Angiospermen*, VEB Gustav Fischer, Jena, Germany, 1959.

Takhtajan, A., *Flowering Plants: Origin and Dispersal*, Smithsonian Institution Press, Washington, DC, 1969.

Takhtajan, A., *Evolutionary Trends in Flowering Plants*, Columbia University Press, New York, 1991.

Takhtajan, A., *Diversity and Classification of Flowering Plants*, Columbia University Press, New York, 1997.

Tassy, P., Présentation; introduction; construction systématique et soumission au test: une forme de connaissance objective, in *L'Ordre et la Diversité du Vivant: Quel Statut Scientifique pour les Classifications Biologiques?*, Tassy, P., Ed., Fondation Diderot/Arthème Fayard, Paris, 1986, pp. 9–13, 15–18, 83–98.

Wagenitz, G., *Wörterbuch der Botanik, Morphologie, Anatomie, Taxonomie, Evolution: Die Termini in Ihrem Historischen Zusammenhang*, Gustav Fischer, Jena, Germany, 1996.

Wagner, W.H., Jr., Origin and philosophy of the groundplan-divergence method of cladistics, *Syst. Bot.*, 5, 173–193, 1980.

Wanntorp, H.-E., Lars Brundin 30 May 1907–17 November 1993, *Cladistics*, 9, 357–367, 1993.

Westoll, T.S., On the evolution of the Dipnoi', in *Genetics, Paleontology, and Evolution*, Jepsen, G.L., Mayr, E., and Simpson, G.G., Eds., Princeton University Press, Princeton, NJ, 1949, pp. 121–84 (Reprint 1963, Atheneum, New York).

White, E.I., Presidential address: a little on lungfishes, Proc. Linn. Soc. London, 177, 1–10, 1966.

Zimmermann, W., *Die Phylogenie der Pflanzen: Ein Überblick über Tatsachen und Probleme*, Gustav Fischer, Jena, Germany, 1930.

Zimmermann, W., Arbeitsweise der botanischen Phylogenetik und anderer Gruppierungswissenschaften, in *Handbuch der Biologischen Arbeitsmethoden*, Vol. 9, Part 3, No. 2, Abderhalden, E., Ed., Urban and Schwarzenberg, Berlin, Germany, 1931, pp. 942–1053.

Zimmermann, W., Research on phylogeny of species and of single characters (Sippenphylogenetik und Merkmalsphylogenetik), *Am. Nat.*, 68, 381–384, 1934.

Zimmermann, W., Die Methoden der Phylogenetik, in *Die Evolution der Organismen: Ergebnisse und Probleme der Abstammungslehre*, Heberer, G., Ed., Gustav Fischer, Jena, Germany, 1943, pp. 20–56.

Zimmermann, W., *Grundfragen der Evolution*, Vittorio Klostermann, Frankfurt, Germany, 1948.

Zimmermann, W., *Geschichte der Pflanzen*, Georg Thieme, Stuttgart, Germany, 1949.

Zimmermann, W., Main results of the "telome theory," *Palaeobotanist*, 1, 456–470, 1952.

Zimmermann, W., *Evolution: Die geschichte ihrer Probleme und Erkenntnisse*, Karl Alber, Freiburg, Germany, 1953.

Zimmermann, W., Methoden der Phylogenetik, in *Die Evolution der Organismen: Ergebnisse und Probleme der Abstammungslehre*. 2. Erweiterte Auflage, Heberer, G., Ed., 1. Lieferung, Gustav Fischer, Stuttgart, Germany, 1954, pp. 25–102.

Zimmermann, W., Phylogenie der Blüten: Vorgetragen auf der Botanikertagung in Hann. Münden, Mai 1956, *Phyton Ann. Rei Bot.*, 7, 162–182, 1957.

Zimmermann, W., *Die Phylogenie der Pflanzen: Ein Überblick über Tatsachen und Probleme. 2. Völlig Neu Bearbeitete Auflage*, Gustav Fischer, Stuttgart, Germany, 1959.

Zimmermann, W., Die Auseinandersetzung mit den Ideen Darwins: Der "Darwinismus" als ideengeschichtliches Phänomen, in *Hundert Jahre Evolutionsforschung: Das Wissenschaftliche Vermächtnis Charles Darwins*, Heberer, G., and Schwanitz, F., Eds., Gustav Fischer, Stuttgart, Germany, 1960a, pp. 290–354.

Zimmermann, W., Geschichte und Methode der Evolutionsforchung, in *Arbeitstagung zu Fragen der Evolution zum Gedenken an Lamarck — Darwin-Haeckel (20. Bis 24. Oktober 1959 in Jena). Herausgegeben von der Biologischen Gesellschaft in der Deutschen Demokratischen Republik*, Gustav Fischer, Jena, Germany, 1960b, pp. 15–36.

Zimmermann, W., Kritische Beiträge zu einige biologischen Problemen IV. Die Ursachen der Evolution, *Acta Biotheoret.*, Ser. A, 14, 121–206, 1962.

Zimmermann, W., Gibt es ausser dem phylogenetischen System "naturliche" Systeme der Organismen? Kritische Bemerkungen zu einigen biologischen Problemen V, *Biol. Zent.bl.*, 82, 525–568, 1963.

Zimmermann, W., *Die Telomtheorie*, Gustav Fischer, Stuttgart, Germany, 1965.

Zimmermann, W., Phänetische und phylogenetische Verwandschaft, *Phyton Ann. Rei Bot.*, 11, 145–163, 1966.

Zimmermann, W., Methoden der Evolutionswissenschaft (=Phylogenetik), in *Die Evolution der Organismen: Ergebnisse und Probleme der Abstammungslehre. 3. Völlig Neu Bearbeitete und Erweiterte Auflage*, Heberer, G., Ed., Band 1, Gustav Fischer, Stuttgart, Germany, 1967, pp. 61–160.

Zimmermann, W., About Mesozoic pteridophylls, *Am. J. Bot.*, 56, 814–819, 1969.

Zimmermann, W. and Schneider, A.M., Mesozoische Pteridophylle, *Phytomorphology*, 17, 336–345, 1967.

Chapter 7

Systematics and Paleontology

Peter L. Forey

Abstract

The relationship between systematics and paleontology throughout the twentieth century was close and tempestuous. It did not begin well. Darwin's theory of evolution had predicted that paleontology would play a central role since the past is the key to the present, and certainly vertebrate paleontologists took up the task. However, at the beginning of the century unanswered questions of the kinds of groups to be recognized and to be reflected in classification led to a general recognition that discovering phylogenetic pathways was an impossible task. The early years of the twentieth century witnessed several attempts to separate two strands of evolutionary theory — genealogy and adaptation. The attempt was cut short because in the 1930s and 1940s there was a concerted effort to try to synthesize the contemporary advances in Mendelian/population genetics and traditional studies of comparative anatomy and paleontology so that a more holistic approach to phylogeny reconstruction may emerge. As a result, the synthesis imposed an adaptationist veneer, which clouded rather than clarified attempts to separate the two strands of evolutionary theory. The adaptationist viewpoint prevailed, and the palaeontological contribution to phylogeny reconstruction was channeled into a self-fulfilling enterprise. Phylogeny was to be reconstructed through a consideration of adaptation blended with inferences about genealogy. Hennig's phylogenetic systematics (later cladistics) challenged paleontology's authority for phylogenetic reconstruction on three grounds: primacy of the Recent world, inability to specify ancestors, and the use of time. General cladistic methods prevailed and were (and are) used extensively in paleontological phylogeny reconstruction. Latterly, time has resurfaced as a potential key element, but discussions as to how and when it is used are ongoing.

Introduction

This chapter attempts to cover the history of the relationship between paleontology and systematics in the twentieth century by identifying milestones. I interpret milestones here as meetings and publications that had significant impact on the ways in which paleontological data were used to infer phylogeny as well as the way in which systematic methods influenced the way we interpret the fossil record. The thesis of this chapter is that the twentieth century began with a complete acceptance that paleontology was crucial to phylogeny reconstruction and there was a mutual influence of paleontology on systematic theory and vice versa. Later, this harmony was broken when a spectrum of distrust of the fossil record and certain systematic methodologies surfaced. The unique nature of paleontological data has directed some investigators to follow certain methods of phylogeny reconstruction not open to neontologists. However, some systematic methodologies have influenced how we interpret fossils and weigh their importance in unraveling the tree of life.

There are a great many strands to the development of paleontology as a discipline, and the history of paleontology has been marked by highly significant advances as well as languishing in periods of stagnation. Many of the successful advances of paleontology are seen in its role in deciphering what George Gaylord Simpson called the "Tempo and Mode" of evolution (Simpson, 1944). Libraries have been filled with literature relating to rates of morphological and taxonomic evolution, the shape of cladogenesis, patterns of extinction and recovery, adaptive radiations, biogeographic history and coevolution. This chapter deals with a tiny fraction of that history — the relationship with systematics — but it is the fraction that underpins most of the rest. What follows is a personal filtering of that history with sins of omission only justified by ignorance or concentration on my areas of prejudice. Most of my examples are drawn from the vertebrate world, but it is likely that readers can substitute invertebrate or paleobotanical examples with little difficulty. Before entering the twentieth century it is worth mentioning some key changes that took place in nineteenth century because they set the scene of the dawn of the twentieth century and the mindset that paleontologists and systematists had inherited and what they thought they had left behind.

Nineteenth Century

It is appropriate to begin this chapter with Louis Agassiz (1807–1873) since he was one of the first to classify Recent and fossil organisms together, thereby validating ideas at the beginning of the nineteenth century that extinct and modern organisms are subject to universal laws. Figure 7.1A shows a classification of the class of fishes (Agassiz, 1844). It plots 44 families of fishes (including 4 exclusively Recent and 5 totally extinct) — (grouped into four orders: Cycloids, Ctenoids, Ganoids and Placoids) recognized by the structure of the scales. Time is shown on the vertical axis, and the relative numbers of species contained within each family is shown by the thickness of the lines. Patterson (1977:584) points out that diagrams such as this are alluringly like that exemplified in Figure 7.1B — this one produced by Romer (1966) a century and a quarter later. Although they look similar, there is a world of difference in the philosophical background used in their drafting. Agassiz was pre-Darwinian and explained the reality of his groups in terms of proof of the existence of divine creation (Lurie, 1960; Rieppel, 1988a). Romer (1966) worked comfortably cocooned within a theory of evolution.

Agassiz's diagram shows a hierarchy of taxa recognized by characters plotted against a time axis. Notice that none of the lines join. That is, there is no implication that one family gave rise, or transformed, to another. The lines do converge, though. They converge on an abstraction of the homologies by which the groups are recognized, and those taxa within each Order registered in larger font were thought by Agassiz to most faithfully demonstrate those homologies. We could view this diagram from above, as circles within circles: the eight families of placoids nested within one circle and placed alongside the nine families of ganoids within another and so on. The four circles would themselves be contained within a more inclusive circle — Pisces, or fishes. The systematic content of both the vertical diagram and the circles would be the same. Agassiz drew it vertically to give the reader the additional information that some of his families occurred at different times of geological history. Agassiz gave considerable importance to the stratigraphic occurrence of organisms as one strand of his threefold parallelism between geological succession, embryological development and the classification of natural groups (Agassiz, 1859).

Romerian spindle trees also show times of occurrence as well as indicating the numbers of species — in this case within families of carnivores — but this time the lines do join, sometimes with more surety than others, and here there is the implication of genetic

A

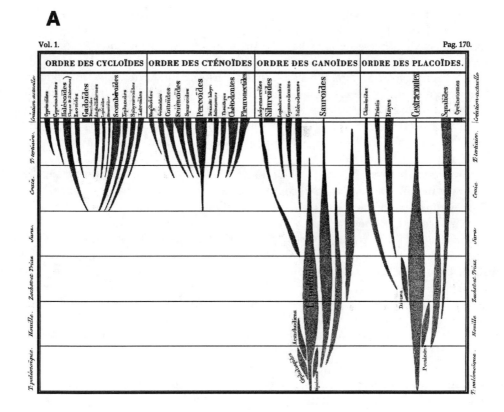

B

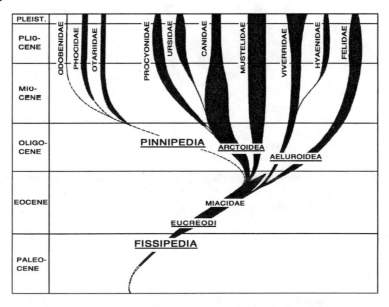

FIGURE 7.1   (A) Agassiz's classification of fishes in which he linked fossil and Recent taxa together (Agassiz 1844:170). (B) The phylogeny of the Carnivora. (From Romer, A.S., *Vertebrate Paleontology,* 3rd ed., University of Chicago Press, Chicago, 1966, Figure 337. With permission.)

relationships where one group gave rise to another. For example, in Figure 7.1B the family Miacidae is indicated to give rise to all the modern families of carnivores. Those connecting lines, like Agassiz's converging lines have to be abstractions, grounded in some theory and discovered by some method. We could not draw the ideas contained in this diagram exclusively as circles within circles because that describing the Miacidae would be incapable of being recognizable by its own individual characters. It would, in effect, be comparable to a circle embracing all the remainder of the carnivores (Miacidae would be synonymous with Carnivora). For Romer's diagram the stratigraphic occurrence is not just a luxurious add-on — (to give additional information); rather, it is essential to the construction of the theory depicted.

The difference between Agassiz's diagram and Romer's diagram is, of course, the influence of Darwin and the desire to recognize ancestor-descendent relationships. Patterson (1977) questions whether Agassiz's depiction of affinity through hierarchies of characters is any more abstract than the ancestor-descendent relationships illustrated by Romer.

Darwin placed great store in fossils as the source of ancestors. They have primitive characters and they are to be found in the fossil record. Shortly after *The Origin* was published, fossils took on a new significance in the eyes of researchers. Figure 7.2 shows two of the earliest attempts to portray a phylogeny through the use of fossils. Albert Gaudry worked with mammalian fossils discovered in the Miocene of Greece, and he believed that by using these discoveries he could plot lines of ancestry and descent. Figure 7.2A (Gaudry, 1862–1867) is one of many that he drew, and it shows a phylogeny of horses from about 15 million years to the present day, depicting familiar lines of descent from fossil ancestors to descendants ending, with the Recent horse and the donkey. Figure 7.2B (after Hilgendorf, 1863 in Reif, 1983) illustrates — literally by gluing fossils on to a card tablet — the evolution of Miocene planorbid snails collected in the Steinheim basin. These diagrams, and others like them, were constructed by attaching paramount importance to the relative stratigraphic position of fossils, the recognition of ancestors and particular patterns of genealogy, some leading from one species to the next, others becoming extinct, others giving rise to more than one descendant and, in the case of Hilgendorf's snails, species arising by hybridization.

There is one big difference between Gaudry's and Hilgendorf's diagrams and Romer's spindle tree of the Carnivora. Gaudry and Hilgendorf linked species with species, or variety with variety, and this is a faithful depiction of evolutionary theory in which species give rise to species. In this they were anticipating what became formalized as stratophenetics (see section on Current Debates). Romer linked family with family; in other words, Romer's diagram shows supraspecific ancestors. However, these ancestors are abstractions or, at best, approximations since Romer hardly believed that a family gave rise to the rest of the carnivores.

Toward the end of the nineteenth century, many diagrams appeared showing graphical tree-like relationships between organisms. Those by Ernst Haeckel (1866; e.g., Figure 7.3) were the earliest, but they essentially took pre-Darwinian classifications and redrafted them to simulate a tree of life. No organisms had been nailed to the trunk or branches. However, by 1874 Haeckel had firmly embraced ancestor-descendent relationships, and groups such as Acrania, Cyclostoma and Selachii are very clearly in ancestor-descendent relationships on the trunk leading to man (Figure 7.4).

Huxley (1864, 1875, 1880) carried the idea of evolutionary lineages much further by suggesting that relationships are better expressed by grades; the lowest grade being a nonmember of the group under study (a presumed ancestor) and the highest being the modern representatives. Membership of the particular grades was assessed by how far from the ancestral type (the nonmember) a particular animal had progressed. Huxley's particular case was that of the horse lineage beginning with *Eohippus,* showing four equally developed toes on the forefoot, coupled with low-crown cheek teeth and ending with modern *Equus,*

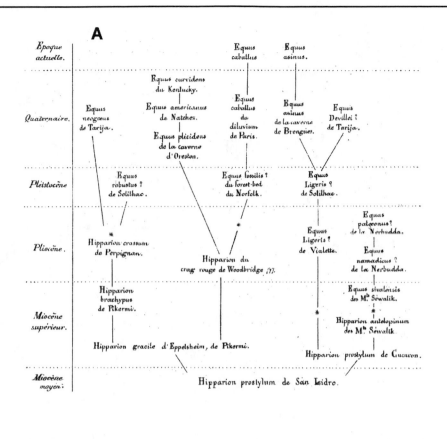

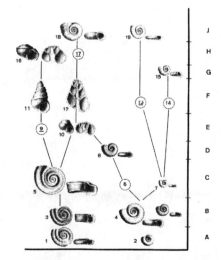

FIGURE 7.2   Two early depictions of phylogeny showing ancestral-descendant lineages and cladogenesis arrived at by combining stratigraphic occurrence with some unstated judgment of morphological similarity. (A) One of the first evolutionary trees plotting inferred lineages from fossils to Recent taxa — the horse and donkey, from Gaudry (1862–1867). (B) Diagram showing the ideas of Hilgendorf (1863) on the evolution of Miocene planorbid snails. Hilgendorf's thesis did not contain any diagrams. This one is reproduced from Reif who derived it from individual cards on which snails were glued and labeled by Hilgendorf. (From Reif, W.E., *Paläontol. Z.*, 57, 7–20, 1983, Figure 3. With permission.)

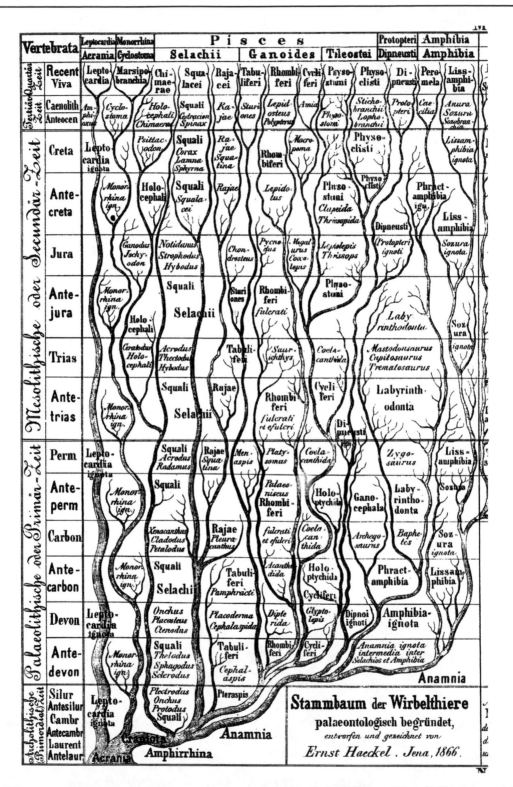

FIGURE 7.3   Phylogenetic tree of lower vertebrates, from Haeckel (1866:Plate 7 [part]). This is essentially a phylogenetic tree based on the ideas of pre-Darwinian classification. Notice that there are no animals of equal rank placed on the branches. There are no specified ancestors. Ideas of ancestry are contained solely in the graphic depiction of branches.

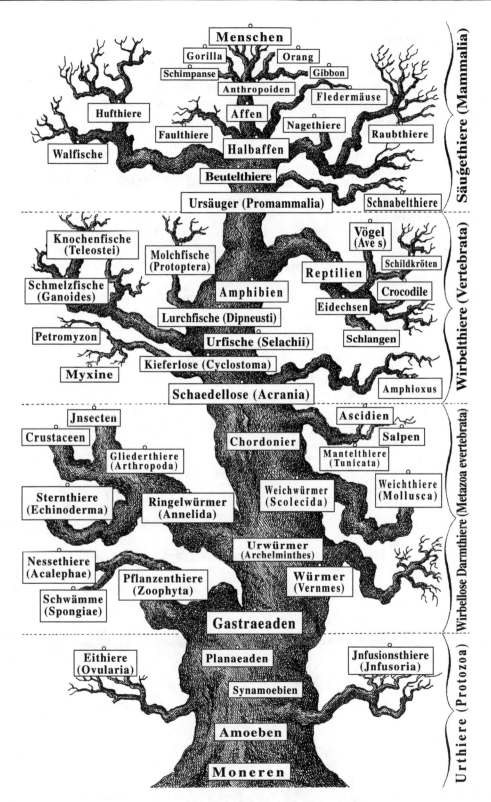

FIGURE 7.4   Phylogenetic tree of the animal kingdom from Haeckel (1874). In comparison with his earlier tree (Figure 7.3) this one shows ancestors on the trunk and branches leading to terminal taxa.

with one prominent digit and high-crowned teeth. Between the Eocene *Eohippus* and *Equus* a series of intermediate forms were already known, their progression harmonizing with their appearance in Cenozoic strata. For Huxley, the palaeontological record was crucial in that he accepted the earlier forms as showing the more general characters. He extended this to all mammalian groups, recognizing that some had progressed further than others had. By dividing up an assumed evolutionary progression into a series of horizontal morphological levels he knew he was flying very near to the discredited *scala naturae* but he nevertheless thought that this was central to classification:

Though no one will pretend to defend Bonnet's *échelle* today, the existence of a *scala animantium* is a necessary consequence of the doctrine of evolution, and its establishment constitutes, I believe, the foundation of scientific taxonomy (Huxley, 1880:652).

It is tempting to equate Huxley's doctrine of evolution as explained by him with Bonnet's attainment of perfection. Huxley would, however, have shied away from equating perfection with Bonnet's angels. Huxley (Figure 7.5) recognized four groups of mammals that were grades of evolutionary progression from the Hypotheria (a hypothetical ancestral group) to Eutheria (modern placental mammals), each one characterized by an increasing complement of eutherian characters.

The idea of grades was also prevalent in invertebrate classifications. Indeed it lingered long. For example, Bulman's (1955: Figures 49, 50) ideas on the phylogeny of some graptolite groups was a direct descendant of Lapworth's (1873) tabular classification of graptolites, which was almost entirely gradual. One striking difference between Huxley and Lapworth, however, was the latter's recognition of the stratigraphic distribution of graptolites.

Cope (1883) continued the use of grade taxa in his classification of mammals in which groups were erected specifically to reflect some estimate of evolutionary level (for details see Polly, 1993).

Some have suggested that classification was unaffected by Darwin's theory of evolution (De Queiroz 1988; De Queiroz and Gauthier 1990, 1992), but, as Polly (1993) rightly points out, this was not so. Huxley's introduction of grade taxa significantly changed the way groups were recognized and given categorical names. Huxley's grades were not quite what we think of as grades today (see below), but the idea that we can establish and name groups that embrace taxa that are thought not necessarily to be genealogical neighbors is a common element. The fact that we use Prototheria, Metatheria and Eutheria today as names of recognized monophyletic taxa is irrelevant to the context in which they were originally established. Richard Fortey (personal communication) sees a direct parallel here with the early classification of graptolites.

One area of Haeckel's phylogenetic trees is of particular interest to me, as it concerns the origin of tetrapods — an event that has been subject to a great deal of theoretical debate relevant to this chapter (Halstead, 1978; Gardiner et al., 1979; Halstead et al., 1979; Patterson, 1980; Rosen et al., 1981). Figure 7.6A is taken from Haeckel (1889) and shows part of a phylogeny of fishes and tetrapods. Within this diagram Haeckel showed a group, Crossopterygii, as giving rise to lungfishes, and this in turn gave rise to Amphibians, which represent the tetrapods or land-dwelling vertebrates. So, here lungfishes were regarded as the ancestor of tetrapods. There is, however, a question mark placed against the Crossopterygii — the presumed ancestors of the lungfishes.

Dollo (1896) gave another diagram (Figure 7.6B), showing that lungfishes were no longer regarded as the ancestors of Amphibia but that they were collateral descendants from a common crossopterygian ancestor.

What happened between these two diagrams? Haeckel's is essentially a pre-Darwinian classification, ultimately derived from examination of a hierarchy of characters, redrawn as a phylogenetic tree. Figure 7.6B contains ideas of ancestry and descent. The lungfishes were removed from the direct line of ancestry because (a) they were living, and (b) they were

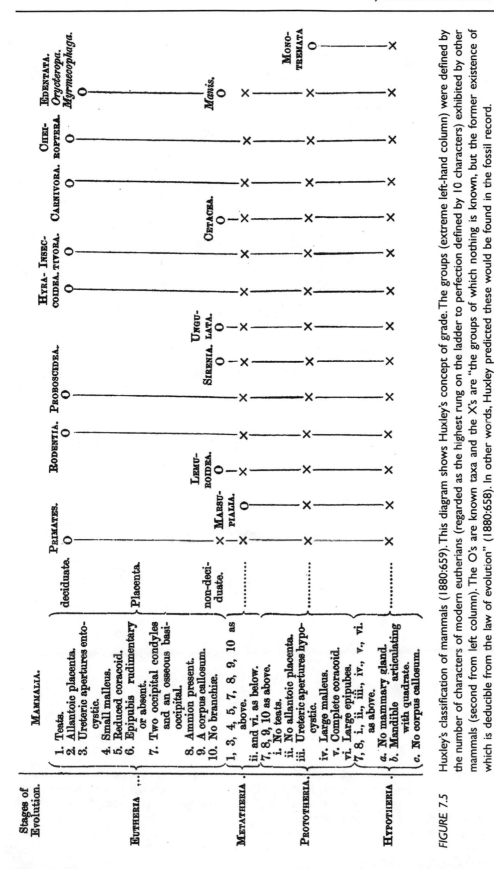

FIGURE 7.5   Huxley's classification of mammals (1880:659). This diagram shows Huxley's concept of grade. The groups (extreme left-hand column) were defined by the number of characters of modern eutherians (regarded as the highest rung on the ladder to perfection defined by 10 characters) exhibited by other mammals (second from left column). The O's are known taxa and the X's are "the groups of which nothing is known, but the former existence of which is deducible from the law of evolution" (1880:658). In other words, Huxley predicted these would be found in the fossil record.

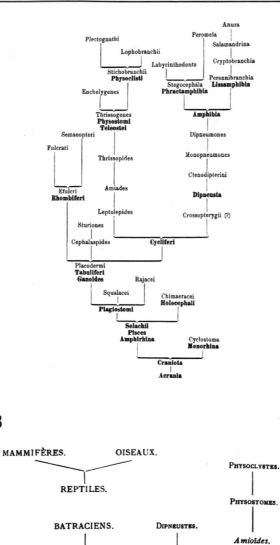

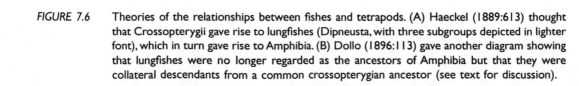

FIGURE 7.6   Theories of the relationships between fishes and tetrapods. (A) Haeckel (1889:613) thought that Crossopterygii gave rise to lungfishes (Dipneusta, with three subgroups depicted in lighter font), which in turn gave rise to Amphibia. (B) Dollo (1896:113) gave another diagram showing that lungfishes were no longer regarded as the ancestors of Amphibia but that they were collateral descendants from a common crossopterygian ancestor (see text for discussion).

thought to be structurally too specialized. In other words, structures within lungfishes could not be imagined to have given rise to those in amphibians: The morphological transitions necessary to draw a connecting line between lungfishes and tetrapods were too extreme to be considered an evolutionary possibility. What about those crossopterygians that Haeckel had designated by a question mark and Dollo had placed so firmly in an ancestral position (just like Romer was to place the Miacidae at the base of the Carnivora)?

Huxley (1861) had erected the group Crossopterygii (Figure 7.7) as a subgroup of Agassiz's Ganoids. He put together a living fish — (*Polypterus*) with six fossil fishes including fossil lungfishes. This group was supposed to contain the ancestors of the tetrapods. Most were Devonian, that is, they were stratigraphically appropriate as ancestors of tetrapods, which in those days were known as dating from the overlying Carboniferous rocks. The problem was that no one could agree on the membership of the Crossopterygii. Cope (1887) included four. Günther (1871) included just three, and so did Watson (1912), but not the same three as Günther, while Cope (1892) changed his mind completely and agreed with Pollard (1891) that only the living *Polypterus* could be included. Doubt over the membership lead Haeckel to place that question mark against the name because he did not really know what was there. The reason no one could agree on membership is that there were no characters for this group, which was supposed to be ancestral to tetrapods: Ancestors do not have characters, yet they needed to be classified.

Early Twentieth Century

Therefore, at the beginning of the twentieth century the attempt to classify vertebrate fossils and Recent animals together using hierarchical distribution of characters had given way to using fossils as ancestors to link groups together with the creation of paraphyletic groups (although the formal recognition of paraphyly was a long way off). The concept of grade was also introduced, but the twentieth century concept of a grade as an adaptive level was yet to surface. The problems of characterizing such ancestral groups were put to one side and so was the uncomfortable dichotomy between those who were trying to recognize groups (e.g., Agassiz and Owen) and those who were trying to recognize ancestors (e.g., Huxley and Dollo). The latter won out, and the relation between vertebrate fossils and systematics was set at the start of the twentieth century. Fossils were useful for linking groups in evolutionary lines because some of them would be ancestors and would display primitive characters.

However, this proved more difficult and more subjective than first realized. Gregory wrote (1910:106), "Phylogeny is essentially an inductive subject, a reasoning by analogy, which is the shifting sand whereon hypotheses and theories are built." He went on to lay out seven principles of good science that may help stabilize the sand; these basically caution against overextrapolation of the evidence. He continued (1910:106), "These principles may indeed seem to be obvious councils of perfection; but so much zoölogical study has been vitiated by the neglect of them that it has come to be scarcely respectable to draw up a phylogenetic tree."

This was by no means an uncommon feeling. Bather (1927:lxxxii) looked back to the late nineteenth century and overextrapolation:

> So early as 1886 Bateson (*Quart. J. Micr. Sci.* xxvi) protested against the multi-
> plication of genealogical trees that involved hypothetical groups as a process
> liable to grave error. There were many other absurdities, such as the designation
> of certain forms as generalised types, or of other forms as linking two distinct
> classes; and these seem to have been due to some uncritical carrying over of old
> transcendental ideas, and a failure to recognise that the new notions demanded

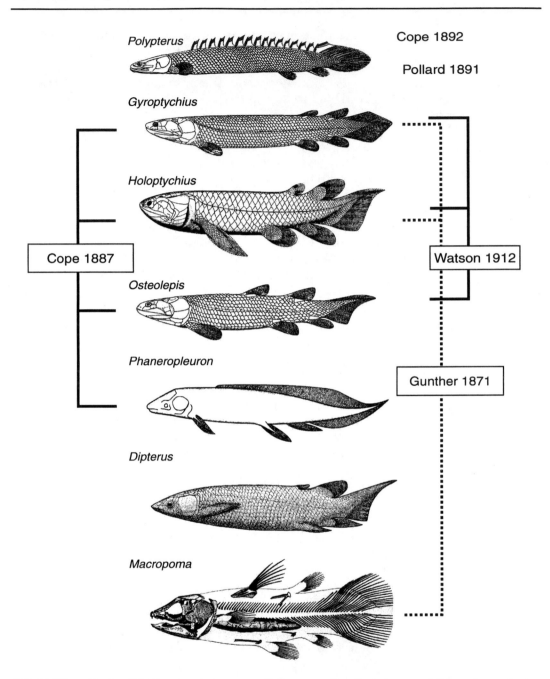

FIGURE 7.7   Huxley (1861) erected the group Crossopterygii to include several fishes that today are scattered among vertebrate classification. Collectively, the Crossopterygii was supposed to contain the tetrapod ancestor. However, different authors disagreed on the membership of this group, and their views are given alongside Huxley's original role call. *Gyroptychius, Osteolepis* and *Holoptychius* are from Jarvik, *Phaneropleuron* from Moy-Thomas and Miles, *Dipterus* from Ahlberg and Trewin and *Macropoma* from Forey. (From Jarvik, E., *Basic Structure and Evolution of Vertebrates*, Academic Press, London, 1980. With permission. Moy-Thomas, J.A. and Miles, R.S., *Palaeozoic Fishes*, 2nd ed., Chapman & Hall, London, 1971. With permission from Kluwer Academic. Ahlberg, P.E. and Trewin, N.H., *Trans. R. Soc. Edinburgh Earth Sci.*, 85, 159–175, 1995. With permission of the Royal Societ of Edinburgh. Forey, P.L., *History of the Coelacanth Fishes*. Chapman & Hall, London, 1998. With permission of the Natural History Museum.)

new lines of attack. The disillusioned biologists consequently transferred their activity to other fields.

In transcendental ideas, Bather could have been speaking of Haeckel's trees.

For paleontology in particular, Gregory (1910) well knew some of the problems that foiled attempts to reconstruct phylogeny and he considered that not enough attention was paid to confront them, leading to the general distrust and disinterest in phylogenetic trees. He listed these as imperfections of the fossil record, which included the difficulty of distinguishing indigenes and immigrants within any one section, and difficulties in separating characters likely to yield phylogenetic signal from those that can be considered adaptive.

Notwithstanding this rather depressing outlook at the turn of the twentieth century, vertebrate paleontologists were defiant and ambitious. Two quotes from highly influential paleontologists encapsulate the accepted role of vertebrate fossils at the beginning of the twentieth century (even though these words were written some years later): "Stratigraphic studies ... could actually trace in succession of strata the progressive evolution of the different races, verifying in specimen after specimen the primitive characters of those from the lower layers" (Matthew, 1926:464).

Romer, certainly the most widely read vertebrate paleontologist of the latter half of the twentieth century, reflected on his early career when he switched from zoology to paleontology: "But increasingly I found myself turning to paleontology. Why content myself with hypothetical ancestral types when actual ancestors may be discovered (Romer, 1949:49).

In fact, so strong was the belief that fossils could do this, that for many workers fossils were essential if we were to understand the phylogenetic history of a group. Here are three quotes — note the years of authorship — to show that the belief is still with us.

"In order to understand the ancestry and connections between living genera and families it is necessary to know the fossil record" (Newell, 1959:275).

"The evolutionary origins and relationships of elasmobranchs are something of a mystery, due mostly to the lack of fossils" (Nash et al., 1976:74).

However, their phylogenetic relationships have not been well established because of limited fossil records" (Lin et al., 2001:252).

Implicit in the last two quotes is the idea that the quality of the fossil record is significant in our attempts to determine phylogenetic relationships, and discussion of this matter has come to dominate the last decade of the twentieth century (see below) with afterburn into the twenty-first. At the turn of the twentieth century, reactions to the quality of the record were diverse. For invertebrate paleontology most workers were impressed by the high quality but, with some exceptions, not inclined to use such data for phylogeny reconstruction. Instead, their efforts were concentrated in stratigraphic correlation and zonation of strata. Animals such as trilobites, brachiopods, graptolites and cephalopod molluscs (goniatites and ammonites in particular) were used highly successfully in biostratigraphic correlation, and this probably focused attention on efforts to compartmentalize groups horizontally (see below). Bather recounted that at the end of the nineteenth century evolutionary theory and attempts to reconstruct phylogeny of the conspicuous invertebrates had not started:

When I began work in London thirty years after the publication of the 'Origin of species', the classification of the Echinoderma was still untouched by an evolutionist; Davidson had just left the Brachiopoda in an order admiral no doubt, but still on the old lines; as for Cephalopoda, Hyatt, the younger Buckman, and a few Continental workers were beginning to appal the collectors of ammonites, but the general arrangement of the class held out for some time to come

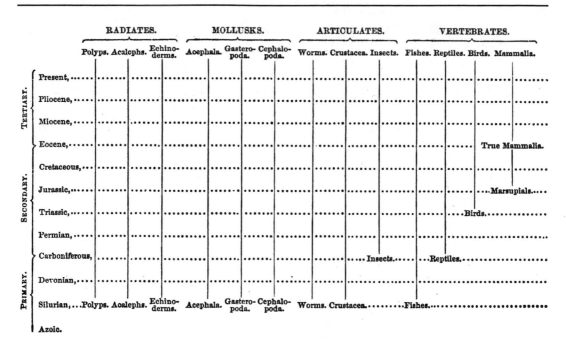

FIGURE 7.8   Chart showing the age of appearance of the principal groups of animals according to Agassiz (1868:95).

against the new school. Many paid lip service to evolution, but few of the established systematists troubled to apply principles which they either did not approve or did not understand. (1927:lxxxi)

For vertebrate paleontologists, the pattern of the fossil record concentrated efforts on finding ancestors (missing links or intercalary types). This may have been a function of the then-known fossil record itself and can be seen embedded in a diagram from Agassiz (1868: figure on p. 95 reproduced here as Figure 7.8). This diagram illustrates the then-known record. It records the major classes of Cuvier's four embranchements of animals, showing that all except insects, reptiles, birds and mammals had a fossil record beginning at the base of Phanerozoic time in the Silurian (the Cambrian and Ordovician had not then been recognized). Agassiz, of course, used such a diagram for a quite different purpose — to suggest that, as all but a few of the classes appeared suddenly, one could not have been derived from another[1]. However, for the evolutionary taxonomist, it also meant that the greatest likelihood of finding a missing link in the fossil record would be within the vertebrate embranchement since there were time gaps in appearance of the subgroups to be potentially filled with fossils.

Another dimension to the recognition of ancestors came to color the relationship between paleontology and systematics in the mid-twentieth century. I have already mentioned the problem faced by late nineteenth century and early twentieth century workers in trying to

---

[1]   Agassiz explained the sequential appearance of vertebrate classes by claiming that the world was not ready for their appearance. For reptiles (then including amphibians), the Carboniferous was a time when "... the extensive marshes afforded means for the half-aquatic, half terrestrial life now characteristic of all our larger Reptiles" (1868:96). For the insects that appeared at the same time, he said, "... while Insects, so dependent on vegetable growth, make their appearance with the first forests; so that we need not infer, because these and other classes come in after the earlier ones, that they are therefore a growth out of them, since it is altogether probable that they would not be created till the conditions necessary for their maintenance on earth were established." (Agassiz, 1868:96).

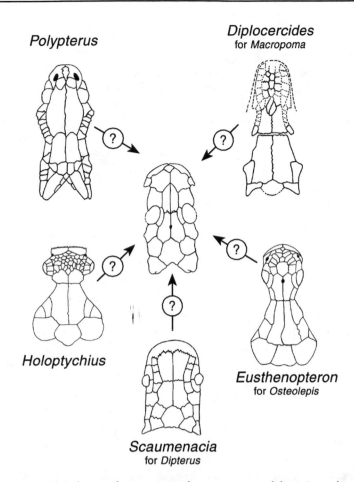

Polypterus

Diplocercides
for *Macropoma*

Holoptychius

Eusthenopteron
for *Osteolepis*

Scaumenacia
for *Dipterus*

FIGURE 7.9   Sometimes the choice of an ancestor from among candidates is made on the assumption of one process or another. The tetrapod skull roof (the descendant) is shown in the center, and different examples of skull roofs of Crossopterygii (sensu Huxley, 1861) are shown alongside. At different times most have been suggested as tetrapod ancestral conditions of the skull roof. Today, that of *Eusthenopteron* is thought the closest. Skull roofs not to scale. (Individual diagrams from Rosen, D.E., Forey, P.L., Gardiner, B.G., and Patterson, C., *Bull. Am. Mus. Nat. Hist.*, 167, 159–276, 1981. With permission of the American Museum of Natural History.)

characterize the assumed ancestral group of tetrapods — the so-called crossopterygians. Another problem revolves around the fact that evolution is a process; it involves change from something that is ancestral to something that is descendent. In trying to find an ancestor, we have to look for something that is close to it but not the same. Workers on crossopterygian fishes and the origin of tetrapods did just that.

The roof of the skull of different crossopterygians (and for the sake of argument I accept Huxley's original formulation for the crossopterygians) is made up of several bones arranged in specific patterns. In Figure 7.9 the descendant is shown in the center, and the various crossopterygians, which are supposed to make up the ancestral group, are arranged around the outside. None of the skull-bone patterns is identical, so the argument as to which is the actual ancestor of the tetrapods — the one Romer would have looked for — depends on what kind of process theory we are going to accept to get from the ancestor to the descendant. A small aside will exemplify many of the arguments over these patterns during the middle of the twentieth century. Säve-Söderbergh (1932) wrote, "In certain characters *Dipterus* seems to be more closely related to the Ichthyostegids than are the Crossopterygians" (by 1932 lungfishes, of which *Dipterus* is one, had been excluded from the crossopterygians).[2]

Those certain characters were the fusion of bones of the skull, but which bones? Different authors had deduced different patterns of phylogenetic bone fusion (and loss, which in terms of pattern recognition among adult animals is the same thing as fusion) and this, in turn depended on the position of the landmarks used (position of the pineal foramen or the paths of the sensory canals were two frequently used) to recognize topographic homology. The arguments are convoluted (Rosen et al., 1981) and/ of secondary interest here, but the common ground is that the recognition of a similar pattern and, hence, identifying the ancestor has been clouded by the assumption of a particular process. Is it going to be bone fusion, fragmentation, bone loss, bone movement, etc.? Our decision is not made on observation but on a theory of transformation. Is this any less abstract than Agassiz's hierarchy of homology?

It is worth backtracking to pick up another thread. At the beginning of the twentieth century two American vertebrate paleontologists began to shape the classificatory method used widely for the next 80 years and generally known as evolutionary taxonomy. William Diller Matthew (Chapter 6, this volume) and William King Gregory, both mammalian paleontologists, stressed the idea that evolutionary change was brought about by response to the changing environment. Therefore, parallel and convergent evolution were to be expected and may well confuse our attempts to reconstruct phylogeny, which they regarded as exclusively an expression of genealogy. They also appreciated the likely confusion between primitive and specialized characters. Although they both considered that our classifications should reflect ancestor-descendent relationships, established as they always had been by superimposing one fossil upon another, they also suggested that the resemblance between the ancestral species should be acknowledged with the establishment of groups. Matthew (1901) recognized two types of groups, vertical and horizontal, terms he attributes to Henry Fairfield Osborn, while he preferred to call them races and groups, respectively. The former contains primitive and specialized members of recognizable and diagnosable taxa. The latter represent the unspecialized species that may be ancestral to many other taxa but that resemble each other more closely in their lack of specializations. As Polly wrote,

> The horizontal grouping described by Matthew is something new. It is not a grade taxon in the sense of Huxley, because only primitive ancestral forms are included, then various descendants grouped in separate clades [the vertical groups], and because membership was contingent on common ancestry, not possession of certain characteristics. (1993:23)

However, characters — or rather lack of them — were used as the price of membership of the horizontal groups — what we would now call paraphyletic groups — because the members had not diverged far in morphology from the common ancestor to all of them (exactly the same argument expressed as genetic similarity that Mayr (1974:102) used: "To him [the evolutionary taxonomist] relationship means the inferred amount of shared genotype, it means gene content rather than purely formalistic kinship."

It is worthwhile to pause and briefly consider what approaches to phylogeny reconstruction and classification were being pursued by invertebrate paleontologists. Early twentieth century invertebrate paleontologists were, with some exceptions (e.g., Rowe, 1899), either

---

[2] For the record, Säve-Söderbergh did not suggest that lungfishes gave rise to all of the tetrapods. He originally (1932) suggested that lungfishes and crossopterygians + tetrapods were derived from a common hypothetical ancestor. Later (1934), he changed his mind to suggest that lungfishes were the ancestors of some tetrapods (urodele amphibians), while crossopterygians were ancestral to the rest of tetrapods.

sceptical of phylogeny reconstruction or disinterested, preferring to invest their energies elsewhere. This changed in the first quarter of the century; classic studies, such as those of ammonite phylogeny, began to appear (e.g., Brinkmann, 1929) and these followed the traditional paleontological method of giving primary consideration to geological succession. This was not universal. Fortey (2001), looking back over 75 years of trilobite systematics, identified three chronologically overlapping phases: evolutionary taxonomy, the stratigraphic paradigm and the cladistic paradigm (which did not appear until the later part of the century). Interestingly, he suggested that initial attempts to reconstruct trilobite phylogeny happened in the 1920s and that methodologists were employing evolutionary taxonomy. Only later, in the 1930s, was the stratigraphic paradigm adopted. Trilobites, unlike many other invertebrate groups, have always been seen as morphologically complex, and it is possible that this, together with the discovery of ontogenetic stages, allowed more precise assessments to be made of primitive versus derived morphologies independently of the stratigraphic record.

Mid-Twentieth Century

The writings of Matthew and Gregory, among others, stimulated George Gaylord Simpson (1944, 1945, 1953, 1961), who carried Matthew's ideas further down the same philosophical road. Simpson drew on additional armament in the rapidly accumulating data of population genetics. Simpson retained the horizontal groups, although he justified them in terms of adaptation rather than as characters of organisms. The explanation of the evolution of species through adaptation could be extrapolated to higher taxa. He also considered it legitimate to separate phylogeny from classification. Therefore, the one need not mirror the other, although there was little guidance offered as to exactly when and by how much classification could, and indeed should, deviate from reconstructed phylogeny. This was the aspect of classification that Simpson characterized as art (1961:130). As Polly wrote,

> The view of the relation of classification that Simpson brought into the synthesis was a complex one. It was a combination of several divergent views on the question that each had advocates in the late nineteenth century. It combined the purely genealogical clades of Kovalevskii, the continuous grades of Huxley, and the character-based classes of Cope. In that it *partially* separated classification from the underlying mechanism producing natural order, it differed in philosophy from all pre-Darwinian methods of classification. (1993: 25)

Looking back through the lens of cladistics Simpson's view of phylogenetic reconstruction and classification seems unstructured and held together only by authority. However, I consider his writings, particularly Simpson 1944, as collectively constituting a milestone, if not in systematics in particular then in paleontology in general, because of the numbers of people influenced and the diversity of work these writings stimulated.

In the 1930s scientists pursuing inquires in the fields of Mendelian genetics, population genetics, systematics and paleontology began talking to each other — or at least occupying the same rooms. There was much debate about the nature of species and the meaning and purpose of classification. The Systematics Association was founded as a result of that debate, and a volume edited by Huxley (1940) synthesized that discussion. Later (Jepsen et al., 1949), much the same subject areas were covered in a meeting organized by The Committee on Common Problems of Genetics, Paleontology, and Systematics of the National Research Council and held in Princeton, NJ, in January 1947. These two publications outlined what became known as the new (modern) synthesis.

From a paleontological perspective, it is remarkable that this synthesis happened at all. Neo-Darwinism emphasized the environmental selection of small nondirectional mutations. Paleontology had been demonstrating directional change with occasional abrupt appearance of species (that is why stratigraphic correlation and zonation works). Nine paleontologists, who were or were to become paleontological household names, contributed to these publications (W.J. Arkell, R.W. Chaney, E.H. Colbert, J.A. Moy-Thomas, B. Patterson, A.S. Romer, D.M.S. Watson, T.S Westoll and H.E. Wood II). Within the essays written by these paleontologists there is much discussion about tracing evolutionary trends through fossils (Westoll, 1949; Watson, 1949; Romer, 1949; Chaney, 1949; Patterson, 1949; Wood, 1949) and the nature of paleontological species (Arkell and Moy-Thomas, 1940) and how they may be classified. A great deal was written about ecology and functional adaptation as portrayed in the fossil record. However, the actual patterns of relationships that functional adaptation was supposed to explain were never questioned, nor was the method by which patterns were to be identified ever discussed. Given the new injection of population genetics and our increased knowledge about the relationship between genes and the environment, it seemed there was no pattern that could not be explained. The skull-roof bones could change in any way you wanted; only authority or stratigraphy would show the way.

Romer's characteristically lucid contribution is interesting from several perspectives. He begins (1949:103), "In the study of evolution it is impossible to overemphasise the importance of the fossil record, the evolutionary lines apparent in this record, and the evidence of evolutionary processes and trends which is revealed by the study of these phyletic lines." He thereby sums up the dominant feeling at the time: namely, that the phyletic lines revealed by the fossil record are the facts that the modern synthesis of the theory of natural selection and Mendelian genetics would explain. Dwight Davis (1949:65), a comparative anatomist, similarly held faith with paleontology: "... the facts of paleontology demonstrate evolutionary progress far more convincingly than do inferences of comparative anatomy."

Exactly how were the "phyletic lines" and "facts of palaeontology" revealed? Watson came closest to explaining an actual method of phylogenetic reconstruction, and he spelled out three stages. These were (1) collection of fossils and their assignment to relative ages, (2) determination of the species present in each age by comparing the morphological variation with that seen in Recent species of assumed relatives, and (3) grouping and classification of those species "based on similarities of structure between two or more allied forms, the similarities being judged according to an order of importance established by the application of morphological principles" (Watson, 1949:45). He went on (p. 45): "The methods of morphology have never been precisely stated and discussed. They are known traditionally and by example to working zoologists."

Here is an interesting juxtaposition. A neontologist (Dwight-Davis) placed responsibility for discovering phylogeny in the hands of paleontologists (most were very keen to accept that task), and a paleontologist (Watson) held faith in traditionally imbibed and, according to him imprecisely specified, morphological principles (not withstanding the nineteenth century efforts of Agassiz [1859, 1868]. If there is anything to be distilled from such quotes, it is that reconstruction of phylogeny is a paleontological enterprise but one grounded in comparative anatomy — the principles of which may or may not have been explained. There was little specific discussion of time — the unique paleontological perspective — or how time was to be used. Perhaps, like morphological principles, time was an attribute known traditionally by working paleontologists.

To be sure, the two publications (Huxley, 1940; Jepson et al., 1949) summarizing the new synthesis must be regarded as milestones, but milestones are not inherently guides to enlightenment; they may simply be markers. My reading of these essays was that consideration of evolutionary processes took hold and that systematics was done a disservice. This particularly affected paleontologists because, remember, they had taken it on themselves, or were charged with the responsibility of discovering the phylogenies, the patterns to be explained.

One of the outcomes of the new synthesis, which pervades the essays, was an emphasis on phylogeny as visualized as successive attainments of new adaptive zones (grades in the sense of Matthew and Gregory rather than those of Huxley), and Simpson had a great part to play in this. Adaptation was the province of neo-Darwinism, and there were many good experimental reasons for this. However, the invocation of a particular adaptation to explain a particular pattern was left rather vague. The mists of adaptation had settled on the facts revealed by paleontology and the possibility that traditionally recognized groups of animals (and plants) could have arisen several times in response to an adaptive drive, perhaps associated with interspecific competition, became a popular idea among paleontologists. Vertebrate paleontologists suggested that groups such as Cyclostomata (hagfishes and lampreys) (Stensiö, 1958), Chondrichthyes (cartilaginous fishes) (Ørvig, 1962), Tetrapoda (Holmgren, 1939; Säve-Söderberg, 1934; Jarvik, 1942), Reptilia (Kuhn-Schnyder, 1962) and Mammalia (Simpson, 1959), among many others, were considered to be polyphyletic (I include diphyly here), the members of these groups having arisen from more than one progenitor within an ancestral group.

In all cases certain morphological attributes were thought to have arisen on more than one occasion because of their adaptive significance. For example, the naked, eel-shaped and jawless cyclostomes (modern hagfishes and lampreys) were thought to have arisen from two separate extinct groups of armored free-swimming, fish-shaped jawless fishes (Stensiö, 1958). The structural similarity between the hagfishes and the lampreys was thought to be due to the adoption of a common semiparasitic way of life induced by competition from superior fishes with jaws and moveable fins. Thus, it was the common way of life that resulted in a similar morphology that should be reflected in classification. As Gould (1987:22) put it, "Ways of making a living (function) are few: lineages are many." It therefore became fashionable among paleontologists to identify these ways of living in classifications. Romer explained the definition of such groups:

> We customarily define a group by the characters which its living members possess in common. But in many cases we suspect, and in certain case we know, that part or all of these characters were not present in the ancestral forms. How, then, may we define such a group? May we not make our definition in terms of common trends, saying in effect that all these have basically similar genetic constitution and hence are liable to the similar modification? (1949:116)

This marries perfectly with Mayr's (1969) definition of relationship — genes in common. It is also the justification for the grouping organisms on the basis of parallelisms in cladistic analysis as advocated by Saether (1979; "underlying synapomorphy") or Mayr (1974; "concealed genotype").

Lest it be thought that this practice was confined to vertebrate paleontologists, more extreme and explicit examples can be found in the world of invertebrate paleontology. Wright (1979) suggested that brachiopods arose on seven occasions from different unknown ancestors predicted to have looked like phoronid worms. Here the dorsal and ventral shell found in all brachiopods is assumed to have developed in parallel seven times. Valentine (1975) justified this by a process explanation, namely, that near the base of the Cambrian (when brachiopods suddenly appear with shells) several separate soft-bodied animals living in the mud moved to an epifaunal position to take advantage of cleaner water and developed a pedicle or stalk for anchorage and a shell for protection since they were now exposed. This is harmless enough as an explanation, but how can it be used to justify the pattern of relationships — or nonrelationships? Valentine, who in the 1970s advocated polyphyletic origins of many of the coelomate phyla, wrote (1973:101), "A functional ecological approach to the origin of higher taxa produces models of adaptive pathways that have much explanatory power and ... . may yet serve to constrain the possibilities so that a well supported model of phylogeny will emerge."

This is more serious because if one can impose a model of evolution onto phylogeny reconstruction then the phylogeny becomes disengaged from observation. In this case the fact that brachiopods all share a particular type of shell — the observation — has been dismissed as irrelevant because a model of an adaptive pathway proclaims this observation to be misleading. Such methods of reconstructing phylogeny also mean that choice between competing hypotheses are choices between the models.

Late Twentieth Century

Paleontology's primacy in being the chief discipline for tracing phylogenetic pathways was seriously challenged at the next milestone identified here: Hennig's phylogenetic systematics (Hennig, 1950, 1965), which gained wide recognition with the publication of an English translation of a manuscript, modified from his 1950 book (Hennig, 1966). I would identify another associated milestone here (Brundin, 1966), one that may well have laid hidden within the brambles of alpha taxonomy were it not for the keen eyes of Gary Nelson. Brundin's work was a 472-page monograph on chironimid midges, probably of interest to only a handful of entomologists. However, it contained an introductory chapter summarizing the theory and methodology of phylogenetic systematics of Hennig (Hennig's 1966 book had not appeared). Brundin's writing was as clear and lucid as Hennig's was dense. Hennig (1966) must stand as the milestone that caused a revolution in systematics, but Brundin (1966, 1968) provided the synopses that spurred many to read Hennig. On Nelson's recommendation, Brundin's introduction (1966) was read during the next year by Niels Bonde in Copenhagen and Colin Patterson in London, both fish paleontologists, who recall their reaction (Patterson, 1995; Bonde, 1999, 2000; Nelson, 2000). The spread of phylogenetic systematics (cladistics) was immediate, and its effect profound (Hull, 1988). Hennig's insight allowed him to redefine the meaning of relationship, to clarify the meaning of monophyletic and polyphyletic groups as well as to introduce the concept of paraphyletic groups, which to my knowledge, had not been done before. Paraphyletic groups were of particular interest to paleontologists as being the ancestral groups. Over the next decade papers by Nelson (1969, 1970, 1971), Bonde (1977), Schaeffer et al. (1972) and Patterson (1977) argued passionately for the adoption of phylogenetic systematics and discussed its relationship with paleontology.

There seem to have been three types of immediate reaction to Hennig's ideas among paleontologists. For some, cladistics was seen as simply adding a complex terminology to what everyone had been doing for years (Boucot, 1979:200: "The cladists seem unfortunately to have swallowed a rhyming dictionary rich in classic roots of all sorts; the resulting deposit has now fertilized a plague of toadstools, sprouting on our beautiful taxonomic lawn.") For others (Mayr, 1969:70, 1974; Charig, 1982) the apparently blinkered view of cladistics as allowing only relative positions of branching points to be expressed in classifications denied the explanatory power of adaptation and different rates of evolution. Others totally accepted cladistics (Schaeffer et al., 1972[3]).

What exactly was the impact of cladistics on the methods previously used by paleontologists to reconstruct phylogeny? There were three body blows. First, as Brundin (1966) spelled out, the establishment of phylogenetic relationships of Recent taxa are a necessary

---

[3]   This paper, published in 1972, I believe was inspired by a talk given by Gary Nelson at the American Museum of Natural History in 1970. The original talk is now in press with some commentary (Williams and Ebach, in press).

prerequisite to the interpretation of systematic position of fossil groups. This immediately stripped paleontology of its position of primary authority in matters phylogenetic. Second, in establishing phylogenies, we can only consider ancestors as being hypothetical. This was anathema to most paleontologists. Would Romer go back to being a zoologist? Authority for phylogeny and recognition of ancestors were the crowning achievements of paleontologists for the previous hundred years. This was now being challenged. Third, time — the unique attribute of fossils — was not the key to unraveling the paths of lineages (Crowson, 1970), although Hennig (1966) did suggest that the stratigraphic occurrence of fossils would allow characters to be polarized (his phylogenetic systematics method suggested that characters be polarized *a priori* [see Chapter 5, this volume]).

The distrust of cladistics by many paleontologists was eventually traced to the distinction between cladograms, which are general atemporal statements of character distribution, and phylogenetic trees, which are precise statements of ancestor-descendant lineages spread out along a time dimension (Platnick, 1979). Since several evolutionary trees are equivalent to one cladogram, then extra information is needed to choose one tree over another. That extra information may be some auxiliary theory of character evolution or the inclusion of the stratigraphic position of fossils with or without recognition of ancestors (see later). In terms of character distribution, however, the only phylogenetic tree that is isomorphic with the cladogram is the one in which the ancestors are hypothetical.

Another paper that deserves the symbolism of a milestone is that of Patterson (1981), "Significance of Fossils in Determining Evolutionary Relationships." In this paper Patterson (1981:218) wrote, "... instances of fossils overturning theories of relationship based on Recent organisms are very rare, and may be non-existent. It follows that the wide-spread belief that fossils are the only, or best, means of determining evolutionary relationships is a myth."

As support, Patterson cites Mayr's comment on the phylogeny of Recent mammals: "Our ideas of their relationships, based on a study of their comparative anatomy, have in no case been refuted by subsequent discoveries in the fossil record" (Mayr, 1974:98). Since Simpson (1961:83) had already written that the classifications of mammals "have come to depend more on fossils than on recent animals," Patterson, along with Mayr, was questioning Simpson and the contribution of paleontology Simpson thought so vital. Patterson's original quote cited above was based on his sample of requests from paleontologists he had asked to supply instances in their own groups of where they thought paleontology had contributed uniquely to ideas of relationships among the Recent taxa. None of the examples stood up to critical examination. However, Patterson did concede that optimization of characters in matrices containing fossils and Recent taxa could overturn original theories of homology amongst the Recent taxa. Indeed, fossils are more likely than Recent taxa to overturn a theory of homology because they tend to have larger numbers of plesiomorphic character states (Huelsenbeck, 1992).

Very few studies have succeeded in including fossils to overturn a theory of relationships based on modern taxa. Some that I know of in the fish world are Arratia (1997), Murray and Wilson (1999) and Wilson (1992). One of the most publicized is the study by Gauthier et al. (1988) who suggested that "Amniote phylogeny seems a perfect test case of the proposition that fossils are so uninformative that they can have little or no impact on a hypothesis derived from the comparatively information-rich extant biota" (p. 106). Their study concerned the relationships among amniote animals. In the initial study of Recent animals only Gauthier et al. used 5 Recent taxa: (turtles, lizards and snakes, crocodiles, birds and mammals) and a total of 109 characters. Within this data matrix 42 of these characters were from the skeleton and 67 were soft anatomical (including physiological characters), which were unavailable in fossils. Gautheir et al. arrived at the hypothesis that mammals are the sister-group of birds + crocodiles. However, when they added an additional

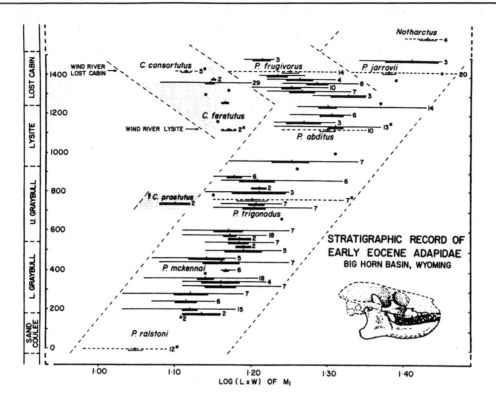

FIGURE 7.10   Stratophenetic tree of primitive primates collected as samples from early Eocene sediments, representing about six million years in the Big Horn Basin, Wyoming. The Y-axis shows the local lithographic subdivisions with the height of the section from the base. The X-axis is the morphological variation of the samples taken at different sampling intervals. See text for discussion. (From Gingerich, P.D. and Simmons, 1977. With permission.)

24 fossil taxa the mutual relationships of the Recent groups changed so that Mammals were now resolved as the sister group to the remainder of the amniotes. Therefore, they concluded that fossils were highly influential and could overturn a theory based on Recent taxa.

However, the study was a lot more complicated than this brief description; for a fuller discussion, see Forey (2001), who identified six kinds of change when translating from the Recent data set to the Recent + fossil data set. Different outgroups were used for some character codings. Different terminal taxa were substituted as representatives of the Recent taxa. A substantial increase occurred in numbers of characters owing to the varied morphology of the fossils; this increase was concentrated on the skeletal morphology, where the 42 Recent skeletal characters increased to 204, thereby introducing bias (Forey 1998b). Introduction of nonapplicable codes for some of the characters of Recent taxa occurred. Introduction of question marks against most of the fossil taxa for the Recent soft anatomical characters occurred. This example shows that straightforward tests of whether fossils can overturn theories are not easy to devise. With the inevitable increase in skeletal characters, increase in number of taxa, recoding using different outgroups, and substitution of taxa, there is a change in the nature of the problem. This is one role that fossils may play: they change the nature of the problem.

There is little point in trying to demonstrate that one or another data set (Recent, fossil or Recent + fossil) yields the best result. Fossils always remain subsidiary to Recent taxa because the identification of primary homology (one of the key stages of what Wägele calls "character analysis" — [Chapter 5, this volume]) depends on interpreting structures in fossils in the light of those in the Recent world. It is more productive to consider the strengths of fossils for phylogenetic reconstruction, which I identify as the following (Forey, 2001; Forey and Fortey, 2001): the ability to break up long branches (dependent on being able to assign a fossil to a Recent taxon on the basis of synapomorphy); stabilization of theories of character evolution; elucidation of the sequential steps involved in the evolution of a complex character (e.g., the evolution of the feather among "dino-birds"); arbitration between equally parsimonious theories based on Recent taxa or molecular data (e.g., Smith and Littlewood, 1994); acting as outgroups for modern taxa (Smith, 1998).

Current Debates

Paleontology has the additional and unique information of time, measured either as absolute age or as relative stratigraphic position. The assumption that time imparts a key role to fossils in phylogeny reconstruction has been with us since the nineteenth century. However, it is only within the last quarter of the twentieth century that the strengths and limitations of time have been discussed in detail. There is an ongoing debate about the impact that consideration of time should have on our ability to reconstruct phylogeny.

The practice of using time or stratigraphy in phylogeny reconstruction has a long history, but it was a practice that, like Watson's view on morphological principles (see above), had been known traditionally and by example to working paleontologists but never spelled out. Gingerich (1976) explained very clearly the method of combining stratigraphic occurrence of fossils with phenetic comparisons between them, which had been in use since Gaudry's and Hilgendorf's work. I identify Gingerich's paper as a milestone because Gingerich gave the method a name — stratophenetics (1976:16) — and thereby legitimized a school.

His explanation was structured around Figure 7.10. This shows a graph where the Y-axis is time as measured in feet starting from the base of the Eocene sediments in the Big Horn Basin of Wyoming. The total time from top to bottom represents about six million years. The X-axis represents morphological variation in the crown area of the first lower molar of the primate *Pelycodus* (now called *Cantius*). Each of the horizontal lines is the range in sample size, with the mean and standard error of the mean as the solid line. Even though the actual morphology that is being measured is the crown area of one particular tooth, this is assumed to reflect morphological difference in the whole animal. Other stratophenetic studies of other organisms have considered much more of the morphology (e.g., Geary, 1990), but the point is that similarity and difference in one or more morphological measurements is seen as reflecting genes in common. These studies follow the Mayrian view of relationships. *Relationship* means genes in common (Mayr, 1969). In Gingerich's study the crown area of the first lower molar is a proxy for genes in common.

The names given on Figure 7.10 are considered species that are recognized on two criteria. The top two species, *P. frugivorus* and *P. jarrowii*, are contemporaneous and sympatric, yet their molar dimensions do not overlap; hence these are distinct species. The different species recognized in the swathe of samples (populations) running from *P. ralstoni* at bottom left to *P. jarrowii* at top right are not so distinguished. These samples have been split into three species by arbitrarily dividing time; hence these species are chronospecies. More complex means to split the samples into species have been done by other workers using stratophenetics (e.g., Geary, 1990; Lazarus, 1986). More important is what happens next — how these species are linked to one another to reconstruct a phylogeny. This is shown in Gingerich's explanatory table reproduced here, which he explained as follows (1979:54):

Level 3	A"	B	C
Level 2	A'	B'	C
Level 1	A	B	C

We can first look at A' in Level 2 and ask which species in level one it most closely resembles (answer A) and then which species in Level 3 it most closely resembles (Answer: A''). Thus A-A'-A'' can be linked together on the basis of their overall similarity. Each successive stage represents a progressive change in the original sample A. Similarly B-B'-B can be linked together, with B' interpreted as a character innovation developed in the original B that was subsequently lost again by the time of Level 3. The clearest case of linking comes when C in Level 1 is linked to C in level 2 and in turn to C in Level 3. This represents an example when the same taxon C appears unchanged /at all three levels.

So here is a clear explanation. But is it? Are all the A's supposed to be the same species, or all the B's the same species since the species were recognized by arbitrarily dividing the time horizontally into Levels 1, 2 and 3. B is said to show an acquisition and subsequent loss of a character innovation not to change species. To return to the real example then, samples of what is called *P. trigonodus* at level 800 are identical to samples of *P. frugivorus* at level 1500, but the intervening samples deviate to the right through what is called *P. abditus*. So B, B' and B in levels 1, 2 and 3 are, respectively, *P. trigonodus*, *P.abditus* and *P. frugivorus*, and these form an ancestor-descendant sequence. Yet genes in common, as measured through the proxy of molar crown area, place *P. trigonodus* and *P. frugivorus* as nearest relatives, if not identical. The phylogeny is not based on overall morphological similarity or difference, nor on any special similarity but on stratigraphy since that is the deciding factor leading to this particular tree.

Gingerich was careful to limit stratophenetics to situations where the sedimentary record appears to be complete and to situations within a closed basin. He acknowledged that the method only works where we have the totality of the biotic and geologic universe. Other people — particularly those working with microfossils — argue that if the recorded sequence of taxa can be repeated in different parts of the world (e.g., Bolli, 1986; Prothero and Lazarus, 1980) then this justifies the use of stratophenetics.

Therefore, the use of time in phylogeny reconstruction is intimately tied to estimations of the completeness of the sedimentary record. Much recent effort has been devoted to estimating of the quality of the record (for a recent review see contributions in Donovan and Paul, 1998 as well as Smith, 2000b). There has also been a great deal of work comparing a phylogeny based on morphological data with the temporal sequence of the included taxa and with the development of a variety of indices devised to express the congruence of morphological phylogenies with stratigraphic occurrence (Norell, 2001; Angielczyk, 2002). Such indices are useful in helping to arbitrate between equally parsimonious phylogenetic trees based on morphological data.

Certainly, stratophenetics is still used, and indeed some workers (e.g., Lazarus and Prothero, 1984) suggest that the stratigraphic order must be used in reconstructing the phylogeny of some groups because there is not enough morphological variation useful for standard phylogenetic analysis. But is it a widely used method? The second edition of the *Encyclopaedia of Palaeobiology* (Briggs and Crowther, 2001) does not include an entry for stratophenetics (it was there in the first edition 11 years previously), and in this 538-page book there is only one index citation to one line. There are many recent stratophenetic studies, and they do seem to have their place. As Fortey (2001:1144) recorded, within the study of trilobites there have been some elegant and successful studies (measured by congruence with purely morphological cladistic studies), "but as a general reference system for trilobite classification the reconstruction of stratigraphy-based trees has not worked." The implication is that the use of stratophenetics is for relatively low taxonomic rank levels (genera and species), and this view has been put forward by practicing stratopheneticists (e.g., Pearson, 1998, who restricts its use to the species-level). Fortey and Jefferies (1982) made a similar point. Does this mean that stratigraphy has a highly restricted use in phylogenetic reconstruction? Some think not.

The use of stratigraphic position as a vital element in phylogeny reconstruction at any taxonomic level has resurfaced under the names *stratocladistics* and *stratolikelihood*. Stratocladistics was devised and named by Dan Fisher (1991, 1992, 1994). Fisher starts from the premise that we never know the true tree based on morphology alone, so we are entitled to use stratigraphic occurrence to add more data — in this case time data — and combine the two to yield a total evidence answer. Time is used as a character to interact with morphological data during parsimony analysis. The method works by comparing the phylogenetic tree based on morphological data with the tree that would be constructed using stratigraphic occurrence alone (that is, as if the taxa are arranged in an ancestor-descendent sequence). The time throughout which the taxa lived is divided into equal horizontal slices. Then the two trees (morphologic and stratigraphic) are compared to each other to see how many steps have to be added to each of the trees. The tree that requires fewer steps is chosen.

Several objections may be raised to stratocladistics. The most serious is that in cladistic analysis we assume that character states are related to each other in a hierarchical way, but time or stratigraphy is not hierarchical; it is linear. So it does not behave as we assume all other morphological (or molecular characters) do. Another criticism is that the observed stratigraphic range may bear 1 to 100% faithfulness to the actual time the taxon lived. Many studies have refined methods of estimating true stratigraphic ranges and some are extremely elegant (see Marshall, 1998 for a recent review). In the end, though, they remain estimates, and just as morphological data can be miscoded, so can the time period be misjudged.

Another criticism of using time as a character is that the method is sensitive to how finely we divide the stratigraphic column (Smith, 2000a). The more finely time is divided, the more likely it is that stratigraphy will overturn a morphological tree simply because there are more time slices lacking the fossils that the morphological tree predicts will be crossed (Figure 7.11). There are no rules or conventions for how finely we divide time. The scale of the stratigraphic resolution varies with the geographic scale of the fossil-bearing bed as well as the taphonomic processes resulting in fossilization (Fisher et al., 2002). So, the arguments over which tree to choose may ultimately depend on arguments about the methods and efficiency of dividing time (see, e.g., Alroy, 2002), not about morphology. There are many parameters by which we may divide time into units for analysis. For local rock sections, there may be good arguments for dividing time very finely (e.g., continuous sedimentation, equally stratified varves [Barton and Wilson, 1999], correlation with Milankovitch cycles [Gale et al., 2000]). However, when dealing with systematic groups of organisms that are geographically widespread, it will be difficult to justify the claim that the time bands as recogniszd in the different geological sections are, in fact, exactly contemporaneous in time and equal in their duration.

An extension to stratocladistics has come to prominence in the past few years. This is stratolikelihood, a method devised by Wagner (1998a). A set of optimal and suboptimal trees is generated from morphological characters alone. Then, this method attempts to take into consideration some probabilistic estimate of the quality of the fossil record. From this the stratolikelihood method derives maximum likelihood scores for extinction rates and the probability of sampling a taxon within one particular time band. The next step is to produce simulated trees using the same size morphological data set and to model changes in character changes and extinction and origination rates using estimations of the fossil ranges. This gives us a likelihood estimate of observing the maximum parsimony tree. We then multiply the probabilities together to get the most parsimonious maximum likelihood tree.

There seem to be three problems here. Maximum likelihood techniques are rarely applied to morphological data because (unlike molecular data) there is little empirical justification for assuming certain patterns or models of evolution. For instance, the arguments of bone fusion or fragmentation of the skull of crossopterygian fishes were basically sterile. Second, likelihood estimates of the fossil record only measure how well we have sampled the available

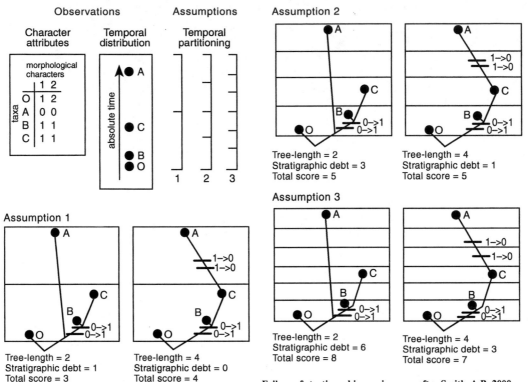

Failure of stratigraphic parsimony – after Smith, A.B. 2000
Jl. Paleont. 74: 764

FIGURE 7.11   The stratocladistic method is sensitive to how finely time is divided. The diagram at the top left gives the morphological and temporal matrix as well as three assumptions as to how finely we divide time. Under assumption one the morphological tree shows two character changes but demands an extra step when compared to the stratigraphic tree because taxon A is not recorded in the lower level. Under assumption two the trees are equally parsimonious, whereas under assumption three (time divided most finely) the stratigraphic tree is more parsimonious. (After Smith, 2000a: fig. 1, with the permission of the author.)

rock sample. How this rock sample relates to what actually lived at any particular time is completely unknown. Third, maximum likelihood methods are really testing the model parameters that were set up in the first place. How this bears on reality, beyond the assumptions of the model of evolution, is unclear.

I mention both stratolikelihood and stratocladistics because these are the current developments influencing theoretical paleontological research into phylogeny reconstruction. In many ways, these methods reflect a desire to put stratigraphy, ancestor-descendent relationships and evolutionary models back into phylogeny reconstruction. We must await the reaction of those practicing paleontologists.

One last challenge may be mentioned. Wagner (1998b, 2000) suggests that during clade history, homoplasy accumulates in morphological characters such that phylogenetic signals become increasingly confused. If true, this would limit the capability of parsimony analysis to recover phylogenetic structure and our ability to deduce patterns of evolutionary change. Stratigraphy may therefore become vital to separate out original from secondary appearances of character states.

Conclusions

Conclusions in a historical essay such as this are somewhat superfluous, since they are simply where we are today. The twentieth century witnessed many discussions about the meanings of relationship and how they were to be expressed in our classifications. That debate is by no means over (see Chapters 6 and 9, this volume) and it is one that is equally applicable to neontology and paleontology. At the beginning of the century paleontological data was considered both necessary and sufficient for phylogeny reconstruction. That was rightly questioned. Through that questioning, the role played by fossils has been critically examined such that the strengths and limitations of the record have been more carefully defined. The past 25 years have seen a great deal of effort put into assessing the quality of the record, and I consider that to be a direct result of that questioning. The spin-offs have been very rewarding. As for the future, that may lie in more careful assessments of exactly under what real (as opposed to simulated) circumstances the fossil record truly documents the history of life and if there are prescribed conditions as to when stratophenetics should be preferred over cladistics.

Acknowledgments

I express my thanks to David Polly, Queen Mary College, London University for discussion and for bringing his thesis to my attention and to Norman Macleod for information on foraminiferal phylogeny and Richard Fortey (both of The Natural History Museum) for reading a preliminary draft.

References

Agassiz, J.L.R., *Recherches sur les Poissons Fossiles,* Vol. 1. *Text,* Petitpierre, Neuchâtel, Switzerland, 1844.

Agassiz, J.L.R., *An Essay on Classification,* Longmans, Roberts and Trübner, London, 1859

Agassiz, J.L.R., *Methods of Study in Natural History,* 2nd ed., Ticknor & Fields, Boston, 1868.

Ahlberg, P.E. and Trewin, N.H., The postcranial skeleton of the Middle Devonian lungfish *Dipterus valenciennesi. Trans. R. Soc. Edinburgh Earth Sci.,* 85, 159–175, 1995.

Alroy, J., Stratigraphy in phylogeny reconstruction — reply to Smith (2000), *J. Paleontol.,* 76, 587–589, 2002.

Angielczyk, K., A character-based method for measuring the fit of a cladogram to the fossil record, *Syst. Biol.,* 51, 176–191, 2002.

Arkell, W.J. and Moy-Thomas, J.A., Palaeontology and the taxonomic problem, in *The New Systematics,* Huxley, J., Ed., Clarendon Press, Oxford, 2002, pp. 395–410, 1940.

Arratia, G., Basal teleosts and teleostean phylogeny, *Palaeoichthyol. München,* 7, 5–168, 1997.

Barton, D.G. and Wilson, M.V.H., Microstratigraphic study of meristic variation in an Eocene fish from a 10,000-year varved interval at Horsefly, British Columbia, *Can. J. Earth Sci.,* 36, 2059–2072, 1999.

Bather, F.A., Biological classification: past and present, *Q. J. Geol. Soc. London,* 83, lxii–civ, 1927.

Bolli, H.M., Evolutionary trends in planktonic foraminifera from early Cretaceous to Recent, with special emphasis on selected Tertiary lineages, *Bull. Centres Rech. Explor.-Prod. Elf-Aquitaine* 10, 565–577, 1986.

Bonde, N., Colin Patterson (1933-1998): a major vertebrate paleontologist of this century, *Geol. Mijnb.,* 78, 255–260, 1999.

Bonde, N., Colin Patterson: the greatest fish palaeobiologist of the 20th Century, *Linnean,* Special issue No. 2, 33–38, 2000.

Bonde, N., Cladistic classification as applied to Vertebrates, in Major Patterns in Vertebrate Evolution, Hecht, M.K., Goody, P.C. and Hecht, B.M., Eds., Plenum Press, New York, 1976, pp. 741–804.

Boucot, A.J., Cladistics: is it really different from classical taxonomy?, in *Phylogenetic Analysis and Paleontology*, Cracraft, J. and Eldredge, N., Eds., Columbia University Press, New York, 1979, pp. 199–210.

Briggs, D.E. and Crowther, P.R., Eds., *Palaeobiology II*. Blackwell Science, Oxford, 2001.

Brinkmann, R., Statistisch-biostratigraphische untersuchengen an mittlejurrassisichen ammoniten uber artbegreff und stammesentwicklung, *Abh. Ges. Wiss. Gött. Math.-Phys. Kl.,* 13, 1–249, 1929.

Brundin, L., Transantarctic relationships and their significance, as evidenced by the chironimid midges, *K. Sven. Vetenskapsakad. Handl.,* 11, 1–472, 1966.

Brundin, L., Application of phylogenetic principles in systematics and evolutionary theory, in *Nobel Symposium 4. Current Problems of Lower Vertebrate Phylogeny*, Ørvig, T., Ed., Almqvist & Wiksell, Stockholm, 1968, pp. 473–496.

Bulman, O.M.B., *Treatise on Invertebrate Paleontology. Part V. Graptolithina with Sections on Enteropneusta and Pterobranchia*, University of Kansas, Lawrence, 1955.

Chaney, R., Evolutionary trends in the angiosperms, in *Genetics, Paleontology and Evolution*, Jepsen, G.L., Mayr, E., and Simpson, G.G., Eds., Princeton University Press, Princeton, NJ, 1949, pp. 190–201.

Charig, A.J., Systematics in biology: a fundamental comparison of some major schools of thought, in *Problems of Phylogenetic Reconstruction, Systematics Association Special Vol. 21*, Joysey, K.A. and Friday, A.E., Eds., Academic Press, London, 1982, pp. 363–440.

Cope, E.D., *The Vertebrata of the Tertiary Formations of the West. United States Geological Survey of the Territories*, Government Printing Office, Washington, DC., 1883.

Cope, E.D., Geology and paleontology, *Am. Nat.,* 21, 1014–1019, 1887.

Cope, E.D., On the phylogeny of the vertebrata, *Proc. Am. Philos. Soc.,* 30, 278–281, 1892.

Crowson, R.A., *Classification and Biology*. Aldine, Chicago, 1970.

De Queiroz, K., Systematics and the Darwinian revolution, *Philos. Sci.,* 55, 238–259, 1988.

De Queiroz, K. and Gauthier, J., Phylogeny as a central principle in taxonomy: phylogenetic definitions of taxon names, *Syst. Zool.,* 39, 307–322, 1990.

De Queiroz, K. and Gauthier, J., Phylogenetic taxonomy, *Annu. Rev. Ecol. Syst.,* 23, 449—480, 1992.

Dollo, L., Sur la phylogénie des dipneustes, *Bull. Soc. Belg. Géol., Paléontol. Hydrol.,* 9, 79–128, 1896.

Donovan, S.K. and Paul, C.R.C., Eds., *The Adequacy of the Fossil Record*. John Wiley & Sons, Chichester, UK, 1998.

Dwight Davis, D., Comparative anatomy and the evolution of vertebrates, in *Genetics, Paleontology and Evolution*, Jepsen, G.L., Mayr, E., and Simpson, G.G., Eds., Princeton University Press, Princeton, NJ, 1949, pp. 64–89.

Fisher, D.C., Phylogenetic analysis and its application in evolutionary paleobiology, in *Short Courses in Paleontology: Analytical Paleobiology*, Gilinsky, N.L. and Signor, P.W., Eds., Paleontological Society, Knoxville, TN, 1991, pp. 103–122.

Fisher, D.C., Stratigraphic parsimony, in *MacClade: Analysis of Phylogeny and Character Evolution*, Version 3, Maddison, W.P. and Maddison, D.R., Eds., Sinauer, Sunderland, MA, 1992, pp. 124–129.

Fisher, D.C., Stratocladistics: morphological and temporal patterns and their relation to phylogenetic process, in *Interpreting the Hierarchy of Nature: From Systematic Patterns to Evolutionary Process Theories*, Grande, L. and Rieppel, O., Eds., Academic Press, San Diego, CA, 1994, pp. 133–171.

Fisher, D.C., Foote, M., Fox, D.L. and Leighton, L.R., Stratigraphy in phylogeny reconstruction: comment on Smith (2000), *J. Paleontol.,* 76, 585–586, 2002.

Forey, P.L., *History of the Coelacanth Fishes*. Chapman & Hall, London, 1998.

Forey, P.L., Les fossiles et la systèmatique, *Biosystema,* 19, 1–28, 2001.

Forey, P.L. and Fortey, R.A., Fossils in the reconstruction of phylogeny, in *Palaeobiology II*, Briggs, D.E.G., and Crowther, P.R., Eds., Blackwell Science, Oxford, 2001, pp. 515–519.

Fortey, R.A., Trilobite systematics: the last 75 years, *J. Paleontol.*, 75, 1141–1151, 2001.

Fortey, R.A. and Jefferies, R.P.S., Fossils and phylogeny: a compromise approach, in *Problems of Phylogenetic Reconstruction*, Systematics Association Special Vol.. 21, Joysey, K.A. and Friday, A.E., Eds., Academic Press, London, 1982, pp. 197–234.

Gale, A.S., Smith, A.B., Monks, N.E.A., Young, J.A., Howard, A., Wray, D.S., and Huggett, J.M., Marine biodiversity through the late Cenomanian-Early Turonian: palaeoceanographic controls and sequence stratigraphic biases, *J. Geol. Soc.*, 157, 745–757, 2000.

Gardiner, B.G., Janvier, P., Patterson, C., Forey, P.L., Greenwood, P.H., Miles, R.S., and Jefferies, R.P.S., The salmon, the lungfish and the cow: a reply, *Nat. London*, 277, 175–176, 1979.

Gaudry, A., *Animaux Fossiles et Géologie de l'Attique d'aprés les Recherches Faites en 1855-56 et 1860 Sous les Auspices de l'Academie des Sciences*, Paris, 1862–1867.

Gauthier, J., Kluge, A.G., and Rowe, T., Amniote phylogeny and the importance of fossils, *Cladistics*, 4, 105–209, 1988.

Geary, D.H., Patterns of evolutionary tempo and mode in the radiation of *Melanopsis* (Gastropoda; Melanopsidae), *Paleobiology*, 16, 492–511, 1990.

Gingerich, P.D., Cranial anatomy and evolution of early Tertiary Plesiadapidae (Mammalia, Primates), *Univ. Mich. Pap. Paleontol.*, 15, 1–141, 1976.

Gingerich, P.D. and Simmons, E., Systematics, phylogeny, and evolution of early Eocene Adapidae (Mammalia, Primates) in North America, Univ. Mich. Contribs., 24, 241-267. 1977.

Gingerich, P.D., The stratophenetic approach to phylogeny reconstruction in vertebrate paleontology, in *Phylogenetic Analysis and Paleontology*, Cracraft, J. and Eldredge, N., Eds., Columbia University Press, New York, 1979, pp. 41–77.

Gould, S.J., Hatracks and theories, *Nat. Hist.*, 1987, 12–23, 1987.

Gregory, W.K., The orders of mammals, *Bull. Am. Mus. Nat. Hist.*, 27, 1–526, 1910.

Günther, A.E., Description of *Ceratodus*, a genus of ganoid fishes, recently discovered in the rivers of Queensland, *Philos. Trans. R. Soc. London*, 161, 511–571, 1871.

Haeckel, E., *Entwickelungseschicte der Organism*. Verlag von Geor Reimer, Berlin, 1866.

Haeckel, E., *Anthropogenie oder Entwickelungsgeschichte des Menschen*, W. Engelmann, Leipzig, Germany, 1874.

Haeckel, E., *Schöpfungs Geschichte: Gemeinverständliche Wissenschaftliche Vorträge über die Entwicklungslehre im Allgemeinen und Diejenige von Darwin, Goethe und Lamarck im Besonderen/von Ernst Haeckel*, 8th ed., G. Reimer, Berlin, 1889.

Halstead, L.B., The cladistic revolution: can it make the grade?, *Nat. London*, 276, 759, 1978.

Halstead, L.B., White, E.I., and Macintyre, G.T., L.B. Halstead and colleagues reply, *Nat. London*, 277, 176, 1979.

Hennig, W., *Grundzüge einer Theorie der Phylogenetischen Systematik*. Deutsche Zentralverlag, Berlin, 1950.

Hennig, W., Phylogenetic systematics, *Annu. Rev. Entomol.*, 10, 97–116, 1965.

Hennig, W., *Phylogenetic Systematics*, University of Illinois Press, Urbana, 1966.

Hilgendorf, F., Beiträge zur Kenntniß des Süßwasserkalkes von Steinheim. thesis, University of Tübingen, Germany, 1863.

Holmgren, N., Contribution to the question of the origin of the tetrapod limb, *Acta Zool. Stockholm*, 20, 89–124, 1939.

Huelsenbeck, J.P., When are fossils better than extant taxa in phylogenetic analysis?, *Syst. Zool.*, 40, 458–469, 1992.

Hull, D.L., *Science as a Process*. Chicago University Press, Chicago, 1988.

Huxley, T.H., Preliminary essay upon the systematic arrangement of the fishes of the Devonian epoch, *Mem. Geol. Surv. U.K. London*, Decade 10, 1–40, 1861.

Huxley, T.H., *Lectures on the Elements of Comparative Anatomy*. Churchill, London, 1864.

Huxley, T.H., On *Stagonolepis robertsoni*, and the evolution of the crocodilia, in *The Scientific Memoirs of Thomas Henry Huxley*, Foster, M. and Lankester, E.R., Eds., Macmillan, London, 1875, pp. 649–662.

Huxley, T.H., On the application of the laws of evolution to the arrangement of the vertebrata, and more particularly of the Mammalia, *Proc. Zool. Soc. London,* 43, 649–662, 1880.

Huxley, J.E., *The New Systematics.* Clarendon Press, Oxford, 1940.

Jarvik, E., On the structure of the snout of crossopterygians and lower gnathostomes in general, *Zool. Bidr. Upps.,* 21, 235–675, 1942.

Jarvik, E., *Basic Structure and Evolution of Vertebrates*, Academic Press, London, 1980.

Jepsen, G.L., Mayr, E., and Simpson, G.G., Eds., *Genetics, Paleontology and Evolution.* Princeton University Press, Princeton, NJ, 1949.

Kuhn-Schnyder, E., La position des nothosauridés dans le système des reptiles, *Colloq. Int. Cent. Nat. Rech. Sci. Paris,* 104, 135–144, 1962.

Lapworth, C., On an improved classification of the Rhabdophora. Part II, *Geol. Mag.,* 10, 555–560, 1873.

Lazarus, D.B., Tempo and mode of morphologic evolution near the origin of the radiolarian lineage *Pterocranium prismatum*, *Paleobiology,* 12, 175–189, 1986.

Lazarus, D.B. and Prothero, D.R., The role of stratigraphic and morphologic data in phylogeny, *J. Paleontol.,* 58, 163–172, 1984.

Lin, Y.-S., Poh, Y.-P., and Tzeng, C.-S., A phylogeny of freshwater eels inferred from mitochondrial genes, *Mol. Phylogenet. Evol.,* 20, 252–261, 2001.

Lurie, E., *Louis Agassiz: A Life in Science.* University of Chicago Press, Chicago, 1960.

Marshall, C.R., Determining stratigraphic ranges, in *The Adequacy of the Fossil Record*, Donovan, S.K. and Paul, C.R.C., Eds., John Wiley & Sons, Chichester, UK, 1998, pp. 23–53.

Matthew, W.D., Additional observations on the Creodonta, *Bull. Am. Mus. Nat. Hist.,* 14, 1–38, 1901.

Matthew, W.D., Early days of fossil hunting in the high plains, *Nat. Hist.,* 26, 449–454, 1926.

Mayr, E., *Principles of Systematic Zoology*, McGraw-Hill, New York, 1969.

Mayr, E., Cladistic analysis or cladistic classification?, *Z. Zool. Syst. Evol.forch.,* 12, 94–128, 1974.

Moy-Thomas, J.A. and Miles, R.S., *Palaeozoic Fishes,* 2nd ed., Chapman & Hall, London, 1971.

Murray, A.M. and Wilson, M.V.H., Contributions of fossils to the phylogenetic relationships of the percopsiform fishes (Teleostei: Paracanthopterygii): order restored, in *Mesozoic Fishes. 2. Systematics and the Fossil Record*, Arratia, G. and Schultze, H.-P., Eds., Freidrich Pfeil, Munich, 1999, pp. 397–411.

Nash, A.R., Fisher, W.K., and Thompson, E.O.P., Haemoglobins of the shark, *Heterodontus portusjacksoni*. II. Amino acid sequence of the α-chain, *Aust. J. Biol. Sci.,* 29, 73–97, 1976.

Nelson, G.J., Origin and diversification of teleostean fishes, *Ann. NY Acad. Sci.,* 167, 18–30, 1969.

Nelson, G.J., Outline of a theory of comparative biology, *Syst. Zool.,* 19, 373–384, 1970.

Nelson, G.J., "Cladism" as a philosophy of classification, *Syst. Zool.,* 20, 373–376, 1971.

Nelson, G., Ancient perspectives and influence in the theoretical systematics of a bold fisherman, *Linnean*, Special issue No. 2, 9–23, 2000.

Newell, N.D., The nature of the fossil record, *Proc. Am. Philos. Soc.,* 103, 264–85, 1959.

Norell, M.A., Stratigraphic tests of cladistic hypotheses, in *Palaeobiology II*, Briggs, D.E. and Crowther, P.R., Eds., Blackwell Science, Oxford, 2001, pp. 519–522.

Ørvig, T., Y a-t-il une relation directe entre les arthrodires ptyctodontides et les holocéphales, *Colloq. Int. Cent. Nat. Rech. Sci. Paris,* 104, 87–101, 1962.

Patterson, B., Rates of evolution in taeniodonts, in *Genetics, Paleontology and Evolution*, Jepsen, G.L., Mayr, E., and Simpson, G.G., Eds., Princeton University Press, Princeton, NJ, 1949, pp. 243–277.

Patterson, C., The contribution of paleontology to teleostean phylogeny, in *Major Patterns in Vertebrate Evolution*, Hecht, M.K., Goody, P.C., and Hecht, B.M., Eds., Plenum Press, New York, 1977, pp. 579–643.

Patterson, C., Origin of Tetrapods: historical introduction to the problem, in *The Terrestrial Environment and the Origin of Land Vertebrates*, Panchen, A.L., Ed., Academic Press, London, 1980, pp. 159–175.

Patterson, C., Significance of fossils in determining evolutionary relationships, *Annu. Rev. Ecol. Syst.*, 12, 195–223, 1981.

Patterson, C., Adventures in the fish trade. Annual Address to the Systematics Association, London, December 6, 1995.

Pearson, P.N., Week 4: Rejoinder, *Nature on line debate Is the fossil record adequate?*, www.nature.com/debates/fossil/fossil_contents.html. Accessed 10/98.

Platnick, N.I., Philosophy and the transformation of cladistics, *Syst. Zool.*, 28, 537–546, 1979.

Pollard, H.B., On the anatomy and phylogenetic position of *Polypterus*, *Anat. Anz.*, 6, 338–344, 1891.

Polly, P.D., Hyaenodontidae (Creodonta, Mammalia) and the Position of Systematics in Evolutionary Biology, Ph.D. thesis, University of Califormia, Berkeley, 1993.

Prothero, D.R. and Lazarus, D.B., Planktonic microfossils and the recognition of ancestors, *Syst. Zool.*, 29, 119–129, 1980.

Reif, W.E., Hilgendorf's (1863) dissertation on Steinheim planorbids (Gasteropoda; Miocene): the development of a phylogenetic research program for palaeontology, *Paläontol. Z.*, 57, 7–20, 1983.

Rieppel, O., Louis Agassiz (1807-1873) and the reality of natural groups, *Biol. Philos.*, 3, 27–47, 1988a.

Rieppel, O., *Fundamentals of Comparative Biology*. Birkhäuser Verlag, Basel, Switzerland, 1988b.

Romer, A.S., Time series and trends in animal evolution, in *Genetics, Paleontology and Evolution*, Jepsen, G.L., Mayr, E., and Simpson, G.G., Eds., Princeton University Press, Princeton, NJ, 1949, pp. 103–120.

Romer, A.S., *Vertebrate Paleontology*, 3rd ed., University of Chicago Press, Chicago, 1966.

Romer, A.S., Vertebrate paleontology and zoology, *Biologist*, 51, 49–53, 1969.

Rosen, D.E., Forey, P.L., Gardiner, B.G., and Patterson, C., Lungfishes, tetrapods, paleontology, and plesiomorphy, *Bull. Am. Mus. Nat. Hist.*, 167, 159–276, 1981.

Rowe, A.W., An analysis of the genus *Micraster* as determined by rigid zonal collection from the zone of *Rhynchonella cuvieri* to that of *Micraster coranguinum*, *Q. J. Geol. Soc. London*, 55, 494–546, 1899.

Saether, O.A., Underlying synapomorphies and anagenetic analysis, *Zool. Scr.*, 8, 305–312, 1979.

Säve-Söderbergh, G., Preliminary note on Devonian stegocephalians from East Greenland, *Meddelser Grønl.*, 94, 1–107, 1932.

Säve-Söderbergh, G., Some points of view concerning the evolution of the vertebrates and classification of this group, *Ark. Zool.*, 26, 1–20, 1934.

Schaeffer, B., Hecht, B.M., and Eldredge, N., Phylogeny and paleontology, *Evol. Biol.*, 6, 31–46, 1972.

Simpson, G.G., *Tempo and Mode in Evolution*, Columbia University Press, New York, 1944.

Simpson, G.G., The principles of classification and a classification of mammals, *Bull. Am. Mus. Nat. Hist.*, 85, 1–350, 1945.

Simpson, G.G., *The Major Features of Evolution*. Columbia University Press, New York, 1953.

Simpson, G.G., Mesozoic mammals and the polyphyletic origin of mammals, *Evolution*, 13, 405–414, 1959.

Simpson, G.G., *Principles of Animal Taxonomy*, Columbia University Press, New York, 1961.

Smith, A. and Littlewood, D.T.L., Paleontological data and molecular phylogenetic analysis, *Paleobiology*, 20, 259–273, 1994.

Smith, A.B., What does palaeontology contribute to systematics in a molecular world, *Mol. Phylogeny Evol.*, 9, 437–447, 1998.

Smith, A.B., Stratigraphy in phylogeny reconstruction, *J. Paleontol.*, 74, 763–766, 2000a.

Smith, A.B., Large-scale heterogeneity of the fossil record: implications for Phanerozoic biodiversity studies, *Philos. Trans. R. Soc. London B*, 356, 351–367, 2000b.

Stensiö, E.A., Les cyclostomes fossiles ou ostracoderms, in *Traité de Zoologié*, Grassé, P.-P., Ed., Masson et Cie, Paris, 1958, pp. 1–229.

Valentine, J.W., Coelomate superphyla, *Syst. Zool.*, 22, 97–102, 1973.

Valentine, J.W., Adaptive strategy and the origin of grades and ground plans, *Am. Zool.*, 15, 391–404, 1975.

Wagner, P.J., A likelihood approach for evaluating estimates of phylogenetic relationships among fossil taxa, *Paleobiology*, 24, 430–449, 1998a.

Wagner, P.J., Week 2: why phylogenetic hypotheses need testing by stratigraphic data. *Nature on line debate. Is the fossil record adequate?* www.nature.com/debates/fossil/fossil_contents.html. Accessed 10/98.

Wagner, P.J., Exhaustion of cladistic character states among fossil taxa, *Evolution*, 54, 365–386, 2000.

Watson, D.M.S., The larger Coal Measure amphibia, *Mem. Proc. Manch. Lit. Philos. Soc.*, 57, 1–14, 1912.

Watson, D.M.S., The evidence afforded by fossil vertebrates on the nature of evolution, in *Genetics, Paleontology and Evolution*, Jepsen, G.L., Mayr, E., and Simpson, G.G., Eds., Princeton University Press, Princeton, NJ, 1949, pp. 45–63.

Westoll, T.S., On the evolution of the Dipnoi, in *Genetics, Paleontology and Evolution*, Jepsen, G.L., Mayr, E., and Simpson, G.G., Eds., Princeton University Press, Princeton, NJ, 1949, pp. 121–184.

Williams, D.M. and Ebach, M.C., The reform of palaeontology and the rise of biogeography: 25 years after "Ontogeny, Phylogeny, Paleontology and the Biogenetic law" (Nelson, 1978), *J. Biogeogr.*, in press.

Wilson, M.V.H., Importance for phylogeny of single and multiple stem-group fossil species with examples from freshwater fishes, *Syst. Biol.*, 41,462–470, 1992.

Wood H.E., II, Evolutionary rates and trends in rhinoceroces, in *Genetics, Paleontology and Evolution*, Jepsen, G.L., Mayr, E., and Simpson, G.G., Eds., Princeton University Press, Princeton, NJ, 1949, pp. 185–189.

Woodward, A.S. and Sherborn, C.D., *A Synopsis of the Vertebrate Fossils of the English Chalk*, Dulau, London, 1890.

Wright, A., Brachiopod radiation, in *The Origin of Major Invertebrate Groups*, House, M.R., Ed., Academic Press, London, 1979, pp. 323–358.

Chapter 8

Parsimony and Computers

A.W.F. Edwards

Abstract

The introduction and use of parsimony methods in phylogenetic estimation is described and discussed against the background of Ockham's razor and *extremum* principles in science generally. It is argued that the method of maximum likelihood applied to realistic models of evolutionary divergence is the optimum procedure, to which parsimony methods are often an acceptable approximation.

Historical Introduction

Parsimony or *maximum parsimony* are terms now in regular use to describe methods of phylogenetic estimation that involve seeking the tree that minimizes some function of evolutionary change such as the total number of mutations or the total length of the branches in a continuous character space (see, for example, the recent books by Weir [1996], Page and Holmes [1998] and Durbin et al. [1998]). The adoption of the word was perhaps unfortunate because it already had a well-established use in mathematics to describe the classical notion (Pappus, fourth century) that geometrical proofs should rely only on ruler-and-compass constructions, and that in general the minimum number of postulates should be used in mathematical proofs. It was also unfortunate because it had a well-established use in philosophy as "multiplicity ought not to be posited unnecessarily" (*"pluralitas non est ponenda sine necessitate"* [William of Ockham, fourteenth century]).

However, these were not the considerations that led Camin and Sokal (1965) to coin the phrase "principle of evolutionary parsimony" when they noticed that on their simulated data "those trees which most closely resembled the true cladistics invariably required for their construction the least number of postulated evolutionary steps for the character studied," since their justification was "the assumption that nature is indeed parsimonious." In a *Scientific American* article Sokal (1966:115) explained what they meant by this: "that nature is fundamentally parsimonious, so that the diversity in character states within a given group was achieved at or close to the minimum number of evolutionary steps." This teleological justification, embodying the idea that nature is trying to achieve something, quickly attracted criticism as being unacceptable (Edwards, 1966; Cavalli-Sforza and Edwards, 1967).

In fact there was already a publication and justification of maximum parsimony in the systematics literature, for at the Systematics Association meeting, Phenetic and Phylogenetic Classification, in Liverpool in 1964 (Heywood and McNeill, 1964:Figure 8.1) Cavalli-Sforza and I had referred to the use of our method of minimum evolution and had commented, "It is probable that this method gives a tree which is approximately the same as the projection

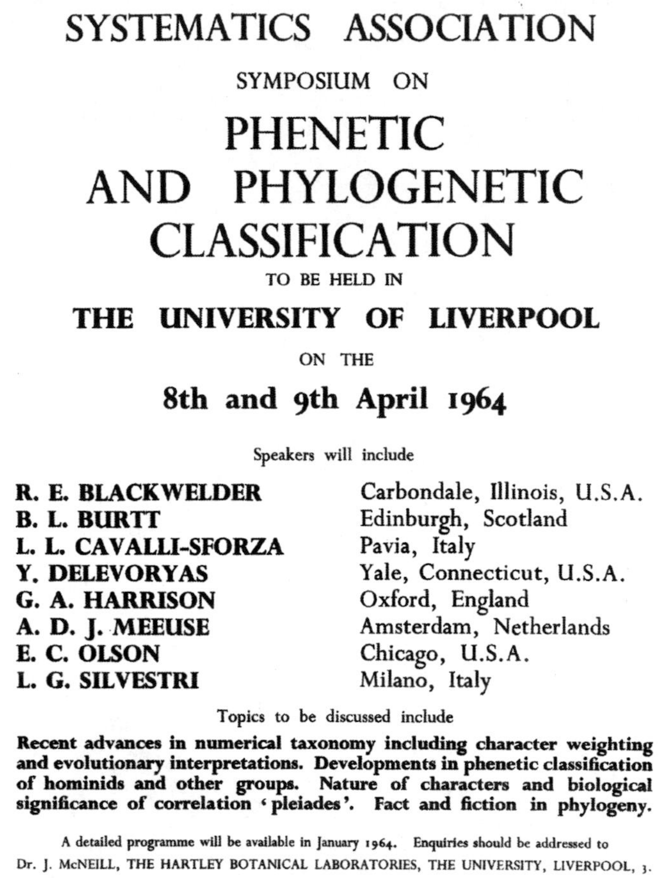

SYSTEMATICS   ASSOCIATION

SYMPOSIUM   ON

PHENETIC
AND   PHYLOGENETIC
CLASSIFICATION

TO BE HELD IN

THE   UNIVERSITY   OF   LIVERPOOL

ON   THE

8th and 9th April 1964

Speakers will include

R. E. BLACKWELDER	Carbondale, Illinois, U.S.A.
B. L. BURTT	Edinburgh, Scotland
L. L. CAVALLI-SFORZA	Pavia, Italy
Y. DELEVORYAS	Yale, Connecticut, U.S.A.
G. A. HARRISON	Oxford, England
A. D. J. MEEUSE	Amsterdam, Netherlands
E. C. OLSON	Chicago, U.S.A.
L. G. SILVESTRI	Milano, Italy

Topics to be discussed include

**Recent advances in numerical taxonomy including character weighting
and evolutionary interpretations. Developments in phenetic classification
of hominids and other groups. Nature of characters and biological
significance of correlation 'pleiades'. Fact and fiction in phylogeny.**

A detailed programme will be available in January 1964.   Enquiries should be addressed to
Dr. J. McNEILL, THE HARTLEY BOTANICAL LABORATORIES, THE UNIVERSITY, LIVERPOOL, 3.

FIGURE 8.1   Poster advertising the Systmatics Association meeting, Phenetic and Phylogenetic Classification, in Liverpool in 1964. (Heywood, V.H. and McNeill, J., Eds. With permission.)

of the maximum-likelihood tree onto the 'now' character space" (Edwards and Cavalli-
Sforza, 1964:75). Maximum-likelihood trees are statistical estimates of trees based on the
criterion of likelihood applied to a specific probability model, and likelihood itself is a well-
known concept in statistics (see, for example, Edwards, 1972, 1992).

Our method of minimum evolution, originally called a principle until we realized that
the word was misleading, was presented in 1963 (Edwards and Cavalli-Sforza 1963a:
104–105), and it is worth recalling its precise formulation:

The Reconstruction of Evolution

Darwin's theory of evolution invites us to believe that closely similar species are
closely related. In cases in which there is no fossil record, or other means of
following evolutionary history, this concept provides the only means of drawing
inferences about the course of evolution, and it is important to see whether it
can be defined sufficiently precisely to be used for estimating the most likely form
of an evolutionary tree solely from data on the similarities and differences
between species. A detailed consideration of this problem leads to the conclusion
that the following principle may be employed:

The Principle of Minimum Evolution

The most plausible estimate of the evolutionary tree is that which invokes the minimum nett amount of evolution.

It is not hindsight to note the careful wording. We wanted to draw an inference by defining *sufficiently precisely* what was meant by similar species being closely related so that we could *estimate* the *most likely* form of an evolutionary tree. The precision was important because we needed an algorithm as the basis for our computer programs, while *estimate* and *most likely* are standard statistical parlance. The computer revolution, in full swing in the early 1960s, had two distinct aspects (Edwards, 1966; Sokal, 1966). Most obviously, it facilitated far larger calculations of all kinds, and in phylogenetic reconstruction this meant calculations that embodied searches through multitudinous tree forms. Less obviously, but just as importantly, the discipline of writing a computer program imposed an absolute requirement for clear thinking and thus for the consideration of basic principles.

It happened that Cavalli-Sforza and I were computer-literate geneticists, but exactly the same phenomenon had been appreciated by the computer-literate authors Sokal and Sneath (1963) in their contemporary book *Principles of Numerical Taxonomy,* to which it is a pleasure to pay tribute. Indeed, the session at the 1964 meeting in which Cavalli-Sforza and I referred to our method of minimum evolution was entitled "Numerical Taxonomy" and introduced by Sneath with some wise words about phylogenies (Sneath, 1964). Sneath (1995) has recently written a historical review.

Ever since Darwin's time phylogeneticists had been implicitly using minimum-evolution or parsimony arguments, but their explicit use had to wait until the requirements of computers compelled algorithmic thinking. When we presented some results of our phylogenetic analysis at the XIth International Congress of Genetics at The Hague in 1963 (Cavalli-Sforza and Edwards, 1964) Professor Silvestri of Milan queried the method of minimum evolution and in reply I said, "It seems to us that our principle has been implicitly used in all evolutionary studies, including cases in which fossil evidence has been available, and all we have done is to admit it explicitly and develop it as the basis for statistical estimation."

Silvestri was also at the Systematics Association meeting, and in his paper with Hill he enlarged on his doubts:

In more general terms, going beyond the limits of microbiology and also those of biology itself, it does not seem possible for us to determine the stages through which a particular system has evolved simply by studying the final state of that system, if historical documentation is not available, just as an analysis of the position of chess pieces at a particular stage of the game does not allow us to reconstruct the moves made to reach that stage. At the most, all that is possible is to determine,.for example, the most economical sequence of moves, or any other sequence which conforms with other principles. In this case, however, the hypothetical reconstruction can only be accepted if the postulated principle, e.g., of economy, can be verified. This is precisely the objection that could be raised against the presentation of the problem by Edwards and Cavalli-Sforza (1963[a]). Using data relative to the present horizon, they reconstruct a possible evolution of races of *Homo sapiens* according to the principle of minimum evolution, which is postulated but not verified. (Silvestri and Hill, 1964:91)

I do not know how good chess enthusiasts are at inferring recent moves from a glance at a mid-game position, but the biological problem was precisely the one that Cavalli-Sforza and I claimed to be able to solve by statistical methods, contrary to these authors' opinion. Their strictures (with which we were already familiar through our contact with Silvestri in Milan) were valuable, however, in forcing us to think about the basis of our methods and to realize that our principle of economy was justified by an appeal to probability models and maximum likelihood.

In order to implement a computer program "which invokes the minimum nett amount of evolution" it is necessary to define *amount of evolution,* and to us geneticists that presented no problem. We happened to be primarily studying human populations characterized by their gene frequencies at the common blood-group loci, and an amount of evolution was obviously a distance measure in some suitable space of gene frequencies, the details of which need not now concern us but which were dictated by the population geneticists' standard model of gene-frequency change through random genetic drift.

With discrete data, to which our principle was also designed to apply, amount of evolution was simply measured by the number of changes, whether of characters or "evolutionary steps" (as in Camin and Sokal, 1965) or mutations in the DNA sequences (as with modern molecular phylogenetics). In a recent paper (Edwards, 1996) I examined, with all the care that I could summon, the literature of the 1960s to see if I could find other explicit statements of the method of minimum evolution, and the earliest rival I unearthed was Eck and Dayhoff (1966), who without offering a justification, sought "the [tree] topology that has the minimum number of mutations [amino acid residue changes to proteins]" (Sneath, 1995 has a reference to the same *Atlas of Protein Sequence and Structure* of the previous year, which I have been unable to consult).[1] Further references can be found in Edwards (1996), to which I have nothing to add except that Dr. E. Zuckerkandl, in commenting on that paper, assures me that he had the principle in mind in late 1964 and regrets not having then stated it explicitly.

It is hardly surprising that geneticists, particularly population geneticists interested in Darwinian evolution, applied their newly won computing skills to the problem of phylogenetic reconstruction. Cavalli-Sforza mentioned in 1991 that he had "started thinking about … the reconstruction of where human populations originated and the paths by which they spread throughout the world" more than 40 years previously while he was in R.A. Fisher's department of genetics at Cambridge. Fisher himself (1958) remarked in a BBC broadcast, the recording of which unfortunately no longer exists, "The reconstruction of racial history, though a distant aim, is not an unreasonable one." We were all thoroughly familiar with the work of A.E. Mourant, which had culminated in his book, *The Distribution of the Human Blood Groups* (1954).

On reflection, if one asks what kind of training and background would be valuable if one wanted to initiate research on how to construct phylogenies based on data from contemporary populations, races, species, etc., whose variability is accepted as being due to splitting, genetic isolation and evolutionary change, the answer would have to be evolutionary biology and statistical inference. Cavalli-Sforza and I had both been in Fisher's department, and Fisher was the author of both *The Genetical Theory of Natural Selection* (1930) and *Statistical Methods for Research Workers* (1925), two of the most famous and influential scientific books of the twentieth century.

---

[1]   According to the Margaret Oakley Dayhoff Memorial website (http://www.dayhoff.cc/modbooks.html), the 1965 atlas is listed as "Atlas of Protein Sequence and Structure, 1965 (Black Cover) by Dayhoff, Eck, Chang, and Sochard, 93 printed one-sided pages."

The Influence of Traditional Procedures

But what about traditional systematics and taxonomy? The 1964 Systematics Association meeting was my introduction to these fields, and much of the discussion at that meeting was about philosophical issues. There were many amusing exchanges. At one point I am recorded as having replied to Sneath, "What we are searching around for is for some taxonomists to say why they are doing taxonomy — as opposed to phylogeny for instance. Mostly they are not engaged in pure information retrieval taxonomy but are undertaking applied taxonomy — but for what purpose? Until this purpose is declared then the required answers cannot be provided."

It all seems less of a mystery nearly 40 years on. Of course a taxonomic classification is, or should be, an efficient information-retrieval system, but it will also possess predictive value through the simple circumstance that characters are correlated (or else no tree structure would have emerged). The predictive value arises because the correlations are not confined to the data actually used. As to why these correlations exist, an underlying tree structure is the obvious answer, which will usually be justified by evolutionary considerations. In this connection, I should like to draw attention to the work of Cavalli-Sforza and Piazza (1975) on what they called "treeness." Although it has not been of much practical importance, their treeness test is an interesting application of the idea that a distance matrix contains within it evidence about any underlying tree structure that is amenable to statistical testing. An accessible account is contained in Cavalli-Sforza et al. (1994).

These remarks on taxonomy bring me to the vexing question of the influence of taxonomic procedures on modern phylogenetic analysis. Here I have little to add to Edwards (1996), where I reported on my efforts to trace the origins of various methods, notably parsimony. Farris and Kluge (1997) disputed my conclusions, preferring to emphasize the contributions of W. Hennig and W.H. Wagner (for references see Edwards, 1996). It is quite difficult to see what the argument is about. I have already quoted my 1964 remark about minimum evolution having been implicitly used in all evolutionary studies "and all we have done is to admit it explicitly and develop it as a basis for statistical estimation." Wagner (1984:115) was of the same opinion with respect to his own groundplan-divergence (GPD) method: "Students of the history of systematic biology will recognize that there is nothing wholly new or original in the elements of the GPD method. Most of its components are based on time-honoured concepts that are as old or nearly as old as evolutionary systematics itself." Perceiving in retrospect that algorithmic methods such as those attributed to Hennig and to Wagner make use of some form of parsimony argument is interesting, but it is not the same thing as starting with an explicit statement of a quantity to be minimized and then deducing the consequences in the manner of Edwards and Cavalli-Sforza (1963a), Camin and Sokal (1965) and Eck and Dayhoff (1966), leading eventually to the perfecting of computer algorithms.

The only reference to Hennig in the 1964 Systematics Association volume was by Sneath (1964:45), who referred to him as having pointed out "that the relation by ancestry of adults and larvae is necessarily the same, while their phenetic relationships need not be so." In the dominant field of modern molecular phylogeny, the three recent books I mentioned at the outset (Weir [1996], Page and Holmes [1998], and Durbin et. al. [1998]) contain hundreds and hundreds of references and indexed names, but not Hennig's.

In this account I have tended to use *parsimony* and *minimum evolution* as synonyms, but it should be mentioned that in recent writing (for example, Nei et al., 1998) there has been a tendency to polarize the terms so they apply to the discrete and the continuous, respectively. There seems to be no good reason for this, though it is a harmless enough convention. A further use of *parsimony* which has recently been noted by Steel and Penny (2000) refers to the preference for assuming as little as possible in the way of an underlying evolutionary model.

Maximum Likelihood

The method of maximum likelihood is the standard estimation procedure in statistical inference. Whether one looks at the inference problem from the point of view of classical repeated-sampling theory or Bayesian theory or straightforward likelihood theory, maximizing the likelihood emerges as the preferred procedure. There really is no dispute about this in regular estimation problems, and phylogenetic inference does seem to be unexceptional from a statistical point of view, even though it took a little while for the initial difficulties in the application of maximum likelihood to be sorted out. This was mainly done by Felsenstein (1968) and Thompson (1974) in their Ph.D. dissertations and subsequent publications.

It became clear quite early that there was no question of minimum evolution providing the maximum-likelihood solution exactly in the case of continuous characters (Cavalli-Sforza and Edwards, 1967; Edwards, 1969), and since there is no difference in principle between the continuous and discrete cases, it came as no surprise that a similar situation was found for discrete models (Felsenstein, 1973). It was also clear on intuitive grounds that the two methods (and, for that matter, almost any sensible algorithmic method that could be devised) were nevertheless going to give quite similar phylogenetic trees on any data with high treeness. The important point is that, as already mentioned in 1964, minimum-evolution algorithms are justified by likelihood considerations. Such considerations may also explain the appeal of Ockham's razor, as Sober (1988) concluded in his book *Reconstructing the Past: Parsimony, Evolution and Inference* and later in "Let's Razor Ockham's Razor" (1990). Edwards 1996 contains a full discussion. For a detailed discussion on the relations between minimum evolution and maximum likelihood the papers by Goldman (1990) and Steel and Penny (2000) should be consulted.

Sometimes attempts have been made to rank methods according to the probability of them giving the right result from simulated data. This is an old canard in statistical inference, and in connection with phylogenetic trees I tried to shoot it down in a letter to *Science* (Edwards, 1995). The approach confuses before-trial evaluation methods that are usually right and after-trial evaluation methods that are credible on each occasion of use. I can do no better than quote the last paragraph of my letter:

> The message from statistical theory is simple: You cannot simulate evolutionary trees without a probability model, and if you possess a probability model you should be performing efficient after-trial evaluation using the method of maximum likelihood. Simulation is valuable in assessing the robustness of a model, but not the suitability of an estimation procedure: modern statistical theory has solved that problem already. (Edwards, 1995)

However, the point has not yet been taken. For example Nei et al. (1998) used computer simulation to compare various methods, one of which, parsimony, they still claim has as its "theoretical basis" William of Ockham's "philosophical idea."

Extremum Principles in Science

Extremum principles, minimizing this or maximizing that, exert a peculiar fascination in science, and biologists have been eager to emulate the success of their physics colleagues in harnessing them. Alas, the situations are quite different. The classic physics example is from optics, where Fermat in the seventeenth century established that when a ray of light passes from a point in one medium through a plane interface to a point in another medium with a different refractive index, its route is the one that minimizes the time it takes for the light to travel. That is, on various assumptions about the speed of light in the different media, he showed that this solution corresponds to Snel's sine law of refraction. Modern books on optics find it necessary to frame Fermat's principle rather differently, not using the notion of a minimum at all, but the point for us is that in its original form it solves at least some optical path problems by minimization. The mathematics works out just right, but that is not to agree with Fermat that "nature is economical."

In the eighteenth and nineteenth centuries mathematical physics unearthed a whole range of *extremum* principles of great practical value because they provided a unifying point of view that gave results that are mathematically identical to the older approaches. Ideas such as potential energy and its minimization led to the notions of potentials and fields of force in gravitational and electromagnetic theory. The great eighteenth century developments in calculus not only provided the necessary mathematical tools but also encouraged thinking in terms of maxima and minima. No one supposed any longer that nature was economical, or quoted Ockham's razor. Rather, there was an implicit appeal to the classical concept of parsimony, that the simplest mathematical approach to any problem should be preferred, and this often involved an *extremum* principle. Ockham's razor cut away the unnecessary mathematics but offered no explanation of nature. The success of mathematical physics has nothing to teach phylogeneticists trying to infer an evolutionary tree, though it did unfortunately beguile evolutionary biologists into thinking they could emulate the physicists by constructing theories of fitness maximization — the survival of the fittest — so that they could use the metaphors of potential theory without troubling themselves with the mathematics. The whole saga of "adaptive surfaces" (or "topographies" or "landscapes") is a warning against such analogical seduction (Edwards, 1994).

A quite different kind of *extremum* principle grew up more or less simultaneously in statistics in the eighteenth and nineteenth centuries, the most famous example being the method of least squares originating in the work of Legendre and Gauss. Far from being a mathematically economical way of solving some problem as is Fermat's principle, it is a method for selecting among probability models the one that most closely fits some observed data (in Gauss's case, the observations of cometary positions). There is an element of arbitrariness in the choice of the criterion to be minimized (here the sum of the squares of the residual errors unexplained by the model), but subsequent research in statistical inference has produced plenty of justifications, and Fisher's method of maximum likelihood is one of them.

These outstandingly successful developments involving *extremum* principles contributed to the idea that one can elevate into a principle the minimization of some quantity or other — in our case a quantity of evolution — in order to infer an evolutionary tree. But the situations are quite different, and an amount of evolution is analogous neither to the potentials of physics nor the errors of statistics.

As we have seen, the true justification of any method of minimum evolution lies in an appeal to a probability model and the primitive notion of likelihood, and to that extent is closer to the method of least squares than to the methods of mathematical physics. A matter of minor historical interest is that the first attempt at a statistical solution of the phylogeny problem did in fact invoke least-squares:

> An additive model may be appropriate, in which the lengths of the segments are so estimated that the observed distance between two points is closely approximated by the expected distance along the tree, formed by the sum of the connecting segments. Least squares provides an elegant estimation procedure, and we are currently engaged in applying this approach to taxonomic data. Our results will be published elsewhere. (Edwards and Cavalli-Sforza, 1963b)

Epilogue

It is exactly 40 years since I joined Professor L.L. Cavalli-Sforza at the University of Pavia to work in his program of statistical human genetics. He introduced me to the phylogenetic trees problem and made the above suggestion about using least-squares on distance data. By the time I left for the United States in September 1964 we had working computer programs for least-squares and for minimum-evolution on distance data, and one for maximum-likelihood estimation that was at the prototype stage, handling only four populations. We were fully aware of contemporary work both in numerical taxonomy and in cluster analysis (through personal contact and from studying Sokal and Sneath, 1963). These programs were, of course, primitive compared to those available today. Personal computers did not exist, and suitable mainframe machines were often not available locally. For my later work in Pavia I had to travel to Milan to use an IBM machine, and even when at the University of Aberdeen from 1965 onwards I had initially to travel to WHO in Geneva to find adequate facilities. Nor could the primitive programs be easily made available to other workers, first because to do so required either the repunching of the program punch cards by the recipient or the mailing of heavy parcels of cards and also because each computer had its own idiosyncrasies that the program had to meet. Nevertheless, the programs, especially EVOMIN, our minimum-evolution program, did travel and were used by others. It was eventually written up as a student project by Thompson (1973).

My overwhelming impression of those years is that our small group, assisted by colleagues such as Professor L. Sylvestri of Milan and Dr. P.A.H. Sneath of London, argued over most of the methodological issues that have since occupied the ensuing four decades. We quickly came to the conclusion that although any reasonable method of cluster analysis would come up with a reasonable tree (and inventing reasonable methods was not difficult), a proper scientific approach required a probability model and maximum-likelihood estimation. Anything less, including the method of minimum evolution, could only be justified as an approximation. I can only repeat once more what we reported at the 1964 Systematics Association meeting: "It is probable that this method gives a tree which is approximately the same as the projection of the maximum-likelihood tree onto the 'now' character space."

At the 1987 Systematics Association meeting Felsenstein (1988:113) remarked that our 1964 paper, "the first to propose a statistical approach to inferring phylogenies," though "extremely influential," was "little known and even less read." He concluded his paper, which is a fine summary of the history, by saying,

> With a bit of luck and much hard work, we will find ourselves in an era in which there is a much more fruitful interaction between systematists and population geneticists. Ancient prejudices of each group against the other will have to be modified. We will need, on the way, to set aside the hypothetico-deductive model of phylogenetic inference, and adopt a statistical one. Whether we can treat statistical models in phylogenetic inference with the appropriate mixture of reverence and scepticism remains to be seen.' (Felsenstein, 1988:124)

This paper is intended as a contribution to the realization of Felsenstein's hopes.

References

Camin, J.H. and Sokal, R.R., A method for deducing branching sequences in phylogeny, *Evolution*, 19, 311–326, 1965 (reprinted in Duncan and Stuessy, 1985).

Cavalli-Sforza, L.L., Genes, peoples and languages, *Sci. Am.*, 265, 104–110, 1991.

Cavalli-Sforza, L.L. and Edwards, A.W.F., Analysis of human evolution, in *Genetics Today, Proceedings of the 11th International Congress of Genetics*, 1963, 3, 923–933, 1964.

Cavalli-Sforza, L.L. and Edwards, A.W.F., Phylogenetic analysis: models and estimation procedures, *Evolution*, 21, 550–570; *Am. J. Hum. Genet.* (Supplement), 19, 233–257, 1967.

Cavalli-Sforza, L.L., Menozzi, P., and Piazza, A., *The History and Geography of Human Genes*, Princeton University Press, Princeton, NJ, 1994 (abridged paperback ed. 1996).

Cavalli-Sforza, L.L. and Piazza, A., Analysis of evolution: evolutionary rates, independence and treeness, *Theoret. Popul. Biol.*, 8, 127–165, 1975.

Duncan, T. and Stuessy, T.F., Eds., *Cladistic Theory and Methodology*, Van Nostrand Reinhold, New York, 1985.

Durbin, R., Eddy, S., Krogh, A., and Mitchison, G., *Biological Sequence Analysis: Probalistic Models of Proteins and Nucleic Acids*, Cambridge University Press, Cambridge, 1998.

Eck, R.V. and Dayhoff, M.O., *Atlas of Protein Sequence and Structure*, National Biomedical Research Foundation, Silver Spring, MD, 1966.

Edwards, A.W.F., Studying human evolution by computer, *New Sci.*, 30, 438–440, 1966.

Edwards, A.W.F., Genetic taxonomy, in *Computer Applications in Genetics*, Morton, N.E., Ed., University of Hawaii Press, Honolulu, 1969, pp. 140–142.

Edwards, A.W.F., *Likelihood*, Cambridge University Press, Cambridge, 1972.

Edwards, A.W.F., *Likelihood*, expanded edition, Johns Hopkins University Press, Baltimore, 1992.

Edwards, A.W.F., The fundamental theorem of natural selection, *Biol. Rev.*, 69, 443–474, 1994.

Edwards, A.W.F., Assessing molecular phylogenies, *Science*, 267, 253, 1995.

Edwards, A.W.F., The origin and early development of the method of minimum evolution for the reconstruction of evolutionary trees, *Syst. Biol.*, 45, 79–91, 1996.

Edwards, A.W.F. and Cavalli-Sforza, L.L., The reconstruction of evolution: abstract, *Heredity*, 18, 553; *Ann. Hum. Genet.*, 27, 104–105, 1963a.

Edwards, A.W.F. and Cavalli-Sforza, L.L., A method for cluster analysis, Preprints 5th International Biometrics Conference, Cambridge, 1963b.

Edwards, A.W.F. and Cavalli-Sforza, L.L., Reconstruction of evolutionary trees, in *Phenetic and Phylogenetic Classification*, Publication 6, Heywood, V.H. and McNeil, J., Eds., Systematics Association, London, 1964, pp. 67–76 (reprinted in Duncan and Stuessy, 1985).

Farris, J.S. and Kluge, A.G., Parsimony and history, *Syst. Biol.*, 46, 215–218, 1997.

Felsenstein, J., Statistical Inference and the Estimation of Phylogenies, PhD dissertation, University of Chicago, 1968.

Felsenstein, J., Maximum-likelihood estimation of evolutionary trees from continuous characters, *Am. J. Hum. Genet.*, 25, 471–492, 1973.

Felsenstein, J., The detection of phylogeny, in *Prospects in Systematics*, Systematics Association Special Vol. 36, Hawksworth, D.L., Ed., Clarendon Press, Oxford, 1988, pp. 112–127.

Fisher, R.A., *Statistical Methods for Research Workers*, Oliver & Boyd, Edinburgh, 1925.

Fisher, R.A., *The Genetical Theory of Natural Selection*, Clarendon Press, Oxford, 1930 (2nd ed., Dover, New York, 1958; variorum ed., Oxford University Press, 1999).

Fisher, R.A., The discontinuous inheritance, *Listener*, 60, 85–87, 1958.

Goldman, N., Maximum likelihood inference of phylogenetic trees, with special reference to a Poisson process model of DNA substitution and to parsimony analyses, *Syst. Zool.*, 39, 345–361, 1990.

Heywood, V.H. and McNeill, J., Eds., *Phenetic and Phylogenetic Classification*, Publication 6, Systematics Association, London, 1964.

Mourant, A.E., *The Distribution of the Human Blood Groups*, Blackwell, Oxford, 1954.

Nei, M., Kumar, S., and Takahashi, K., The optimization principle in phylogenetic analysis tends to give incorrect topologies when the number of nucleotides or amino acids used is small, *Proc. Natl. Acad. Sci. Washington*, 95, 12390–12397, 1998.

Page, R.D.M. and Holmes, E.C., *Molecular Evolution: A Phylogenetic Approach*, Blackwell, Oxford, 1998.

Silvestri, L.G. and Hill, L.R., Some problems of the taxometric approach, in *Phenetic and Phylogenetic Classification*, Publication 6, Heywood, V.H. and McNeil, J., Eds., Systematics Association, London, 1964, pp. 87–103.

Sneath, P.H.A., Introduction, in *Phenetic and Phylogenetic Classification*, Publication 6, Heywood, V.H. and McNeil, J., Eds., Systematics Association, London, 1964, pp. 43–45.

Sneath, P.H.A., Thirty years of numerical taxonomy, *Syst. Biol.*, 44, 281–298, 1995.

Sober, E., *Reconstructing the Past: Parsimony, Evolution, and Inference*, MIT Press, Cambridge MA, 1988.

Sober, E., Let's razor Ockham's razor, in *Explanation and its Limits*, Knowles, D., Ed., Cambridge University Press, Cambridge, 1990, pp. 73–93.

Sokal, R.R., Numerical taxonomy, *Sci. Am.*, 215(6), 106–116, 1966.

Sokal, R.R. and Sneath, P.H.A., *Principles of Numerical Taxonomy*, Freeman, San Francisco, 1963.

Steel, M. and Penny, D., Parsimony, likelihood and the role of models in molecular phylogenetics, *Mol. Biol. Evol.*, 17, 839–850, 2000.

Thompson, E.A., The method of minimum evolution, *Ann. Hum. Genet.*, 36, 333–340, 1973.

Thompson, E.A., Mathematical Analysis of Human Evolution and Population Structure, PhD dissertation, University of Cambridge, 1974.

Wagner, W.H., Jr., Applications of the concepts of groundplan-divergence, in *Cladistics: Perspectives on the Reconstruction of Evolutionary History*, Duncan, T. and Stuessy, T.F., Eds., Columbia University Press, New York, 1984, pp. 95–118.

Weir, B.S., *Genetic Data Analysis II*, Sinauer, Sunderland, MA, 1996.

Chapter 9

Homologues and Homology, Phenetics and Cladistics: 150 Years of Progress

David M. Williams

Abstract

Homology is seen by most, if not all, comparative biologists as the central concept, although its history and development is understood as long and disputatious. Richard Owen usually receives credit for clarifying the concept, at least from the perspective of the nineteenth century. Owen clarified not only homology and analogy but also homologue and analogue. That distinction — homologue as part, homology as a relation — has been lost from general understanding. A revival of this distinction places developments of the past, as well as developments in the recent numerical assessments of phylogeny, in a more exacting light. Homologue and homology relate in the same way as phenetics (similarity) and cladistics (relationships), both notions having a history prior to the numerical approach to phylogeny. The distinction relates to Colin Patterson's transformational and taxic homology, where *transformation* and *hierarchy* relate to phenetics and cladistics, respectively. Patterson's distinction is no longer relevant, as cladistic homology should properly include both transformation (homologues) and hierarchy (taxa), captured by the cladistic parameter of relationship.

Introduction

> *I am toiling over my chapter about Owen, and I believe his ghost in Hades is grinning over my difficulties. The thing that strikes me most is, how he and I and all the things we fought about belong to antiquity. It is almost impertinent to trouble the modern world with such antiquarian business.*
> *(T.H. Huxley to J.D. Hooker, 4th February, 1894 in L. Huxley, 1913:321)*

After a long and distinguished career, Richard Owen died in 1892, at 89 years old. Owen left behind an impressive body of anatomical studies, the first British Museum of Natural History, and a host of enemies. Ironically, the task of summarizing Owen's achievements fell to Thomas Henry Huxley, one of Owen's longest and most pernicious foes (Desmond, 1994, 1997). Huxley was invited to summarize Owen's scientific contributions, which he did in a 59-page review (Huxley, 1894). For the most part Huxley reviewed the history of anatomy and morphology, not mentioning Owen, or his achievements, for nearly 30 pages until he discussed the concepts of homology and analogy, perhaps Owen's major contribution to comparative anatomy. Owen's homological anatomy produced many significant works in comparative anatomy and "the appearance of Owen's homological work met with considerable enthusiasm, and 'homologizing' was all the rage for a decade or so following Owen's [1847] report" (Rupke, 1994:183).

By the time of his death, Owen's reputation had dwindled to such an extent that he was all but written out of history by the new generation of Darwinian biologists (Rupke, 1994). In more recent times Owen's work has been reexamined and his contribution to science reevaluated (Ospovat, 1981; Padian, 1985, 1995, 1997; Richards, 1987; Rupke, 1993, 1994; Camardi, 2001; for a dissenting view on Owen's rehabilitation see Ghiselin, 1997, 2000). Of all the topics Owen addressed during his career that might very well "belong to antiquity," homology has remained to "trouble the modern world," its "tortuous and disputatious history" (Nelson and Platnick, 1981:151) matched only by that of the species concept (De Pinna, 1999; Wheeler and Meier, 2000). Nevertheless, it seems that homology simply cannot be ignored, as it is "... without question the most important principle in comparative biology" (Bock, 1974:386; Patterson, 1982:22).

In keeping with the theme of this book, I have attempted to identify certain milestones from the past 150 years. Three immediately come to mind: (1) Owen's (1843) clarification (my starting point), (2) Patterson's (1982, 1988a) distinction between taxic and transformational homology, and (3) Nelson's (1994) introduction of the concept of relationship into studies of homology. Central to Nelson's formulation is the contrast between homologue as "a part of an organism" (Nelson, 1994:104) and homology as the "phylogenetic relationship between parts, or homologues, of different organisms" (Nelson, 1994:104; see also Nelson, 1985, 1989a). I will use this distinction to orient the following discussion, bearing in mind that the terminology of the past, as well as the present, weaves a tangled web (Haas and Simpson, 1946).

Owen's Milestone: Homologues and Homology, Analogues and Analogy

Homologues and Analogues

Richard Owen was certainly not the first to differentiate kinds of comparisons among organisms, but his clarification of the terms *homology* and *analogy* is usually regarded as the beginning of the modern era of comparative biology, and in this sense is rightly considered a milestone and a convenient point with which to begin. Owen contrasted homologue and homology with analogue and analogy. He first offered concise definitions of *homologue* and *analogue* in the glossary appended to his *Lectures on Comparative Anatomy* (1843):

> Analogue. A part or organ in one animal which has the same function as another part or organ in a different animal. (Owen, 1843:374, 1847:175, 1848:7, 1866: xii)

> Homologue. The same organ in different animals under every variety of form and function. (Owen, 1843:379, 1847:175, 1848:7, 1866:xii)

When comparing the parts of different animals' skeletons, bones that correspond are the homologues, as they are considered to be the same thing or, in Owen's words, "namesakes": "A 'homologue' is a part or organ in one organism so answering to that in another as to require the same name ..." (Owen, 1866:xii).

Owen (1849) provided an excellent example with the illustration included as the frontispiece to his book *On the Nature of Limbs* — an illustration remarkably similar to those produced three centuries earlier by Pierre Belon and Leonardo da Vinci (McMurrich, 1930:Figure 12; Tassy, 2000:34, Figure 2; see also Owen's comments in Sloan, 1992:110). Belon's famous figure illustrated comparable parts of a human and bird skeleton, which immediately conveys to the reader which parts are to be considered the same even though

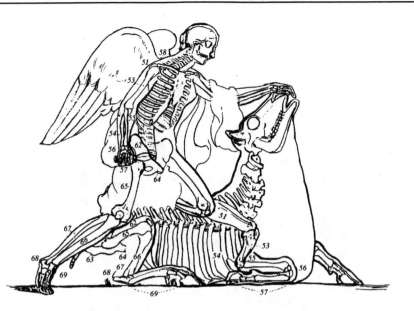

FIGURE 9.1   Reproduction of the illustration used for the frontispiece of Richard Owen's *On the Nature of Limbs*. (From Owen, R., *On the Nature of Limbs*, John van Voorst, London, 184.)

they superficially differ (Belon, 1555:Plates 40 and 41; reprinted in Cole, 1944:8; Zimmermann, 1967:68, abb. 3; Kluge, 1971:4, Figure 1-1, 1977:4, Figure 1.1; Bonde and Hoffmeyer, 1985[1996]:92; Ellenius, 1993:381, Figure 2; Panchen, 1994:43, Figure 8, 1999:6, Figure 1; Zunino and Colomba, 1997:67, Figure 7.2; Minelli, 1999:88; Goujet, 2000:46; Rieppel and Kearney, 2002:71, Figure 2; redrawn in Singer, 1931:87 [1950:87; 1959:91]; Boyden, 1973:Figure 4.1; Papavero et al., 1995:145, for a small sample). To assist the reader, Belon labeled the comparable parts of each skeleton with the same symbols, preceding Owen by three centuries. Owen's frontispiece offered an illustration of a transparent angel and bull standing side by side revealing the skeleton of each animal (Figure 9.1; reproduced in Stevens, 1998, cover illustration).

The parts of each skeleton are numbered such that the equivalent bones in the angel receive the same number as those in the bull. The same parts may be found in one form or another in all vertebrate animals (Figure 9.2).

Almost every modern account of the morphology of any group of plants or animals contains these kinds of illustrations (e.g., Ridley, 1996: Figure 17.7), allowing immediate comparison of different organisms. Owen's "homological anatomy" was an attempt to discover and christen all the namesakes of all vertebrates. The idea can be represented symbolically.

For instance, in Figure 9.1 there is skeletal element 51, or the scapula (see also Figure 9.2). In Figure 9.3 each rectangle represents the particular bone from the bull (left) and the angel (right). Both homologues are enclosed within a larger stippled rectangle, which represents their sameness, or their correspondence (Ghiselin, 1976, 1997). The idea may be further generalized where A' and A'' represent any two features proposed as homologues (Figure 9.4).

Owen's analogues are simpler to understand, as they refer to features that simply share the same function, for example the legs of man and the walking legs of a crayfish (Boyden, 1947).

Homology and Analogy

Owen's detailed descriptions of homology and analogy, as opposed to the stark definitions of homologue and analogue, appeared in his *Archetype and Homologies of the Vertebrate*

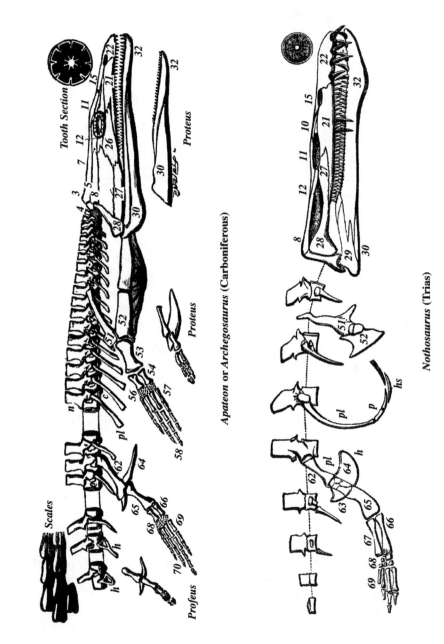

FIGURE 9.2   Reproduction from Owen (1859:Figure 65, "Apateon or Archegosaurus (Carboniferous)," and Figure 70, "Nothosaurus (Trias)").

'Namesakes'

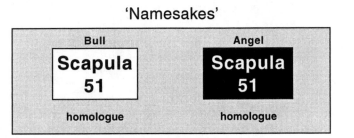

FIGURE 9.3   Diagrammatic representation of two homologues (scapula, part 51) from the bull and angel (after Owen 1849: frontispiece, numbering after Owen). Shaded area represents the similarity; the homologues represent the differences.

Homologues

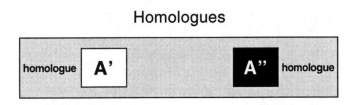

FIGURE 9.4   Diagrammatic representation of two homologues, A' and A", shaded area represents the similarity, the homologues A' and A" represent the differences.

Skeleton (Owen, 1847, 1848). In these accounts (and in an earlier short summary [Owen, 1846]) Owen described three kinds of homology, special, general and serial:

> Relations of homology are of three kinds: the first is that above defined [special homology], viz. The correspondency of a part or organ, determined by its relative position and connections, with a part or organ in a different animal; the determination of which homology indicates that such animals are constructed on a common type ... . (Owen, 1847:175; see also Owen, 1848:7)

> A higher relation of homology is that in which a part or series of parts stands to the fundamental or general type, and its enunciation involves and implies a knowledge of the type on which a natural group of animals, the vertebrate for example, is constructed. (Owen, 1847:175; see also Owen, 1848:7)

> ... any given part of one segment may be repeated in the rest of the series, just as one bone may be reproduced in the skeletons of different species, and this kind of repetition or representative relation in the segments of the same skeleton I call 'serial homology'. (Owen, 1847:175; see also Owen, 1848:7)

Owen's special homology ("the correspondency of a part or organ") could "be determined by the relative position and connection of the parts, and may exist independently of form, proportion, substance, function, and similarity of development" (Owen, 1847:174; see also 1848:6,1866:vii). While Owen stressed that homologies were "mainly, if not wholly" dependent upon these criteria, their meaning was somewhat more complex, as "the determination of homology indicates that such animals are constructed on a common type ...":

> I take for granted that it is generally known, as it is universally admitted by competent anatomists and naturalists, that these limbs or locomotive members,

Special homology

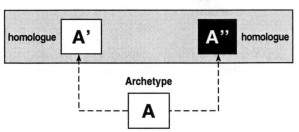

FIGURE 9.5   Diagrammatic representation of two homologues, A' and A", derived from the equivalent structure, A, in the archetype. Shaded area represents the similarity, the homologues A' and A" represent the differences.

which, according to their speciality of form, have received the above special names, are answerable or 'homologous' parts: that the arm of the Man is the fore-leg of the Beast, the wing of the Bird, and the pectoral fin of the fish. This special homology has been long discerned and accepted; but the general homology of the parts or their relations to the vertebrate Archetype, in short their 'Bedeu-tung' or essential nature, is not generally known. (Owen, 1849:3)

Owen refers to the special homology of forelimbs as being well known, but their explanation ("*Bedeutung*" (signification); Owen, 1849:1) not generally understood. By well known, Owen appears to be referring to the identities of particular homologues ("homol-ogous parts"), the bones themselves (as in Figures 9.1 to 9.3). To explain homology, Owen goes on, one needs to consider their general homology. Owen's explanation of special homology was intimately connected to his concept of the archetype, the point of reference from which all parts are seen as modifications in a variety of different organisms (Padian, 1985; Rupke, 1993, 1994; Sloan, 2003). Thus, while it appears that Owen directly equates homologue with special homology, he considered homologues to be empty of meaning without reference to, and interpretation from, the archetype (Owen, 1848:175–203, 1849:87–119, Plate 1 [Owen's Figure 1 is the archetype; the entire Plate 1 is reproduced in, for example, Panchen, 1994:31, 1999:15 and, partly, in Gould, 2002:320). Owen's special homology might be better illustrated by Figure 9.5 (compare with Figure 9.4).

Here, the two homologues A' and A" are related through their general homology, derived from a consideration of the hypothetical (and imaginary) general homologue A of the archetype (Panchen, 1994). Thus, Owen's special homology is not just the homologues themselves but a concept derived from that knowledge (Rudwick, 1976:210: "The determi-nation of homologies was not, however, a simple empirical exercise …"). Figure 9.5 may be compared with some recent diagrammatic representations of Owen's homology (Wagner, 1994:277, Figure 1, upper diagram; Staton, 2000:Figure 1a). Wagner's illustration appears incomplete, as he depicts only the relation between homologues (Figure 9.6). Interestingly, Wagner represents the relation with a line interconnecting the two homologues. Wagner's line may be understood as equivalent to the stippled rectangles of Figures 9.3 to 9.5.

Figures 9.5 and 9.6 may go some way toward explaining the differences in the devel-opment of the concept of homology. Figure 9.6 depicts two homologues, A' and A", con-nected by a line. Both homologues A' and A" are based on observable (but abstracted) features, such as the scapula of an angel and the scapula of a bull (Figure 9.1). Wagner's connecting line is present to indicate that they are indeed the same — the line is not particularly necessary; it is purely symbolic (Figure 9.6); the name (scapula) or even the number (part 51) works just as well (Figure 9.3). Figure 9.5 differs from Figure 9.6 in that

Homologues

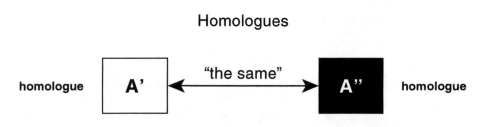

FIGURE 9.6   Diagrammatic representation of Wagner's views of Owen's homology (after Wagner 1994; 277, Figure 1, upper diagram; also Staton, 2000:Figure 1). The line joining the two homologues follows Wagner. (From Wagner, G.P., in *Homology: The Hierarchical Basis of Comparative Biology*, Hall, B.K., Ed., Academic Press, San Diego, 1994, p. 277, Figure 1, upper diagram.)

it includes the additional (but hypothetical) homologue A, which in Owen's case is said to represent the scapula of the archetype, also marked as part 51 in Owen's diagram (Owen, 1848:Plate 2, Figure 1; 1849:Plate 1, Figure 1[1]). In a similar fashion, Belon appears to have regarded his diagrams as generalizations ("La section d'un oyeseau seulement de tous oyseaux l'intérieur demonstre") (Steiner, 1954:2; Dupuis 2000:8), suggesting an early version of a common plan or archetype, an abstraction used as a guide to the essential similarities seen in all birds as well as all vertebrates and equivalent to the stippled rectangles in Figures 9.3 to 9.5. Thus, in Figure 9.6 the two bones are considered comparable by some property or properties common to both, whereas in Figure 9.5 the two bones are explained with reference to their derivation from homologue A. In other words, both diagrams include an abstracted similarity, the principle that unites the homologues (represented by the stippled area in Figures 9.3 to 9.5) and gives rise to the notion that homology can be somewhat paradoxically defined as features that are the same but different. In a general sense, this paradox becomes understandable if the homologues are viewed as modifications, regardless of how that modification came about: "The 'sameness' that constitutes the character (the homology) is thus the *unmodified* state, which all the organisms that show the character share, either in its original or some modified form" (Platnick, 1979:542; see also Nelson and Platnick, 1981:301).

To demonstrate the meaning of analogy, Owen referred to the dragon, *Draco volans*:

Its forelimbs being composed of essentially the same parts as the wings of a bird are homologous with them; but the parachute being composed of different parts, yet performing the same function as the wings of a bird, is analogous to them … . But homologous parts may be, and often are, also analogous parts in a fuller sense, viz. As performing the same functions … the pectoral fin of the flying fish is analogous to the wing of the Bird, but, unlike the wing of the Dragon, it is also homologous with it. (Owen, 1847:175; see also Owen, 1846:526, Owen, 1848:7, 1866:xii)

---

[1] The illustration of the archetype, as opposed to Owen's entire diagram, has often been reproduced (see, for example, Carpenter, 1854:73, Figure 61 [reproduced in Richards, 1992:130, Figure 25]; Ospovat 1981:134; Gould 1986:20; Rupke (1993:241, Figure 3) and Rupke (1994:195, Figure 27). Farber 2000:43). Owen's original pencil sketch is reproduced in Rupke, 1993:241, Figure 3 and Rupke, 1994:195, Figure 27. Rupke placed Owen's figure beside a similar diagram from Carus (1828), noting that "Owen may have been more indebted to Carus than he acknowledged" (Rupke, 1994:241). Russell (1916:105) reproduced Owen's drawing of the archetype but omitted the numbering of the parts. Many later reproductions follow Russell rather than Owen; hence the diagram often appears without the numbered parts, for example Macleod (1965:265, Figure 1), Rudwick (1976:211), Ruse (1979:120) and Patterson (1978:122, 1999:101).

TABLE 9.1   Relations of analogy and homology with reference to the human forelimb.

	Homologous	Nonhomologous
Analogous	Forelimb of man	Forelimb of man
	Forelimb of monkey	"Legs" of crayfish
Nonanalogous	Forelimb of man	Forelimb of man
	Wing of bird	Gill of crayfish

After Boyden, A., *Am. Midl. Nat.*, 37:648–669, 1947.

Analogy is thus defined solely in terms of function: Legs are for walking, wings are for flying. Yet homology and analogy, as relations, can be combined in four permutations. Boyden (1947:231), for instance, gave examples of each relative to the forelimb of man (Table 9.1).

The four permutations illustrate that while both homologues and analogues may be based on the same criteria (similarity in structure and relative position), the difference between analogous and nonanalogous relations are ultimately governed by function, while the differences between homologous and nonhomologous relations are ultimately governed by form.

Homology and analogy were different kinds of relations not opposites; Owen never intended them to be "antonyms," or "antitheses" (Strickland, 1846:35; Panchen, 1994:44). Nevertheless, there has been a tendency to apply the two concepts as if they were indeed opposites rather than different relations (e.g., Ghiselin, 1976, 1984:109). To consider homology and analogy as opposites implies that one is true and the other false, under the general understanding that homology is an indicator of some true relationship (however conceived) among organisms based on hypotheses concerning their parts — their homologues. That dichotomy was a later invention, one that Lankester began (but did not intend) by dissecting the term homology and substituting two terms of his own invention: *homogeny* for similarities that truly indicate common ancestry (the true relationship) and *homoplasy* for similarities otherwise explained.

Homoplasy

In an effort to explain Owen's namesakes, Lankester focused on the possibilities offered by Darwin's new evolutionary perspective:

> ... how can the sameness (if we may use the word) of an organ under every variety of form and function be established and investigated? This is, and always has been, the stumbling-block in the study of homologies without the light of evolutionism ... (Lankester 1870a:35)

While often understood as an attack on Richard Owen's archetypes, Lankester was also targeting embryology and its use as a direct guide for establishing sameness: "... before the appearance of Mr. Darwin's theory many zoologists were turning to embryology as a surer guide than ideal archetypes in tracing the identities of structures in organisms" (Lankester, 1870a:35). The "light of evolutionism" inspired Lankester to redefine homology in terms of "community of descent":

> Without doubt the majority of evolutionists would agree that by asserting an organ A in an animal α to be homologous with an organ B in animal β, they mean that in some common ancestor ε the organs A and B were represented by

Homogeny

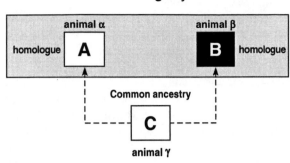

FIGURE 9.7   Diagrammatic representation of two homologues, A and B, derived from the equivalent structure, C, in the common ancestor. Shaded area represents the "similarity," the homologues A and B represent the "differences." Compare with Figure 9.5.

an organ C, and that α and β have inherited their organs A and B from ε. (Lankester, 1870a:36)

Lankester's views can be represented diagrammatically (Figure 9.7). Apart from the obvious metaphysics, Lankester's proposal differs little from Owen's (compare Figure 9.5 with Figure 9.7), the nonmaterial archetype becoming, potentially at least, a material reality (Russell, 1916; Naef, 1919:35; Nelson, 1989a:280). "If we suppose that the ancient progenitor, the archetype as it may be called, of all mammals, had its limbs constructed on the existing general pattern ..." (Darwin, 1859:435) then, of course, it could represent an ancestor instead. That viewpoint helped with another of Lankester's motives, as he also wished to rid biology of the word homology because of the apparent Platonic associations of Owen's archetype (Russell, 1916:267; Hall, 1992:57; Rupke, 1994): "It is less likely to cause confusion if we have a new term than if we amend an old one, which is my reason for not retaining 'homology'" (Lankester, 1870a:42). Lankester proposed the term *homogeny* to indicate "... structures which are genetically related, in so far as they have a single representative in a common ancestor ..." (Lankester, 1870a:36; Figure 9.7).

Lankester recognized that not all namesakes could be directly attributed to or explained by common ancestry. To cover these other instances of sameness he proposed another term, *homoplasy*. Homoplasy would be recognized "when identical or nearly similar forces, or environments, act on two or more parts of an organism which are exactly or nearly alike, the resulting modifications of the various parts will be nearly or exactly alike" (Lankester, 1870a:39). Lankester thought that all similarities might eventually be explained by some kind of cause effectively substituting a process (homogeny or homoplasy) for a structure (homologue).

St. George Mivart (1893), a defender of Owen (Mivart, 1893), quickly replied to Lankester, suggesting that in his view it would be unwise, or at least premature, to dismiss the term homology, as structures that showed "a close resemblance of parts" would remain so until such times as a particular cause was established (Mivart, 1870:115). Mivart, then, equated Owen's homologues with homology as a generalized term for useful similarities, divorcing it from any explanation or meaning ("It will be useful to speak of such undetermined parts simply as 'homologues'" [Mivart, 1870:116[2]]). However, Mivart did not deny a role for causes. Instead, he recognized two kinds of common origin subdividing homogeny into ancestral and developmental homogeny, the former directly equivalent to Lankester's homog-

---

[2] That is not to say Mivart excluded Owen's types; he clearly thought they were of some significance. What he did do was encourage the view that homology could indeed be subdivided, contrary to Lankester's intentions (see Desmond, 1982). Mivart gives a sympathetic account of Lankester's views in Mivart (1871).

eny, the latter for "parts which go through a process of development which is similar in the two individuals" (Mivart, 1870:116). Lankester (1870b:342) felt the need to respond, stressing that homogeny meant only what Mivart referred to as ancestral homogeny; although he allowed that developmental and ancestral homogeny might indeed correspond, but without some common ancestral origin, developmental homogeny, if anything, would be an example of homoplasy.

Lankester's homogeny never did achieved widespread use (but see Haas and Simpson, 1946:322; Simpson, 1959:290; Kaplan, 1984:53; Sattler, 1984:391; Rieppel, 1988:58). Instead, the historical meaning Lankester intended for homogeny was simply applied to the older term, *homology*, used to designate similarities explained by common ancestry, attaching a cause that was in some ways as difficult to establish as the meaning Owen attached to homology. While Lankester appeared to create homogeny and homoplasy as a way of subdividing the older, more generalized homology (Osborn, 1902:264; Patterson, 1982:68; Gould, 2002:1074), Mivart's usage persisted, as Lankester saw no further role for a category called *homology*. As Mivart put it "'homology' will be, as it were, a generic term with homogeny and homoplasy as two species under it" (Mivart, 1870:115). Of course, that interpretation is only possible if homology was retained and directly equated with essential similarity (or homologue), devoid of explanation (Figure 9.4). Simply put, while criteria could assist in the recognition of homologues, causes might provide some absolute explanation as well as a means for establishing homology.

Homoplasy and Evolution

Conflict among homologies has given rise to a renewed preoccupation with Lankester's homoplasy. This interest has been accompanied by a renewed preoccupation with parallelism and its explanation (Roth, 1984:14; Sluys, 1989; Wagner, 1989:55, 66; Brooks, 1996; DeSalle et al., 1996; Gould, 2002). More recently causes have been invoked, as "the significance of this similarity is thus dependent on the existence of a relevant underlying process" (Sanderson and Hufford, 1996:328; cf. Spemann, 1915).

As noted above, Mivart suggested that homology was "... a generic term with homogeny and homoplasy as two species under it" (Mivart, 1870:115), which requires equating homology with similarity (i.e., homologues, the undetermined parts). Henry Fairfield Osborn (1902:264), from his reading of Lankester (1870a), understood homology in the same way: "Homology thus includes homoplasy [and] homogeny" (Osborn, 1902:264). He further noted,

> It follows that subsequent writers, including myself, have misused the term "homoplasy," confusing it with "parallelism" and "convergence" which, as we have seen, may affect absolutely non-homologous structures. *Homoplasy should be confined to structures in which there is an element of* homology. (Osborn, 1902:264–265)

What is the element of homology? Here Osborn meant some degree of common ancestry rather than some degree of similarity. Osborn coined the term *latent homology* to explain cases of what he called "independent evolution" (Osborn, 1902:264), but as he noted, "Prof. Edmund B. Wilson, pointed out to me that such cases were almost exactly covered by the *original definition* of the word 'homoplasy' by Lankester ..." (Osborn, 1902:264). Osborn's latent homology was an orthogenetic explanation of the origin of certain homologues, those often described as parallelisms. Osborn's motivation was to propose an evolutionary mechanism (as opposed to a developmental mechanism) to explain similarities other than those due to immediate common ancestry. Osborn included his 1902 article in *Evolution of Mammalian Molar Teeth* (1907), and a response from Lankester: "I recognized (as hitherto combined under one term 'homology') *only* homogeny — or hereditary quality — and homoplasy or moulded non-hereditary quality. The likeness due 'to other reasons' than homogeny ... cannot be homogeny" (Lankester in Osborn, 1907:238).

There is a paradox here as, in Lankester's terminology, Osborn probably meant "latent homogeny," but as "the likeness due 'to other reasons' than homogeny ... cannot be homogeny," it is, as Osborn later recognized, homoplasy. It seems that, regardless of explanation, homoplasy came to mean unexplained similarities — almost equivalent to homologues, unexplained similarity.

Latent homology resurfaced with De Beer (1971:10) as an explanation for similarities with "hidden causes," Lankester's "other reasons." Latent homology has returned again, accompanied by more modern explanations and modern causes ("Latent homology might be explained by reiterated recruitment of retained developmental genetic pathways" [Hawkins, 2002:46].). Independent homologies have also returned, as rampant parallelism (Sluys, 1989), underlying synapomorphies (Saether, 1979; Brooks, 1996:16), parallelophyly (Mayr and Ashlock, 1991:255, 424; Mayr and Bock, 2002:179) and many others (for a compilation see Nelson, 1996:144). Each version reflects, if not exactly, then in spirit, Osborn's orthogenetic views, which explain "the recurrence of similarity in evolution" (Sanderson and Hufford, 1996).

Current usage of the term *homoplasy* appears to suggest, like Osborn, that homoplasy does involve some homology ("Homoplasy is derived similarity that is not synapomorphy [homology]," [Wake, 1996:xvii; see also Mayr and Ashlock, 1991:418]). That seems to suggest that *homoplasy* is equivalent to *homologue*. A more accurate viewpoint, underlined by Osborn's semantic puzzle, is that homoplasy is simply a term used today to explain away character conflict (Nelson and Platnick, 1981:342), a problem that has yet to find any real solution (cf. Nelson, 1996; Nelson et al., 2003) and looks increasingly likely to become redundant.

Homoplasy and Nonevolution

While common ancestry was considered the explanation for homology, the search for the causes of similarity continued, and a complex terminology evolved, many new terms arising and going extinct with equal rapidity (Haas and Simpson, 1946). Moment, for instance, suggested that "the basic distinctions among the similarities of plants and animals would rather seem to be those based on (1) developmental, (2) genetical, and (3) evolutionary differences" (Moment, 1945:453). Moment's general argument was identical to Lankester's, almost word perfect: "The word homology had perhaps best be discarded as too charged with metaphysical overtones and a neutral descriptive term like similarity substituted" (Moment, 1945:455). Moment's words are interesting, as he did state that, in his understanding, homology could indeed be equated with similarity and by inference with homologue. Like Mivart, Moment suggested a complex terminology to deal with the many different causes: for development homology (similarities) there was *homodynamy*, for evolutionary homology (similarities) there was *homophyly*, and for genetic homology (similarities) there was *homogeny*.[3]

Some 50 years later, Wagner created the biological homology concept, contrasting it with historical homology, those similarities explained by common ancestry. Biological homology was defined as: "Structures from two individuals or from the same individual are homologous if they share a set of developmental constraints, caused by locally acting self-

---

[3] Moment also used heterodynamy, heterogenetic and heterophyletic for dissimilar features caused by developmental, genetic and evolutionary processes, respectively. Moment's terminology is somewhat idiosyncratic relative to the use to which the same words had previously been put (see Haas and Simpson, 1946:324 and Hubbs, 1944, for a review of the original use of these terms).

regulating mechanisms of organ differentiation" (Wagner, 1989:62; see also Wagner, 1994:275). As Rieppel pointed out " ... homodynamy would correspond to Wagner's (1989) concept of biological homology ... " (Rieppel, 1992:711) and, of course, Lankester's homogeny would correspond to "historical homology."

Even more recently, Butler and Saidel (2000) proposed generative homology, which is said to account for the creation of similarity. *Syngeny* (generative homology) is defined as "the relationship of a given character in different taxa that is produced by shared generative pathways" (Butler and Saidel, 2000:849). Syngeny is accompanied by a complementary term, *allogeny* (generative homoplasy), defined as "the relationship of a given character in different taxa that is produced by different generative pathways" (Butler and Saidel, 2000:849).

One might be tempted to draw a general conclusion from these 150 years of effort: Development homogeny (Mivart) = homodynamy (Moment) = biological homology (Wagner) = sygeny (Butler and Saidel) = homoplasy (Lankester). That equation might appear logical as far as terminology goes, but as Lankester (1870b:342) pointed out, developmental and ancestral homogeny, for instance, might correspond and indicate the same relation (a position vigorously developed by Haeckel and his biogenetic law). In that case, development homogeny = ancestral homogeny (Mivart) = homodynamy = homophyly (Moment) = biological homology = historical homology (Wagner) = sygeny (Butler and Saidel) = homogeny (Lankester). In effect, causes, however determined, refer only to homologues and the efforts to establish their sameness with some degree of certainty. In each case the issue is sameness.

Patterson's Milestone: Taxic and Transformational Homology

Recognizing Homologues and Transformational Homology

To recognize homologues is to discover the useful or essential similarities that provide evidence for christening the parts of organisms (discovering the namesakes), a task often understood as relatively straightforward (Owen, 1849:71; Patterson, 1988a:605). T.H. Huxley, for instance, commented on Belon's illustration in his account of Owen's contributions to science:

> The old French naturalist Belon, who must have been a good deal of an artist, and illustrated his book 'L'Histoire de la Nature des Oyseaux', with many 'naifs portraicts', initiated this way of dealing with anatomy [... the artistic fashion ... , Huxley, 1894:287]. The skeleton of a bird is set beside that of a man, and the reader is left to draw the obvious conclusion as to their 'unity of organisation'. A child may see that skull 'answers' to skull; spinal column to spinal column; ribs to ribs; breast bone to breast bone; wings to arms; and legs to legs, in the two. (Huxley, 1894:287–288)

Nearly 100 years later, Patterson wrote,

> When a child learns to recognise birds, the criteria used are, at root, the same as those used by the scientist: the feathers, beak and wings, and so on are 'the same' in a sparrow and a swan, whereas the wings of a beetle or the beak of a turtle do not make those creatures birds — the 'sameness' is different or inessential ... (Patterson, 1987:235)

For Huxley and Patterson, roughly 100 years apart, recognizing homology is child's play. If the task really was simple, then the business of systematics should have come to a happy conclusion many years ago, effectively relegating morphology to the status of "antiquarian business." That appears not to be so (Rieppel and Kearney, 2001, 2002). One might ask what, if anything, led Belon to make his particular comparisons?

> This [Belon's diagram] was a remarkable achievement at the time, and the question must be raised whether it was wholly intuitive, or whether there are some underlying assumptions or operational criteria which guided the comparison of the different types of organization. (Rieppel, 1988:35; see also Rieppel, 1987:416, 1994:70; Rieppel and Kearney, 2002:70)

Owen suggested that homologues were "determined by the relative position and connection of the parts" (Owen, 1847:174; see also Owen, 1848:6, 1866:vii; Boyden, 1943; Panchen, 1994), restating Geoffroy-St. Hilaire's "Principle of Connections" and "Principle of Composition." The most recent full discussion of criteria used for recognizing homologues is in Adolf Remane's book *Die Grundlagen des Natürlichen Systems, der Vergleichenden Anatomie und der Phylogenetik* (1952; see also Remane, 1955, 1961). Remane's criteria have been widely discussed from a variety of different viewpoints (for example Hennig, 1950:172, 1953:11, 1966:94; Kiriakoff, 1954; Simpson, 1961:93; Sokal, 1962; Eckhart, 1964; Sattler, 1964, 1984:338; Bock, 1969, 1974, 1977, 1989; Hagemann, 1975; Riedl, 1978:34, 1983:211, 1989:xiv; Wiley, 1981:131; Froebe, 1982:25; Kaplan, 1984:54, 2001; Remane, 1985:169–170, 1989; Rieppel, 1986:198; Ax, 1987:160; Sluys, 1989; Mayr and Ashlock, 1991:143; Donoghue, 1992:172; Schmitt, 1995:428; Haszprunar 1998:335). According to Simpson, for example, Remane published an "... exceptionally valuable book on taxonomy in general" (Simpson, 1961:93), and Riedl noted that for homology "everything essential was said by Goethe, among the early morphologists, and by Remane, among the modern ones" (Riedl, 1978:23). Oddly, this "absolutely decisive contribution [Remane, 1952], however, was never translated. The fault must have lain in the mistrust that people in 'modern countries' bought to the 'old morphology'" (Riedl, 1983:211). Be that as it may, Remane's book remains untranslated. However, according to others, Remane, like Owen, simply "... reiterate[ed] Geoffroy-St Hilaire's Principle of Connections and Principle of Composition" (Brower, 2000a:14; see Stevens, 2000:84 and Rieppel, 1988:37). Remane did more than that.

Remane proposed three principle criteria (*Hauptkriterien*; Remane, 1952:63) and three auxiliary criteria (*Hilfskriterien*; Remane, 1952:64), of which only the principle criteria may be understood as a restatement of Geoffroy-St Hilaire's "Principle of Connections" and "Principle of Composition"[4] (the auxiliary criteria are discussed further below). Many have suggested that the principle criteria are equivalent to Patterson's (1982) similarity test, which "... does not test the hypothesis [of homology] but validates it as worthy of testing, or evaluates its internal consistency" (Patterson, 1982:38; see also Cracraft, 1981:25). Patterson noted that "topographic correspondence and ontogenetic transformations are the usual criteria, and correspondences that pass such tests merit the same name ..." (Patterson, 1988a:605). That statement seems to apply to the recognition of homologues, the namesakes (the black and white rectangles of Figure 9.6).

The similarity test has also been equated with Patterson's transformational homology (Donoghue, 1992:173; Brower and Schawaroch, 1996; Sluys, 1996; Haszpruner, 1998; Donoghue and Ree, 2000:760). According to Patterson, transformational homology deals with change, usually (but necessarily) interpreted as the literal, material transformation of one homologue into another, representing an actual process, rather than being hypothetical or imaginary (the line connecting the black and white rectangles of Figure 9.6).[5] If homology is understood in this context, then its discovery is a combination of first establishing the homologues and second establishing their connection. To explain the connecting line requires the identification of the process or processes that cause one homologue to change (transform) into another (see above). The principle criteria alone might seem insufficient, as they only establish the similarity of the parts, hence Patterson's inclusion of ontogenetic transformations (see below). Establishing the connection between homologues has been called by some "character phylogeny" ("... [t]he notion of character phylogeny is therefore one of transformation" [Patterson, 1983:21]) and traditionally involves auxiliary evidence, conventionally gained from ontogeny (Holland, Chapter 11, this volume), paleontology (Forey, Chapter 7, this volume) or biogeography (Humphries, Chapter 10, this volume), or even some combination of the three. The use of paleontology and biogeography as evidence for character phylogeny was severely criticized and is no longer practiced for that purpose (Nelson, Chapter 6, this volume; Humphries, Chapter 10, this volume). Today character phylogeny is seen by many as largely irrelevant (but see Wägele, Chapter 5, this volume), being superseded by various computer algorithms, designed to sort homologues into their appropriate sequences. Transformational homology is thus simply a string of homologues that require some analytical approach (a method) to organize them into the correct, or most appropriate sequence; in effect, numerical taxonomy has become the new paleontology (Nelson, Chapter 6, this volume).

Remane's three auxiliary criteria (see above) are concerned with the "coincidence of characters" (Wagner, 1986:152), considered related to, or even synonymous with, Patterson's congruence test (Patterson, 1982; Bock, 1989:336; Schmitt, 1995; Haszpruner, 1998) and often associated with the discovery of taxic homology (Donoghue, 1992:173; Brower and Schawaroch, 1996; Sluys, 1996; Haszpruner, 1998; Donoghue and Ree, 2000:760). That viewpoint presupposes that taxic homology is discovered from data organized prior to analysis.

Recognizing Taxa and Taxic Homology

Patterson's taxic homology (1982) is represented in one of Hennig's early diagrams (Figure 9.8).

In the lower part of Hennig's diagram (Figure 9.8) there are two homologues, α and α' (cf. Figures 9.4 and 9.6), comparable to transformational homology. In the upper part of the diagram, homologue α' (represented by the black boxes) occurs in taxon B and C but

---

[4] The three principle criteria are as follows:
1. *The positional criterion:* "Homology can be recognised by similar position in comparable systems of features" (Remane, 1952:33; translation in Riedl, 1978:34 and Ax, 1987:160).
2. *The structural criterion:* "Similar structures can be homologized, without reference to similar position, when they agree in numerous special features. Certainty increases with the degree of complication and of agreement in the structures compared" (Remane, 1952:46; translation in Riedl, 1978:34 and Ax, 1987:160).
3. *The transitional criterion, Stetigkeitskritium:* "Even dissimilar structures of different position can be regarded as homologous if transitional forms between them can be proved so that, in considering two neighbouring forms, the conditions under headings (1) and (2) are fulfilled. The transitional forms can be taken from the ontogeny of the structure or can be true systematically intermediate forms" (Remane, 1952:49; translation in Riedl, 1978:34 and Ax, 1987:160).

The three auxiliary criteria are as follows:
1. *The general conjunctional criterion:* "Even simple structures can be regarded as homologous when they occur in a great number of adjacent species" (Remane, 1952:64; translation in Riedl, 1978:36 and Ax, 1987:160).
2. *The special conjunction criterion:* "The probability of the homology of simple structures increases with the presence of other similarities, with the same distribution among closely similar species" (Remane, 1952:64; translation in Riedl, 1978:36 and Ax, 1987:160).
3. *The negative conjunctional criterion:* "The probability of the homology of features decreases with the commonness of occurrence of this feature among species which are not certainly related" (Remane, 1952:64; translation in Riedl, 1978:36 and Ax, 1987:160).

The descriptions of Remane's criteria are taken from Riedl (1978:34): "I shall quote him [Remane] verbatim, modifying only the nomenclature" (see also Ax, 1987:160).

[5] That transformational homology need not imply real processes is illustrated by Patterson in his use of Owen's explanation of the derivation of all vertebrate homologues from the vertebrate archetype (Patterson, 1982:34).

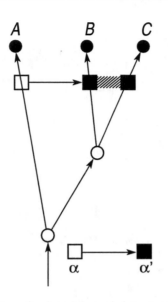

FIGURE 9.8   Diagrammatic illustration of a shared derived character. (Reproduced from Hennig, (1957:66, abt. 8.)

'Homology'

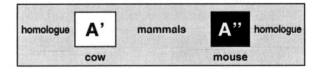

FIGURE 9.9   Diagrammatic representation of two homologues, A' and A", which define (diagnose) the taxon Mammals. Shaded area represents the similarity; the homologues A' and A" represent the differences.

not A. Thus, homologue α' is considered to be a synapomorphy for the group B + C; the homologue and the taxon are the same thing, supporting the monophyly of the group B + C: "The taxic approach (the one I am advocating) is concerned with the monophyly of groups" (Patterson, 1982:34). Figure 9.4 can be simply modified by adding the name of the taxon implied by the homologue (Figure 9.9, cf. Figure 9.3). It is worth noting that in Hennig's diagram homologue α also picks out a group, A + B + C. It too is a synapomorphy. The implication is that synapomorphy equals homology. Platnick clarified the issues:

> ... if alternative character states [homologues] are in some sense the same, how can they be different? There seems to be only two possibilities: either one state [homologue] is a modified form of the other, or both are modified forms of a third state [homologue]. (Platnick, 1979:542; see also Nelson and Platnick, 1981:301; Platnick and Nelson, 1981:123)

Figure 9.10, similar to Hennig's figure (Figure 9.8, lower portion), is a diagrammatic representation of two homologues, A and B, with homologue A present in taxa *a* and *b* and homologue B present in taxa *c* and *d* (Figure 9.10a). In this example, it is assumed that homologue B is a modification of homologue A, Platnick's first possibility. Therefore homologue A is really represented by the larger white rectangle uniting taxa *a* to *d,* and it is

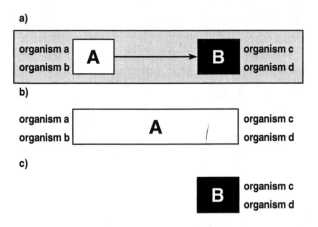

FIGURE 9.10   (a) Diagrammatic representation of two homologues, A and B. Homologue A is possessed by taxa *a* and *b*, and homologue B is possessed by taxa *c* and *d*. (b) Homologue B is a transformation of homologue A; hence A can be represented by the larger white box that unites taxa *a–d*. (c) Homologue B specifies the group *c* + *d*.

irrelevant at the level that discriminates taxa *a* to *d* (Figure 9.10b). Homologue A, when associated with taxon *a* and *b*, would also be called a symplesiomorphy. Homologue B, however, does specify a group, *c* + *d* (Figure 9.10c). While the notion of modification (transformation), however conceived, is significant in giving substance to the idea that the two homologues A and B really are the same but different, the relevant information is supplied by homologue B for the proposed group *c* and *d* and by A for the group *a* to *d*. The taxic approach deals only with the relevant homologue and its associated group.

In Figure 9.11a it is assumed that both homologues A and B are modified forms of a third but unknown homologue, Platnick's second possibility. As in Figure 9.10, the sameness is captured by the large white rectangle, which includes all four taxa *a* to *d* (Figure 9.11b). Even though both homologues A and B are thought to be modifications of some third unknown homologue, the larger white rectangle uniting all four taxa remains relevant only to the entire group. Homologue A, on its own, and homologue B, on its own, are the relevant data (Figure 9.11c). Here also, the notion of modification (transformation) is significant in giving substance to the idea that these two homologues are the same but different, the relevant data are supplied by homologue A and the group *a* and *b* and by homologue B and the group *c* with *d*. Symplesiomorphy is really synapomorphy at a different level, and the two stand in relation to one another (Patterson, 1982:33; De Pinna, 1991:376), with symplesiomorphy becoming an irrelevant concept. Taxic homology is, then, hierarchical rather than linear.

The Numerical Representation of Homology

Here I am interested in what homology might be thought of prior to any analysis. In the early 1960s Hunter coined the term paralogy (Hunter, 1964; see Inglis, 1966; Van Valen, 1982:308, Kaplan, 1984; De Pinna, 1991:389), a concept with a meaning different from its current use in molecular systematics (Fitch, 1970) and biogeography (Nelson and Ladiges, 1996; Ladiges, 1998. For Hunter, a paralogue[6] is

> ... a part or organ in one animal similar in anatomical structure or microana-
> tomical structure to a part or organ in a different animal. Paralogy, then, refers
> only to anatomical similarity and has no phylogenetic or functional implications.
> (Hunter, 1964:604)

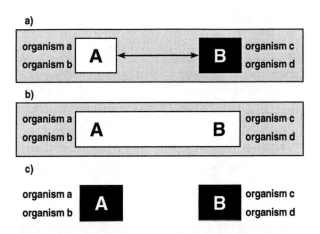

FIGURE 9.11   (a) Diagrammatic representation of two homologues, A and B. Homologue A is possessed by taxa *a* and *b*, and homologue B is possessed by taxa *c* and *d*. (b) Homologues A and B are supposed transformations of an unknown third homologue; hence, together they can be represented by the larger white box that unites taxa *a–d*. (c) Homologue A specifies the group *a + b*. Homologue B specifies the group *c + d*.

Of Hunter's paralogy, De Pinna wrote that "... when translated to current [cladistic] concepts, [it] is equivalent to ... 'shared derived character' or 'putative synapomorphy', i.e., primary homology" (De Pinna, 1991:389). This would suggest that "shared derived character," "putative synapomorphy" and "primary homology" simply imply "anatomical similarity." That equation appears too simplistic. Nevertheless, early in the debate concerning Hennig's phylogenetic systematics, Tuomikoski sensed a difference between Hennig's (1966) and Brundin's (1966) use of the term *synapomorphy*. Tuomikoski noted that for Brundin "... synapomorphy means agreement in apomorphic characters between two groups and thus also covers the similarity which is due to parallelism, chance similarity, etc." (Tuomikoski, 1967:139). Brundin (1966:26), and Hennig on occasion (see Hennig, 1965:605), spoke of true synapomorphy to distinguish it from the more operational definition of shared derived character, with only the former picking out natural or monophyletic groups. Tuomikoski suggested that "... a change in the original sense of the term synapomorphy would be confusing and a term for 'true synapomorphy' would at any rate be needed" (Tuomikoski, 1967:139; see also Bonde, 1977).

The issue was that although synapomorphies (shared derived characters) might be successfully identified, proposed or hypothesised, not all would ultimately indicate monophyletic groups; a certain number will mislead. Once identified, true synapomorphy might readily be called homology:

I emphasise the concept of synapomorphy instead of homology, the former being viewed as an estimator, the latter the parameter. I do so because synapomorphy

---

[6] In the revised edition of *Numerical Taxonomy*, Sneath and Sokal (1973) included the term *isology*, a word first coined by Florkin (1962:820) for the comparison of chemical similarities: "The biochemical compounds, molecules or macromolecules, which show signs of chemical kinship, we shall call *isologues*" (Florkin, 1966:6; a translation of Florkin, 1962:820). Florkin contrasted isology with homology, noting that the latter is explained by having "... a common origin and a common line of descent" and that a "high degree of isology in the primary structure of proteins ... would be taken as a sign of homology" (Florkin, 1966:43). Sneath later wrote, "It should be noted that reference has been made to 'homologies'. The quotation marks indicate that it is not evolutionary homology that is meant here. A better word is *isology*, merely implying sameness. This is operational homology as used in phenetic work ..." (Sneath, 1983:27, 29; see also Sneath, 1982:209, 1988:266). Isology seems to be equivalent to Hunter's paralogy.

TABLE 9.2   The Nature of Resemblances. The neutral term is listed as "unknown" cause, the known cause is attributed to either "common ancestry" or "independent similarities." For the latter kind of resemblances, homoplasy has been universally adopted (see also Schmitt 1995:429). For a similar table see Schuh (2000:71, Table 4.1).

	Unknown Cause	Known Cause	
	Similarity	Common ancestry, natural group	Independent similarities
Lankester (1870a)	—	Homogeny	Homoplasy
Mivart (1870)	Homology	Ancestral homogeny[a]	Developmental homogeny
Florkin (1962), Sneath and Sokal (1973)	Isology	—	—
Hunter (1964)	Paralogy	Homology	—
Brundin (1966)	Synapomorphy	"True synapomorphy"	Homoplasy
Bonde (1977); Kluge (1991); McKitrick (1994)	Synapomorphy	Homology	Homoplasy
Gaffney (1979)	Putative synapomorphy	Synapomorphy	Homoplasy
Farris and Kluge (1979)	Shared derived character	Synapomorphy	Homoplasy
Rieppel (1980)	Topographical homology	Phylogenetic homology	Homoplasy
Rieppel (1988)	homology = topographic correspondence	homogeny = homology	Homoplasy
De Pinna (1991)	Primary homology	Secondary homology	Homoplasy
Rieppel (1992, 1993, 1994)	Topological homology	Phylogenetic homology	Homoplasy
Minelli and Schram (1994); Minelli (1996)	Positional homology	Phylogenetic homology	Homoplasy
Brower and Schawaroch (1996)	"topographic identity" [= transformational homology]	Homology	Homoplasy

[a] Mivart introduces 25 kinds of explanations for resemblances. Ancestral homogeny is the only one significant to this table.

does not presuppose common ancestry whereas homology does. Homology is dealt with only indirectly by maximum character congruence, the ultimate arbiter of character history (see also Kluge, 1993:9; Kluge and Strauss, 1985:258).

"Synapomorphy does not presuppose common ancestry." What, if anything, does it presuppose? In an operational context, is synapomorphy simply just a synonym for similarity, for the homologues? Some have understood it that way, as "matchings, not homologies ... provide evidence pertinent to the inference problem at hand" (Sober, 1988:212; see also Minelli, 1996; Minelli and Schram, 1994; Brower and Schawaroch, 1996), matchings (similarities) devoid of any interfering theory (Cartmill, 1994:119). Table 9.2 documents a selection of viewpoints on the relation between homology and synapomorphy. These views (with the exception of Lankester and Mivart) might collectively be seen as charting the developments of the homology concept in a "post-Hennigian" or, more accurately, a "post-numerical" era. In general the schemes recognize three terms or categories. The first term represents data prior to analysis, those data entered into conventional character X taxon matrices. The remaining two terms are applied to similarities eventually rated as due to common ancestry (homology, secondary homology, phylogenetic homology, etc.) and those that do not,[7] homology emerging from analysis, "dealt with only indirectly" (Kluge, 1993:3). The various tripartite schemes are restatements of Mivart's ideas, where "'homology' [synapomorphy] will be, as it were, a generic term with homogeny [homology] and homoplasy as two species under it" (Mivart, 1870:115). They indicate just the similarities, just the homologues ("If not presumed homologs, what exactly are those things that match?" [Nelson, 1989b:294]).

It might be going too far to suggest that Mivart (or even Osborn) was responsible for the first phenetic approach to systematic data, but the comparison does seem apt. Whatever the case, it is clear that the usual binary data entered into modern data matrices are phenetic, inasmuch as they represent simple similarities, apparently theory free or theory independent. The last time "… theory independent operationalism …" was actively entertained was the phenetic approach to systematics, an episode "… which seems closed at least for the time being in systematics" (Nelson, 1994:138; see also Nelson, 1985:39, but see Williams, 2002). While the data may appear theory free, the methods of analysis are certainly not. Many commentators on the development of cladistics have suggested that the marriage of Hennig's ideas with the data matrix was a major breakthrough (Anderson 1999, 2001; Mitter, 1999). Alternatively, one might view the data matrix as harboring the last vestiges of phenetics, preventing progress in systematics rather than enhancing it (Nelson, Chapter 6, this volume).

Not all agree with the simple matching point of view, as synapomorphy is said to presuppose a group, or taxon (Patterson, 1982; De Pinna, 1991; Kluge, 1991; Rieppel, 1992, 1993, 1994). It is of significance that Patterson (1982) reserved no special term for similarities proposed prior to analysis, as taxic homologies were hypotheses to be tested, some eventually being found true, others false. There appears no need for any special preanalytical term, and the entries in the second column of Table 9.2 are redundant, as they are either explicit (topographic correspondence, topographic identity) or implicit (shared derived character, primary homology, topological homology, topographical homology) synonyms for homologues, or parts. Taxic homology differs simply by acknowledging the equivalence of taxon and homology.

The Analysis of Homologues

Transformational Homology

Transformational homology might be summarized as follows:

1. Transformational homology = symplesiomorphy + synapomorphy. If the task is to distinguish synapomorphy from sympleisomorphy, then the primary data for transformational homology is:
2. Transformational homology = homologue A + homologue B + homologue C, etc.

The series is rather like a multistate character: "… an initial hypothesis of homology among character states, all states are assumed to be homologous, but no particular relationships are specified" (Mabee, 2000:93). Some years ago Griffiths commented on multistate characters (see also Patterson, 1982:68):

> The only requirement for arranging character statements in sequences is that their meaning must be exclusive, in others words that it be impossible for any object to satisfy more than one of the character statements in the sequence at the same time. (Griffiths, 1974:108–109)

---

[7] Homoplasy has been consistently used for similarities that are not explained by common ancestry, a slightly different interpretation to that Lankester had in mind.

The very notion of modification (transformation) requires inclusive definitions, a hierarchy. Exclusive definitions of homology (or characters) seem to suggest that when characters do transform, the original disappears entirely. The notion of transformation, when applied to a series of homologues, seems entirely irrational, and attempts to discover a hierarchical relation becomes one of imposition. If phylogenetic reconstruction (parsimony, compatibility, maximum likelihood, etc.) is about method, then it is about imposition. As Patterson said of the analysis of nucleotides sequences, "... [the trees] don't pop out of the data, so I suppose they come from massaging the data with a theory or with a computer programme based on a theory ..." (Patterson, 2002:27; also Patterson, 1983:26), by using an "algorithm [that] imposes the hierarchy ..." (Kluge, 2001:203). Of course, to justify an algorithm that finds hierarchical relationships, one may assume that phylogeny will conform to a hierarchical, branching model of evolution (Lee, 2002) or one may simply assume hierarchy as an "epistemological axiom" (Brower, 2000b:147; Brower, 2002). Neither seems completely satisfactory.

Alternatively, one might ignore methods altogether and seek justification in known processes (causes), as Patterson did when he suggested that ontogeny (development) "imposes hierarchical organisation on linear [morphological] sequences" (Patterson, 1983:26). Ontogeny is usually seen as decisive in homology decisions. Løvtrup (1978:350) developed the idea calling it Naef's theorem after the Swiss zoologist Adolf Naef (1919, 1921–1923 [for an English translation see Naef, 1972], Naef, 1928 [for an English translation see Naef, 2000]; biographical details in Reif, 1983, 1998; von Boletsky, 1999, 2000). A more succinct account was given by Naef himself: "Parts 'that have been given a stage of development, with regard to a particular morphotype, remain, with regard to the latter, homologous throughout all subsequent transformations" (Zangerl, 1948:368; translated from Naef, 1919).

Naef's comments relate to individual parts such that for morphology it is considered the only direct source of evidence for observing one homologue transforming directly into another so "... the mode of development itself is the most important criterion of [homologues] homology" (Nelson, 1978:335; Patterson, 1981:199, 1982:38, 1983, 1988a:605, 1994:177, 2002:24; Nelson, 1994:108). Recall, however, Patterson's words, cited above: "Topographic correspondence and ontogenetic transformations are the usual criteria [of homology], and correspondences that pass such tests merit the same name ..." (Patterson, 1988a: 605). Once again, the criteria simply refer to homologues.

Taxic Homologues

Taxic homology might be summarized as:

1. Taxic homology = synapomorphy = taxon. If homology = synapomorphy, then
2. Taxic homology = taxon. That equation is possible only to the extent that the homologue indicates the taxon (i.e., feathers = birds), so really
3. Homologue = taxon

Does homologue equal homology? I have argued above that to do so simply equates similarity with homology and takes no account of the transformation or modification of homologues. Thus, in this sense, taxic homology also represents the homologues, and the sorting procedure (method) is simply seen as a special form of phenetics ("A cladogram ... summarizes information on shared features ... and expresses a special case of phenetic relationship, scoring positive occurrences only" [Patterson, 1988b:78; Platnick, 1985:88]).

Nelson's Milestone: Homology as Relationship

I have rather labored the point about homologues and homology, but that issue seems central and requires close examination before any progress can be made. In fact, Panchen, perhaps unwittingly, was far more succinct:

> However unsatisfying, homology should be defined simply as 'structural [and positional] similarity' rather than as 'any structural similarity due to common ancestry' (Boyden, 1973, p. 82). Or in Owen's words, 'HOMOLOGUE. The same organ in different animals under every variety of form and function.' (Panchen, 1994:56)

Panchen's words may be simplified: However unsatisfying, homology should be defined as *homologue*.

That is unsatisfying. As Nelson noted, "It should be clear, however, that homology, in general or in particular, is a relationship between homologues ... and not simply the homologues themselves" (Nelson, 1994:120). Systematics might make progress if a clear distinction is made between *homologue*, the parts of organisms and *homology*, the parts of taxa, the phylogenetic relations of organisms (Nelson, 1989a). Homologues are established by observation of processes or parts — themselves subject to appropriate theory (Rieppel and Kearney, 2002) — but will remain more or less equivalent to what has previously been called similarity or correspondence. Homology is the relationship specified by the homologues (Nelson, 1985:30, 42–44, 1989a:276, 279, 280, 282, 1994:109, 120, 131; see also Rieppel, 1991). Hennig clarified the meaning of phylogenetic relationship and framed the problem in terms of what might be called the cladistic parameter (Nelson and Platnick, 1981:318; cf. Kluge, 1993:9, cited above):

> "Phylogenetic relationship" is thus a relative concept. It is pointless (since it is self-evident) to say, as is often said, that a species or species-group is "phylogenetically related" to another. The question is rather one of knowing whether a species or species-group is more or less closely related to another than to a third. (Hennig, 1965:97)

But what does relationship mean in the context of homology? Hennig's words might be suitably changed: It is pointless to say, as is often said, that a homologue is "phylogenetically related" to another. The question is rather one of knowing whether a homologue is more or less closely related to another than to a third.

This presents another problem: where to make the comparisons, where to find the third item for comparison, as Hennig's statement above, reworded for homologues, simply states a rather imprecise version of taxic homology. Echoing Hennig's words, Bock also noted that "... to say that two features are homologous or nonhomologous conveys no information" (1969:415). The statement, "The pectoral flipper of whales is homologous to the pectoral fin of sharks" is meaningless (Figure 9.12), or to use another popular example, to say, "The wings of birds are homologous" is equally meaningless (Ghiselin, 1966:129). To compensate, Bock introduced the conditional phrase.

Addition of a conditional phrase was said to allow focus and precision so that one might say, "The pectoral flipper of whales and the pectoral fin of sharks are homologous as vertebrate limbs" (Bock, 1969:415; Figure 9.13), and "... the wings of birds and the wings of bats are homologous as the forelimbs of tetrapods" (Bock, 1974:387).

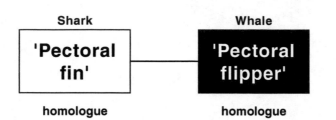

FIGURE 9.12   Diagrammatic representation of two homologues, the pectoral fin of a shark and the pectoral flipper of a whale (cf. Figure 9.5).

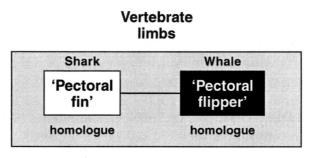

FIGURE 9.13   Diagrammatic representation of two homologues, the pectoral fin of a shark and the pectoral flipper of a whale. The addition of the conditional phrase adds the element of sameness, the shaded area = Vertebrate limb (cf. Figure 9.5).

The conditional phrase notes the hierarchical nature of homologues by distinguishing the common aspect, the element of sameness. The conditional phrase seems to be yet another explanatory device to identify or acknowledge the sameness of different homologues. This may be extended and, differing philosophies aside, allow the following equation: archetype = ancestor = symplesiomorphy = the conditional phrase = sameness. In most of the figures herein, that aspect has been represented by the stippled rectangle (Figures 9.3 to 9.5). From a Cladistic perspective, the conditional phrase remains insufficient as, in Bock's example above, it simply "affirms that wings are arms ..." (Nelson, 1994:129; see also Bang et al., 2002:180), restating the original homologues in a different form (Figures 9.12 and 9.13). The relative nature of homology is the interplay between apomorphy and plesiomorphy, its sameness and difference (Nelson, 1989a:282, 1994).

Figure 9.14a,b is a diagrammatic representation of two homologues, A and B (cf. Figure 9.11). The relevant data are supplied by homologue B and the group *c* and *d*. To capture the relation of homology, to include both apomorphy and plesiomorphy (difference and similarity), a relationship may be specified that includes the taxa within the group (*c* and *d*) and those outside (*a* and *b*). In this case, homologue B implies two specific relationships: Taxon *c* and *d* are more closely related to each other than they are to taxon *a*, and taxon *c* and *d* are more closely related to each other than they are to taxon *b*. Each individual statement of relationship includes the sameness (i.e., plesiomorphy) (taxon *a, b, c* and *d* all share the sameness of the unmodified homologue A) and the difference (i.e., apomorphy) (taxon *c* and *d* share the modified homologue B), hierarchy (taxon *c* and *d* are nested within taxa *a* to *d*) and transformation (homologue B is a modified form of homologue A). In short, each statement is a hypothesis that includes all the data relevant to the relation of homology and unites both aspects of transformational (modification) and taxic (hierarchic) homology. With a suite of specified relationships the only analytical task is to sort true statements from false.

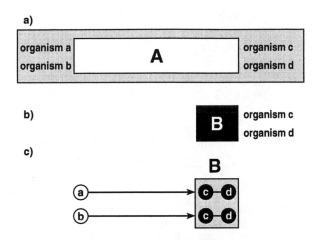

FIGURE 9.14   Diagrammatic representation of two homologues, A and B. Homologue A is possessed by taxa *a* and *b* and homologue B is possessed by taxa *c* and *d*. (a) Homologue A specifies the group *a–d* (b) Homologue B specifies the group *c* and *d*. (c) Homologue B specifies the relationship of *c* + *d* with *a*, and *c* + *d* with *b*. + *d* with *b*.

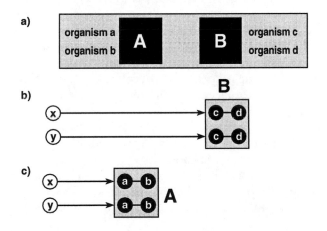

FIGURE 9.15   Diagrammatic representation of two homologues, A and B. (a) Homologue A is possessed by taxa *a* and *b* and homologue B is possessed by taxa *c* and *d*. (b) Homologue B specifies the relationships of *c* + *d* with *x* and *y* (organisms outside the area of immediate concern) (c) Homologue A specifies the relationship of *a* + *b* with *x* and *y* (organisms outside the area of immediate concern).

In a second example (Figure 9.15), there are two homologues, A and B (cf. Figure 9.12). Homologue B is possessed by taxon *c* and *d*, and homologue A is possessed by taxon *a* and *b*. The assumption here is that both homologues A and B are modified forms of some third unknown state. In this example, as homologues A and B are united by a common sameness (plesiomorphy), their relationships are specified beyond the immediate data. Nevertheless, both homologues A and B make precise statements of relationships relative to the taxon they correspond to.

This approach to the understanding of homology has recently been implemented using parsimony analysis of three-item statements (Nelson and Platnick, 1991; Nelson, 1996; Kitching et al., 1998). This has been mistakenly considered to be simply a method of analysis, and criticism has so far dwelt only on technical matters. Little has been said about the possibilities offered by specifying statements of relationship, the problems presented by equating homologues with homology or, more surprisingly, the recognition that current binary data matrices are essentially phenetic, no more than mere lists of things that are similar. Perhaps this is not too surprising, since the equation has only recently been made clear (Nelson, 1989a, 1994).

Discussion

I identified three milestones in the development of the concept of homology: (1) Owen's distinction and clarification of homology and analogy (and homologue and analogue), (2) Patterson's distinction between taxic and transformational homology, and (3) Nelson's introduction of the concept of relationship into studies of homology.

Owen said of his own work, "I have always felt and stated that I was merely making known the meaning of a term introduced into comparative anatomy long ago, and habitually used in the writings of the philosophical anatomists of Germany and France" (Owen, 1846:526). He did, however, suggest that he had advanced things a little by distinguishing different kinds of homology and, more importantly, giving the correspondences some meaning: "I have gone perhaps a little further than Oken and Geoffroy [the philosophical anatomists of Germany and France referred to above] in defining kinds of 'homology,' which appeared to me to be three, viz. 'general,' serial,' and 'special'" (Owen, 1846:526).

Owen's focus was the vertebrate archetype, from which all explanation flowed. Homological anatomy developed into a highly successful research program, which yielded new data and new interpretations of many kinds of organisms. In one sense it provided the most convincing evidence for theories concerning the origin of organisms and, ultimately, the most convincing evidence for the general idea of evolution (Amundson, 1998). The explanatory switch from archetypes to ancestors achieved a change in focus and, stimulated by Ernst Haeckel's genealogical trees, provided a way of representing and explaining all the "principal groups of the natural system" (Haeckel, 1866). Yet a more subtle change occurred, equating homologue (the parts, the namesakes) with homology (the relationship specified by the namesakes). As a consequence, homology was equated with similarity (or even correspondence; Ghiselin, 1997), which paved the way for several decades dedicated to discovering the meaning of *sameness,* either by invoking abstract entities or searching for some absolute process (cause), with neither approach ultimately having much success (De Beer, 1971).

The second milestone I identify is Patterson's distinction between taxic and transformational homology. With hindsight, it seems that transformational homology might also really refer to homologue, albeit with a cause linking the namesakes. However, that insight was only made possible once the idea was contrasted with taxic homology. To offer any meaningful hypotheses, each homologue implies a group, such that homologue and taxon are considered one and the same thing. Yet even the general idea of modification (transformation) suggests that it is not possible to divorce the transformational aspect of homology from the taxic aspect, as all homologues are assumed to be modifications of other homologues (Platnick, 1979). Cladistic analysis, in its numerical incarnation, unwittingly separated the taxic from the transformational aspect and deals only with arrays of homologues (similarity), as if the relationship of homology can only be discovered by imposing some theory of change, some method. Imposition suggests that there is no order to be discovered, and if phylogenetic reconstruction (irrespective of method) is really about imposition, then it seems to represent yet another series of approaches devoted to perfecting artificial classifications, entirely dependent upon the usefulness or otherwise of the particular algorithm of choice.

The third milestone I identify, and one yet to have made its proper impact, is Nelson's (1994) argument for extending the idea of cladistic relationship to the data themselves. It is possible to view Nelson's proposal as a reunion or marriage of the taxic and transformational aspects of homology: the transformational aspect dealing with homologues and their modification, the taxic aspect dealing with their implied relationships. Understanding homology as a relation instead of a structure should satisfy most people. The causes of homology simply become the causes of homologues (the switch is philosophical), and such investigations will certainly continue to produce useful and valuable studies, clarifying the characters (homologues) of organisms. However, with the inclusion of the specific cladistic parameter, methods for arranging and sorting homologues seem no longer necessary, and arguments concerning the superiority of parsimony, compatibility, maximum likelihood, etc. may simply evaporate and disappear (cf. Patterson, 1988b). Nelson's contribution to the homology problem is best seen as a continuation of the long dialogue concerning homology and its meaning; admittedly, that dialogue has been obscured by the desire to develop either the perfect method or find the perfect data, neither of which really do seem to exist (Nelson, 1979:21).

In summary, it might seem that confusion about the concept of homology stemmed from various understandings of the nature of data and what similarities might mean in relation to the organisms that bear them — in short, the science of systematics. Over the past 150 years one might have a certain amount of sympathy for those who have suggested abandoning the term homology altogether, regarding it as insoluble or even irrelevant (e.g., Lankester 1870a; Osborn, 1902; Spemann, 1916; Moment, 1945; Sattler, 1984; Dover, 1987; Maynard-Smith in Tautz, 1998; Stevens, 2000:83): "It is sufficient to 'know' that homology, like truth, exists, and to proceed to use, or coin, more appropriate terms for specifying what we mean in a modern scientific context" (Wake, 1994:268; see also Wake, 1999; Stevens, 2000:83). The problem, of course, might be the "modern scientific context." Over the years many "more appropriate terms" have indeed been coined: isology and paralogy (for morphology), homogeny and homophyly, homomorphy and homonomy, syngeny and allogeny (Haas and Simpson, 1946; Hall, 1992; Butler and Saidel, 2000); molecular data has donated its fair share of new words: orthology and paralogy (for molecules), xenology and paraxenology, synology and gametology, partial homology, pro-orthology, semi-orthology and trans-homology, and no doubt many others (Holland, 1999; Mindell and Meyer, 2001). Stevens noted of the early nineteenth century that "... there was a complex semantic web around words signifying relationships and resemblances — words such as 'primitive,' 'type,' 'essence,' 'organization,' 'symmetry,' and 'analogy.' From this web emerged the term 'homology' ... ." (Stevens, 1984:78)

The new "complex semantic web" (Mindell and Meyer, 2001:Figure 2) may also be understood as a series of synonyms for relationships and resemblances. The idea of resemblance remains useful, if not essential, but most probably will never be exact: "[Similarity] has so far resisted quantification (and is unlikely to submit), and it cannot be precisely defined" (Patterson, 1988a:605). Relationship is another matter. Hennig provided a solution for dealing with taxa in such a way that there was little ambiguity. Phrases such as "A is more closely related to B rather than C" became the stock phrase for cladistics in particular and systematics in general. Statements of relationship are best thought of as a representation of the cladistic parameter. If taxon and homologue are one and the same, then there seems no reason for not extending that precise expression of relationship directly to homologues.

Acknowledgments

For reading various drafts of this paper, I am grateful to Malte Ebach, Peter Forey, Chris Humphries, Gary Nelson, Olivier Rieppel and Ole Seberg.

References

Amundson, R., Typology revisited: two doctrines on the history of evolutionary biology, *Biol. Philos.*, 13, 153–177, 1998.

Andersen, N.M., Quantitative cladistics and the reconciliation of morphological and molecular systematics, in *Phylogenetik und Moleküle*, Schmitt, M., Ed., Edition Archaea, pp. 121–144.

Andersen, N.M., The impact of W. Hennig's 'phylogenetic systematics' on contemporary entomology, *Eur. J. Entomol.*, 98, 133–150, Gelsenkirchen/Schwelm, 2001.

Ax, P., *The Phylogenetic System: The Systematization of Organisms on the Basis of their Phylogenesis* (trans. Jefferies, R.P.S.), Wiley, Chichester, 1987.

Bang, R., Schulz, T.R., and DeSalle, R., Development, homology and systematics, in *Molecular Systematics and Evolution: Theory and Practice*, DeSalle, R., Giribet, G., and Wheeler, W., Eds., Birkhäuser Verlag, Basel, Switzerland, 2002, pp. 17–86.

Belon, P., *L'Histoire de la Nature des Oyseaux avec leurs Descriptions, & Naifs Portraits Retirez du Naturel Escrite en Sept Livres, etc.*, Gilles Corrozet, Paris, 1555.

Bock, W.J., Comparative morphology in systematics, in *Systematic Biology*, Publication 1692, National Academy of Sciences, Washington, DC, 1969, pp. 411–448.

Bock, W.J., Philosophical foundations of classical evolutionary classification, *Syst. Zool.*, 22, 375–392, 1974.

Bock, W.J., Foundations and methods of evolutionary classification, in *Major Patterns in Vertebrate Evolution,* Hecht, M.K., Goody, P.C., and Hecht, B.M., Eds., NATO Advanced Study Institute, Series A, 14, Plenum Press, New York, 1977, pp. 852–895.

Bock, W.J., The homology concept: its philosophical and practical methodology, *Zool. Beitr.*, NF 32, 327–353, 1989.

Bonde, N., Cladistic classification as applied to Vertebrates, in *Major Patterns in Vertebrate Evolution,* Hecht, M.K., Goody, P.C., and Hecht, B.M., Eds., Plenum Press, New York, 1977, pp. 741–804.

Bonde, N. and Hoffmeyer, J., Eds., *Udviklingsideens Historie: Fra Darwins Synthese til Nutidens Krise. Naturens Historie Fortaellere*, Vol. 2, G.E.C. Gads, Copenhagen, 1985[1996], pp. 469–495.

Boyden, A., Homology and analogy: a century after the definitions of "homologue" and "analogue" of Richard Owen, *Quarterly Rev. Biol.,* 18, 228–241, 1943.

Boyden, A., Homology and analogy, *Am. Midl. Nat.*, 37, 648–669, 1947.

Boyden, A., *Perspectives in Zoology*, Pergammon Press, Oxford, 1973.

Brooks, D., Explanation of homoplasy at different levels of biological organisation, in *Homoplasy. The Recurrence of Similarity in Evolution*, Sanderson, M.J. and Hufford, L., Eds., Academic Press, San Diego, CA, 1996, pp. 3–36.

Brower, A.V.Z., Homology and the inference of systematic relationships: some historical and philosophical perspectives, in *Homology and Systematics: Coding Characters for Phylogenetic Analysis*, Scotland, R.W. and Pennington, T., Eds., Taylor and Francis, London, 2000a, pp. 10–21.

Brower, A.V.Z., Evolution is not a necessary assumption of cladistics, *Cladistics*, 16, 143–154, 2000b.

Brower, A.V.Z., Cladistics, phylogeny, evidence and explanation a reply to Lee, *Zool. Scr.*, 31, 221–223, 2002.

Brower, A.V.Z. and Schawaroch, V., Three steps of homology assessment, *Cladistics,* 12, 265–272, 1996.

Brundin, L., Transantarctic relationships and their significance, as evidenced by chironomid midges, with a monograph of the subfamilies Podonominae and Aphroteniinae and the austral Heptagyiae, *K. Sven. Vetenskapsakad. Handl.*, 4(11), 1–472, 1966.

Butler, A.B. and Saidel, W.M., Defining sameness: historical, biological, and generative homology, *BioEssays,* 22, 846–853, 2000.

Camardi, G., Richard Owen, morphology and evolution, *J. Hist. Biol.*, 34, 481–515, 2001.

Carpenter, W.B., *Principles of Comparative Physiology*, 4th ed., London, 1854.

Cartmill, M., A critique of homology as a comparative concept, *Am. J. Phys. Anthropol.*, 94, 115–123, 1994.

Carus, C.G., *Von den Ur-Theilen des Knochen- und* Schalengerüstes, Leipzig, 1828.

Cole, F.J., *A History of Comparative Anatomy from Aristotle to the Eighteenth Century*, London, 1944.

Cracraft, J., The use of functional and adaptive criteria in phylogenetic systematics, *Am. Zool.*, 21, 21–46, 1981.

Darwin, C., *On the Origin of Species by Means of Natural Selection, or the Preservation of Favoured Races in the Struggle for Life*, John Murray, London, 1859.

De Beer, G., *Homology, an Unsolved Problem*, Oxford Biology Readers No. 11, Oxford University Press, London, 1971.

De Pinna, M.C.C., Concepts and tests of homology in the cladistic paradigm, *Cladistics*, 7, 367–394, 1991.

De Pinna, M.C.C., Species concepts and phylogenetics, *Rev. Fish Biol. Fish.*, 9, 353–373, 1999.

DeSalle, R., Agosti, D., Whiting, M., Perez-Sweeney, B., Renson, J., Baker, R., Bonacum, J., and Bang, R., Cross-roads, milestones, and landmarks in insect development and evolution: implications for systematics, *Aliso*, 14, 305–321, 1996.

Desmond, A., *Archetypes and Ancestors: Palaeontology in Victorian London 1850-1875*, Blond & Briggs, London, 1982.

Desmond, A., *Huxley: The Devil's Disciple*, Michael Joseph, London, 1994.

Desmond, A., *Huxley: Evolution's High Priest*, Michael Joseph, London, 1997.

Donoghue, M.J., Homology, in *Keywords in Evolutionary Biology*, Keller, E.F. and Lloyd, E.A., Eds., Harvard University Press, Cambridge, MA, 1992, pp. 170–179.

Donoghue, M.J. and Ree, R.H., Homoplasy and developmental constraints: a model and an example from plants, *Am. Zool.*, 40, 759–769, 2000.

Dover, G., Nonhomologous views of a terminological muddle, *Cell*, 51, 515, 1987.

Dupuis, C., Homologie et caractères: quelques aspects biologiques, *Biosyst.*, 18, 1–19, 2000.

Eckhart, T., Das Homologieproblem und Fälle strittiger Homologien, *Phytomorphology*, 14, 79–92, 1964.

Ellenius, A., Ornithological imagery as a source of scientific information, in *Non-Verbal Communication in Science Prior to 1900*, Mazzolini, R.G., Ed., L.S. Olschki, Florence, Italy, 1993, pp. 375–390.

Farber, P.L., *Finding Order in Nature: The Naturalist Tradition from Linnaeus to E.O. Wilson*, Johns Hopkins University Press, Baltimore, MD, 2000.

Farris, J.S. and Kluge, A.J., A botanical clique, *Syst. Zool.*, 28, 400–411, 1979.

Fitch, W.M., Distinguishing homologous from analogous proteins, *Syst. Zool.*, 19, 99–113, 1970.

Florkin, M., Isologie, homologie, analogie et convergence en biochimie compareé, *Bull. Acad. R. Belg., Cl. Sci.*, Series 5, 48, 819–824, 1962.

Florkin, M., *A Molecular Approach to Phylogeny*, Elsevier, Amsterdam, 1966.

Froebe, H.A., Homologiekriterien oder Argumentationsverfahren, *Ber. Dtsch. Bot. Ges.*, 95, 19–34, 1982.

Gaffney, E.S., An introduction to the logic of phylogeny reconstruction, in *Phylogenetic Analysis and Paleontology*, Cracraft, J. and Eldredge, N., Eds., Columbia University Press, New York, 1979, pp. 79–111.

Ghiselin, M.T., An application of the theory of definitions to systematic principles, *Syst. Zool.*, 15, 127–130, 1966.

Ghiselin, M.T., The nomenclature of correspondence: a new look at "homology" and "analogy," in *Evolution, Brain, and Behavior: Persistent Problems*, Masterton, R.B., Hodos, W., and Jerison, H., Eds., Lawrence Erlbaum Associates, Hillsdale, NJ, 1976, pp. 129–42.

Ghiselin, M.T., "Definition," "character," and other equivocal terms, *Syst. Zool.*, 33, 104–110, 1984.

Ghiselin, M.T., *Metaphysics and the Origin of Species,* State University of New York Press, Albany, 1997.

Ghiselin, M.T., The founders of morphology as alchemists, in *Cultures and Institutions of Natural History, Essays in the History and Philosophy of Science,* Ghiselin, M.T. and Levinton, A.E., Eds., California Academy of Sciences, San Francisco, 2000, pp. 39–49.

Goujet, D., Caractères et homologie: homologie primaire, homologie secondaire', in *Caractères, Biosyst.,* 18, 45–52, 2000.

Gould, S.J., Archetypes and adaptation, *Nat. Hist.,* 95(10), 16–27, 1986.

Gould, S.J., *The Structure of Evolutionary Theory,* Harvard University Press, Cambridge, MA, 2002.

Griffiths, G.C.D., On the foundations of biological systematics, *Acta Biotheoret.,* 23, 85–131, 1974.

Haas, O. and Simpson, G.G., Analysis of some phylogenetic terms, with attempts at redefinition, *Proc. Am. Philos. Soc.,* 90, 319–349, 1946.

Haeckel, E., *Generelle Morphologie der Organismen,* G. Reimer, Berlin, 1866, 2 vols.

Hagemann, W.J., Eine mögliche Strategie der vergleichenden Morphologie zur phylogenetischen Rekonstruktion, *Bot. Jahrb. Syst., Pflanzengesch. Pflanzengeogr.,* 96, 107–124, 1975.

Hall, B.K., *Evolutionary Developmental Biology,* Chapman and Hall, London, 1992.

Haszprunar, G., Parsimony analysis as a specific kind of homology estimation and the implications for character weighting, *Mol. Phylogenet. Evol.,* 9, 333–339, 1998.

Hawkins, J., Evolutionary developmental biology: impact on systematic theory and practice, and the contribution of systematics, in *Developmental Genetics and Plant Evolution,* Cronk, Q.C.B., Bateman, R.M., and Hawkins, J.A., Eds., Taylor Francis, London, pp. 32–51.

Hennig, W., *Grundzüge einer Theorie der Phylogenetischen Systematik,* Deutsche Zentralverlag, Berlin, 1950.

Hennig, W., Kritische Bermerkungen zum phylogenetischen System der Insekten, *Beitr. Entomol.,* 3, 1–85, 1953.

Hennig, W., Systematik und Phylogenese, *Bericht über die Hundertjahrfeier der Deutschen Entomologischen Gesellschaft, Berlin. 30 September bis 5 Oktober 1956,* von Hannemann, H., Ed., Akademie-Verlag, Berlin, 1957, pp. 50–71.

Hennig, W., Phylogenetic systematics, *Annu. Rev. Entomol.,* 10, 97–116, 1965.

Hennig, W., *Phylogenetic Systematics,* University of Illinois Press, Urbana, 1966 (reprinted in 1979, 1999).

Holland, P.W.H., The effect of gene duplication on homology, in *Homology,* Novartis Foundation Symposium 222, John Wiley and Sons, Chichester, UK, 1999, pp. 226–236.

Hubbs, C.L., Concepts of homology and analogy, *Am. Nat.,* 78, 289–307, 1944.

Hunter, I.J., Paralogy, a concept complementary to homology and analogy, *Nature,* 204, 604, 1964.

Huxley, L., *Life and Letters of Thomas Henry Huxley, &c.,* 2nd ed., Macmillan, London, 1913.

Huxley, T.H., Owen's position in the history of anatomical science, in *The Life of Richard Owen by his Grandson R.* Owen, London, 1894, pp. 658–689.

Inglis, W.G., The observational basis of homology, *Syst. Zool.,* 15, 219–228, 1966.

Kaplan, D.R., The concept of homology and its central role in the elucidation of plant systematic relationships, in *Cladistics: Perspectives on the Reconstruction of Evolutionary History,* Duncan, T. and Stuessy, T.F., Eds., Columbia University Press, New York, 1984, pp. 51–70.

Kaplan, D.R., The science of plant morphology: definition, history, and the role of morphology, *Am. J. Bot.,* 88, 1711–1741, 2001.

Kiriakoff, S.G., [Review of] Adolf Remane: *Die Grundlagen des Natürlichen Systems, der Vergleichenden Anatomie und der Phylogenetik. Theoretische Morphologie und Systematik I.* Akademische Verlagsgesellschaft, Geest and Portig, K.-G. Leipzig, *Bull. Ann. Soc. R. Entomol. Belg.,* 90, 157, 1954.

Kitching, I.J., Forey, P.L., Humphries, C.J., and Williams, D.M., *Cladistics: The Theory and Practice of Parsimony Analysis,* 2nd ed., Oxford University Press, Oxford, 1998.

Kluge, A.G., Concepts and principles of morphologic and functional studies, in *Chordate Structure and Function,* Waterman, A.J., Ed., MacMillan, New York, 1971, pp. 3–41.

Kluge, A.G., Concepts and principles of morphologic and functional studies, in *Chordate Structure and Function,* 2nd ed., Kluge, A.G., Ed., MacMillan, New York, 1977, pp. 3–41.

Kluge, A.G., Boine Snake Phylogeny and Research Cycles, *Misc. Publ., Mus. Zool. Univ. Mich.,* 178, 1–58, 1991.

Kluge, A.G., *Aspidites* and the phylogeny of pythonine snakes, *Rec. Aust. Mus., Suppl.,* 19, 1–77, 1993.

Kluge, A.G., Parsimony with and without scientific justification, *Cladistics,* 17, 199–210, 2001.

Kluge, A.G. and Strauss, R.E., Ontogeny and systematics, *Annu. Rev. Ecol. Syst.,* 16, 247–268, 1985.

Ladiges, P.Y., Biogeography after Burbidge, *Aust. Syst. Bot.,* 11, 231–242, 1998.

Lankester, E.R., On the use of the term homology in modern zoology, and the distinction between homogenetic and homoplastic agreements, *Ann. Mag. Nat. Hist.,* Ser. 4, 6, 34–43, 1870a.

Lankester, E.R., On the use of the term homology, *Ann. Mag. Nat. Hist.,* Ser. 4, 6, 342, 1870b.

Lee, M.S.Y., Divergent evolution, hierarchy and cladistics, *Zool. Scr.,* 31, 217–219, 2002.

Løvtrup, S., On von Baerian and Haeckelian Recapitulation, *Syst. Zool.,* 27, 348–352, 1978.

Mabee, P., The usefulness of ontogeny in interpreting morphological characters in *Phylogenetic Analysis of Morphological Data,* Wiens, J., Ed., Smithsonian Institution Press, Washington, DC, 2000, pp. 84–114.

MacLeod, R.M., Evolutionism and Richard Owen, 1830-1868: an episode in Darwin's century, *Isis,* 56, 259–280, 1965.

Mayr, E. and Ashlock, P.D., *Principles of Systematic Zoology,* 2nd ed., McGraw-Hill, New York, 1991.

Mayr, E and Bock, W.J., Classifications and other ordering systems, *J. Zool. Syst. Evol. Res.,* 40, 169–194, 2002.

McKitrick, M.C., On homology and the ontological relationship of parts, *Syst. Biol.,* 43:1–10, 1994.

McMurrich, J.P., *Leonardo da Vinci the Anatomist (1452-1519),* Williams and Wilkins, Baltimore, MD, 1930.

Mindell, D.P. and Meyer, A., Homology evolving, *Trends Ecol. Evol.,* 16, 434–440, 2001.

Minelli, A., Some thoughts on homology 150 years after Owen's definition, *Mem. Soc. Ital. Sci. Nat. Mus. Civ. Stor. Nat. Milano,* 27, 71–79, 1996.

Minelli, A., Uccello Ugual Uomo, Pipistrello Uguale Uomo, in *Volatilia: Animali dell'Aria nella Storia della Scienza da Aristole ai Giorni Nostri,* Longo, O., Ed., Generoso Procaccini, Naples, 1999, pp. 87–100.

Minelli, A. and Schram, F.R., Owen revisited: a reappraisal of morphology in evolutionary biology, *Bijdr. Dierkd.,* 64, 65–74, 1994.

Mitter, C., Sketches of U.S. systematic entomology, circa 1850-2000: return of a golden age, *Ann. Entomol. Soc. Am.,* 92, 798–811, 1999.

Mivart, St. G., On the use of the term "homology," *Ann. Mag. Nat. Hist.,* ser. 4, 6, 113–121, 1870.

Mivart, St. G., Sir Richard Owen's hypotheses, *Nat. Sci.,* 2, 18–23, 1893.

Moment, G.B., The relationship between serial and special homology and organic similarities, *Am. Nat.,* 79, 445–455, 1945.

Naef, A., *Idealistische Morphologie und Phylogenetik (zur Methodik der systematischen),* Jena, Verlag von Gustav Fischer, 1919.

Naef, A., Die Cephalopoden (Systematik), in *Fauna e Flora del Golfo di Napoli,* Monograph 35, Pubblicata dalla Stazione Zoologica di Napoli. R. Friedländer and Sohn, Berlin, 1921–1923.

Naef, A., Die Cephalopoden (Embryologie), in *Fauna e Flora del Golfo di Napoli,* Monograph 35, Pubblicata dalla Stazione Zoologica di Napoli. R. Friedländer and Sohn, Berlin, 1928.

Naef, A., Cephalopoda (systematics), in *Fauna and Flora of the Bay of Naples (Fauna e Flora del Golfo di Napoli)*, Monograph 35, Part I, Vol. I, Fascicle II [End of Vol. I]. Smithsonian Institute Libraries, Washington, DC, 1972 (trans. of Naef 1921–1923).

Naef, A., Cephalopoda: embryology, in *Fauna and Flora of the Bay of Naples [Fauna und Flora des Golfes von Naepel]*. Monograph 35. Part I, Vol. II [Final part of Monograph 35], Smithsonian Institute Libraries, Washington, DC, 2000, pp. 3–461 (trans. of Naef 1928).

Nelson, G.J., Ontogeny, phylogeny, paleontology, and the biogenetic law, *Syst. Zool.*, 27, 324–345, 1978.

Nelson, G.J., Cladistic analysis and synthesis: principles and definitions, with a historical note on Adanson's *Familles des Plantes* (1763-1764), *Syst. Zool.*, 28, 1–21, 1979.

Nelson, G.J., Outgroups and ontogeny, *Cladistics*, 1, 29–45, 1985.

Nelson, G.J., Cladistics and evolutionary models, *Cladistics*, 5, 275–289, 1989a.

Nelson, G.J., [Review of] Reconstructing the Past: Parsimony, Evolution, and Inference, E. Sober, *Syst. Zool.*, 38, 293–294, 1989b.

Nelson, G.J., Homology and systematics, in *Homology: The Hierarchical Basis of Comparative Biology*, Hall, B.K., Ed., Academic Press, London, 1994, pp. 101–149.

Nelson, G.J., *Nullius in verba, J. Comp. Biol.*, 1, 141–152, 1996.

Nelson, G.J. and Ladiges, P.Y., Paralogy in cladistic biogeography and analysis of paralogy-free subtrees, *Am. Mus. Novit.*, 3167, 1–58, 1996.

Nelson, G.J. and Platnick, N.I., *Systematics and Biogeography: Cladistics and Vicariance*, Columbia University Press, New York, 1981.

Nelson, G.J. and Platnick, N.I., Three-taxon statements: a more precise use of parsimony?, *Cladistics*, 7, 351–366, 1991.

Nelson, G.J., Williams, D.M., and Ebach, M.C., A question of conflict: three-items and standard parsimony compared, *Syst. Biodivers.*, 1, 145–149, 2003.

Osborn, H.F., Homoplasy as a law of latent or potential homology, *Am. Nat.*, 36, 259–271, 1902.

Osborn, H.F., *Evolution of Mammalian Molar Teeth to and from the Triangular Type*, New York, 1907.

Ospovat, D., *The Development of Darwin's Theory: Natural History, Natural Theology and Natural Selection 1838-1859*, Cambridge University Press, Cambridge, 1981.

Owen, R., *Lectures on the Comparative Anatomy and Physiology of the Invertebrate Animals, Delivered at the Royal College of Surgeons in 1843, from Notes taken by W. W. Cooper ... and Revised by Prof. Owen, &c.*, Longman, Brown, Green and Longman, London, 1843.

Owen, R., Observation on Mr. Strickland's article on the structural relations of organized beings, *Philos. Mag.*, 28, 525–527, 1846.

Owen, R., Report on the archetype and homologies of the vertebrate skeleton, *Report of the 16th Meeting of the British Association for the Advancement of Science*, Murray, London, 1847, pp. 169–340.

Owen, R., *On the Archetype and Homologies of the Vertebrate Skeleton*, London, 1848.

Owen, R., *On the Nature of Limbs*, John van Voorst, London, 1849.

Owen, R., Palaeontology, in *Encyclopedia Britannica*, Vol. 17, 8th ed., Adam and Charles Black, Edinburgh, 1859, pp. 91–176.

Owen, R., *On the Anatomy of Vertebrates*, Vol. 1. *Fishes and Reptiles*, Longmans, Green, London, 1866

Padian, K., On Richard Owen's archetype, homology, and the vertebral theory: interrelations and implications, in *XVII International Congress on the History of Science, Abstracts Scientific Sections*, Bj 5.P, Acts 1, 1985.

Padian, K., A missing Hunterian lecture on vertebrae by Richard Owen, *J. Hist. Biol.*, 28, 333–368, 1995.

Padian, K., The rehabilitation of Sir Richard Owen, *BioScience*, 47, 446–453, 1997.

Panchen, A.L., Richard Owen and the concept of homology, in *Homology: The Hierarchical Basis of Comparative Biology*, Hall, B.K., Ed., Academic Press, San Diego, CA, 1994, pp. 21–62.

Panchen, A.L., Homology: history of a concept, in *Homology*, Novartis Foundation Symposium 222, John Wiley and Sons, Chichester, UK, 1999, pp. 5–18.

Papavero, N., Llorente-Bousquets, J., and Espinosa-Organista, D., *Historia de la Biología Comparada*, Vol. III. *De Nicolás de Cusa a Francis Bacon*, D.R. Universidad Nacional Autonoma de Mexico, Mexico, 1995.

Patterson, C., *Evolution*, British Museum (Natural History), London, 1978.

Patterson, C., Significance of fossils in determining evolutionary relationships, *Annu. Rev. Ecol. Syst.*, 12:195–223, 1981.

Patterson, C., Morphological characters and homology, in *Problems of Phylogenetic Reconstruction*, Joysey, K.A. and Friday, A.E., Eds., Academic Press, London, 1982, pp. 21–74.

Patterson, C., How does phylogeny differ from ontogeny?, in *Development and Evolution*, Goodwin, B.C., Holder, H., and Wylie, C.C., Eds., Cambridge University Press, Cambridge, 1983, pp. 1–31.

Patterson, C., Evolution: neo-Darwinian theory, in *The Oxford Companion to the Mind*, Gregory, R.L., Ed., Oxford University Press, Oxford, 1987, pp. 234–244.

Patterson, C., Homology in classical and molecular biology, *Mol. Biol. Evol.*, 5, 603–625, 1988a.

Patterson, C., The impact of evolutionary theories on systematics, in *Prospects in Systematics*, Hawksworth, D.L., Ed., Clarendon Press, Oxford, 1988b, pp. 59–91.

Patterson, C., Null or minimal models, in *Models in Phylogeny Reconstruction*, Scotland, R.W., Siebert, D.J., and Williams, D.M., Eds., Clarendon Press, Oxford, 1994, pp. 173–192.

Patterson, C., *Evolution*, 2nd ed., The Natural History Museum, London, 1999.

Patterson, C., Evolutionism and creationism, *Linnean*, 18, 15–32, 2002.

Platnick, N.I., Philosophy and the transformation of cladistics, *Syst. Zool.*, 28, 537–546, 1979.

Platnick, N.I., Philosophy and the transformation of cladistics revisited, *Cladistics*, 1, 87–94, 1985.

Platnick, N.I. and Nelson, G.J., The purpose of biological classification, *Proceedings of the 1978 Biennial Meeting of the Philosophy of Science Association*, 2, 117–129, 1981.

Reif, W.-E., Evolutionary theory in German paleontology, in *Dimensions of Darwinism*, Grene, M., Ed., Cambridge University Press, Cambridge, 1983, pp. 173–203.

Reif, W.-E., Adolf Naefs Idelistiche Morphologie und das Paradigma typologischer Makroevolutionstheorien, *Verh. Gesch. Theor. Biol.*, 1, 411–424, 1998.

Remane, A., *Die Grundlagen des natürlichen Systems, der vergleichenden Anatomie und der Phylogenetik, Theoretische Morphologie und Systematik I.*, Akademische Verlagsgesellschaft, Leipzig, Geest and Portig, K.-G., 1952 (2nd ed. 1956; reprint of 2nd ed. 1971).

Remane, A., Morphologie als Homologienforschung, *Verh. Dtsch. Zool. Gesellschaft*, 18(Suppl.), 159–183, 1955.

Remane, A., Gedanken zum Problem: Homologie und analogie, praeadaption und parallelität, *Zool. Anz.*, 166, 447–465, 1961.

Remane, J., Grenzen der Anwendung von Maschinenanalogie und Okonomieprinzip in der stammesgeschichtlichen Forschung, *Aufs. Reden Senckenb. Naturforsch. Ges.*, 35, 155–177, 1985.

Remane, J., Die Entwicklung des Homologie-Begriffs seit Adolf Remane, *Zool. Beitr.*, 32 (NF), 497–501, 1989.

Richards, E., A question of property rights: Richard Owen's evolutionism reassessed, *Br. J. Hist. Sci.*, 20, 129–171, 1987.

Richards, R.J., *The Meaning of Evolution: The Morphological Construction and Ideological Reconstruction of Darwin's Theory*, University of Chicago Press, Chicago, 1992.

Ridley, M., *Evolution*, 2nd ed., Blackwell Scientific, Cambridge, MA, 1996.

Riedl, R., *Order in Living Organisms: A Systems Analysis of Evolution*, Wiley, Chichester, 1978.

Riedl, R., The role of morphology in the theory of evolution, in *Dimensions of Darwinism*, Grene, M., Ed., Cambridge University Press, Cambridge, 1983, pp. 205–238.

Riedl, R., Opening address and introduction, *Fortschr. Zool. (Progr. Zool.)*, 35, vii–xvi, 1989.

Rieppel, O., Homology, a deductive concept?, *Z. Zool. Syst. Evol.forsch.*, 18, 315–319, 1980.

Rieppel, O., [Review of] *Die Spaltung des Weltbildes: Biologische Grundlagen des Erklärens und Verstehens*. Rupert Riedl, *Cladistics,* 2, 196–200, 1986.

Rieppel, O., Pattern and process: the early classification of snakes, *Biol. J. Linn. Soc.,* 31, 405–420, 1987.

Rieppel, O., *Fundamentals of Comparative Biology,* Birkhäuser, Basel, Switzerland, 1988.

Rieppel, O., Things, taxa and relationships, *Cladistics,* 7, 93–100, 1991.

Rieppel, O., Homology and logical fallacy, *J. Evol. Biol.,* 5, 701–715, 1992.

Rieppel, O., The conceptual relationship of ontogeny, phylogeny, and classification: the taxic approach, *Evol. Biol.,* 27, 1–32, 1993.

Rieppel, O., Homology, topology, and typology: the history of the modern debates, in *Homology: The Hierarchical Basis of Comparative Biology,* Hall, B.K., Ed., Academic Press, London, 1994, pp. 63–100.

Rieppel, O. and Kearney, M., The origin of snakes: limits of a scientific debate, *Biologist,* 48, 110–114, 2001.

Rieppel, O. and Kearney, M., Similarity, *Biol. J. Linn. Soc.,* 75, 59–82, 2002.

Roth, V., On homology, *Biol. J. Linn. Soc.,* 22, 13–29, 1984.

Rudwick, M., *The Meaning of Fossils: Episodes in the History of Palaeontology,* Macdonald, London, 1976.

Ruse, M., *The Darwinian Revolution: Science Red in Tooth and Claw,* University of Chicago Press, Chicago, 1979 (reprinted 1999).

Russell, E.S., *Form and Function: A Contribution to the History of Animal Morphology,* John Murray, London, 1916.

Rupke, N., Richard Owen's vertebrate archetype, *Isis,* 84, 231–251, 1993.

Rupke, N., *Richard Owen: Victorian Naturalist,* Yale University Press, New Haven, CT, 1994.

Saether, O.A., Underlying synapomorphies and anagenetic analysis, *Zool. Scr.,* 8, 305–312, 1979.

Sanderson, M.J. and Hufford, L., Eds., *Homoplasy: The Recurrence of Similarity in Evolution,* Academic Press, San Diego, CA, 1996.

Sattler, R., Methodological problems in taxonomy, *Syst. Zool.,* 13, 19–27, 1964.

Sattler, R., Homology: a continuing challenge, *Syst. Bot.,* 9, 382–384, 1984.

Schmitt, M., The homology concept: still alive, in *The Nervous System of Invertebrates: An Evolutionary and Comparative Approach,* Breidbach, O. and Kutsch, W., Eds., Birkhäuser, Basel, Switzerland, 1995, pp. 425–38.

Schuh, R.T., *Biological Systematics: Principles and Applications,* Comstock, Ithaca, NY, 2000.

Simpson, G.G., Anatomy and morphology: classification and evolution: 1859 and 1959, *Proc. Am. Philos. Soc.,* 103, 286–306, 1959.

Simpson, G.G., *Principles of Animal Taxonomy,* Columbia University Press, New York, 1961.

Singer, C.J., *A Short History of Biology: A General Introduction to the Study of Living Things,* Clarendon Press, Oxford, 1931.

Singer, C.J., *A History of Biology: A General Introduction to the Study of Living Things,* rev. ed., H. K. Lewis, London, 1950.

Singer, C.J., *A History of Biology to about the Year 1900: A General Introduction to the Study of Living Things,* 3rd ed., Abelard-Schuman, London, 1959.

Sloan, P.R., On the edge of evolution, in *R. Owen: The Hunterian Lectures in Comparative Anatomy, May and June 1837, with an Introductory Essay and Commentary by P.R. Sloan,* Natural History Museum, London, 1992, pp. 3–72.

Sloan, P.R., Whewell's philosophy of discovery and the archetype of the vertebrate skeleton: the role of German philosophy of science in Richard Owen's biology, *Ann. Sci.,* 60, 39–61, 2003.

Sluys, R., Rampant parallelism: an appraisal of the use of nonuniversal derived character states in phylogenetic reconstruction, *Syst. Zool.,* 38, 350–370, 1989.

Sluys, R., The notion of homology in current comparative biology, *Z. Zool. Syst. Evolutions-forsch.,* 34, 145–152, 1996.

Sneath, P.H.A., [Review of Nelson, G.J. and Platnick, N.I. Systematics and biogeography, 1981], *Syst. Zool.,* 31, 208–217, 1982.

Sneath, P.H.A., Philosophy and method in biological classification, in *Numerical Taxonomy*, Felsenstein, J., Ed., Springer, Berlin, 1983, pp. 22–37.

Sneath, P.H.A., The phenetic and cladistic approaches, in *Prospects in Systematics*, Hawksworth, D.L., Ed., Clarendon Press, Oxford, 1988, pp. 252–73.

Sneath, P.H.A. and Sokal, R.R., *Numerical Taxonomy: The Principles and Practice of Numerical Classification*, W.H. Freeman, San Francisco, 1973.

Sober, E., *Reconstructing the Past: Parsimony, Evolution, and Inference*, Massachusetts Institute of Technology, Cambridge, MA, 1988.

Sokal, R.R., Typology and empiricism in taxonomy, *J. Theoret. Biol.*, 3, 230–267, 1962.

Spemann, H., Zur Geschichte und Kritik des Begriffes der Homologie, *Kult. Ggw.: Allg. Biol.*, 3, Teil, Abteilungen 4, Bd. 1, 63–86, 1915.

Staton, J.L., Homology in character evolution, in *Encyclopedia of Life Sciences*, London, Nature Publishing Group, http:/www/elsnet, 2000.

Steiner, H., Die Bedeutung des Homologiebegriffes für die Biologie, *Vierteljahrsschr. Nat.forsch. Ges. Zürich*, 99, 1–19, 1954.

Stevens, P.F., Haüy and A.-P. Candolle: crystallography, botanical systematics, and comparative morphology, 1780-1840, *J. Hist. Biol.*, 17, 49–82, 1984.

Stevens, P.F., *Plants and Animals, Form and Relationships: An Exhibition at Houghton Library, June 1–August 21*, 1998, Harvard University, pp. 1–21.

Stevens, P.F., On characters and character states: do overlapping and non-overlapping variation, morphology and molecules all yield data of the same value?, in *Homology and Systematics: Coding Characters for Phylogenetic Analysis*, Scotland, R.W. and Pennington, T., Eds., Taylor and Francis, London, 2000, pp. 81–105.

Strickland, H.E., On the use of the word homology in comparative anatomy, *Philos. Mag.*, 28, 35, 1846.

Tassy, P., La théorisation du charactère en cladistique, *Caractères Biosyst.*, 18, 33–43, 2000.

Tautz, D., Debatable homologies, *Nature*, 395, 17–18, 1998.

Tuomikoski, R., Notes on some principles of phylogenetic systematics, *Ann. Entomol. Fenn.*, 33, 137–147, 1967.

Van Valen, L.M., Homology and causes, *J. Morphol.*, 173, 305–312, 1982.

von Boletsky, S., Systematische Morphologie und Phylogenetik: zur Bedeutung des Werkes von Adolf Naef (1883-1949), *Vierteljahrsschr. Nat.forsch. Ges. Zürich*, 144, 73–82, 1999.

von Boletsky, S., Adolf Naef (1883-1949): A biographical note, in *Fauna and Flora of the Bay of Naples (Fauna und Flora des Golfes von Naepel)*, Monograph 35, Cephalopoda. Embryology. Part I, Vol. II (Final part of Monograph No. 35), pp. ix–xiii. Smithsonian Institute Libraries, Washington, DC, 2000.

Wagner, G.P., The systems approach: an interface between developmental and populational genetic aspects of evolution, in *Patterns and Processes in the History of Life*, Raup, D.M. and Jablonski, D., Eds., Springer-Verlag, Berlin, 1986, pp. 149–165.

Wagner, G.P., The biological homology concept, *Annu. Rev. Ecol. Syst.*, 20, 51–69, 1989.

Wagner, G.P., Homology and the mechanisms of development, in *Homology: The Hierarchical Basis of Comparative Biology*, Hall, B.K., Ed., Academic Press, San Diego, 1994, pp. 273–299.

Wake, D.B., Comparative terminology, *Science*, 265, 268–269, 1994.

Wake, D.B., Introduction, in *Homoplasy: The Recurrence of Similarity in Evolution*, Sanderson, M.J. and Hufford, L., Eds., Academic Press, San Diego, CA, 1996, pp. xvii–xxv.

Wake, D.B., Homoplasy, homology and the problem of 'sameness' in biology, in *Homology*, Novartis Foundation Symposium 222, John Wiley and Sons, Chichester, UK, 1999, pp. 24–33.

Wheeler, Q.D. and Meier, R., *Species Concepts and Phylogenetic Theory*, Columbia University Press, New York, 2000.

Wiley, E.O., *Phylogenetics: The Theory and Practice of Phylogenetic Systematics*, Wiley and Sons, Interscience, New York, 1981.

Williams, D.M., Parsimony and precision, *Taxon,* 51, 143–149, 2002.

Zangerl, R., The methods of comparative anatomy and its contribution to the study of evolution, *Evolution,* 2, 351–374, 1948.

Zimmermann, W., Methoden der Evolutionswissenschaft (=Phylogenetik), in *Die Evolution der Organismen: Ergebnisse und Probleme der Abstammungslehre. 3. Völlig neu Bearbeitete und Erweiterte Auflage,* Heberer, G., Ed., Band 1, 1967, Gustav Fischer, Stuttgart, pp. 61–160.

Zunino, M. and Colomba, M.S., *Ordinando la Natura: Elementi di Storia del Pensiero Sistematico in Biologia,* Medical Books, Palmero, 1997.

Chapter 10

From Dispersal to Geographic Congruence: Comments on Cladistic Biogeography in the Twentieth Century

Christopher J. Humphries

Abstract

Historical biogeography stems from over several hundred years ago, although it is generally acknowledged that its intellectual origins were during the time of Linnaeus, when he started classifying all of the organisms of the globe into a general classification. Linnaeus considered that all organisms originated from a mountainous, tropical island and dispersed to the remainder of the Earth. His dispersalist approach endures to the present day, even though, as shown from major achievements during the past 50 years or more, that the Earth is a dynamic system wherein the evolution of organisms is intimately tied up with tectonic changes through time. The biogeographic literature is cluttered with a range of perspectives and methods around this modern threefold parallelism of space, time and form and whether biogeography determines general patterns of individual species histories. The purpose of this chapter is to examine the key developments in analytical approaches, especially since the publication in 1964 of *Space, Time, Form: The Biological Synthesis* by Léon Croizat to the very recent considerations of cladistic biogeography in the meaning of geographical congruence.

Introduction

By now one might have thought that historical biogeography should have made considerable material progress since Linnaeus (1781) first pronounced that species originated in Paradise (Ball, 1976; Nelson and Platnick, 1981; Humphries and Parenti, 1986, 1999). There have been many developments since that time, and the subject has meandered along with a gradual accumulation of insights in systematics, geology and biogeography. However, at the same time there are some persistent hoary old chestnuts that continue to crop up under different theoretical and methodological guises. It is safe to say that the subject has reached the point that the nature of allopatric differentiation has been "generally agreed upon to be the history of the natural world," although the historical factor underlying allopatry "has been differently conceived by Buffon as *Dispersal*, de Candolle as *Earth History*, by Lyell as *Creation*, by Darwin as *Improbable Dispersal* ('occasional means of transport') and Croizat as *Plate Tectonics*" (Nelson, 1978:302, including quotation above).

The problem is that all of these paradigms persist[1] and have accumulated over the past 150 years, with considerable proliferation in the past 50 years or more. It now seems pertinent to ask what constitutes progress, a topic that has already been touched upon by Crisci (2001). The subject has moved in fits and starts with periodic revivals of enthusiasm in new books and the occasional symposium (Croizat, 1964; Forey, 1981; Nelson and Platnick, 1981; Nelson and Rosen, 1981; Cracraft, 1982, 1985, 1989; Leviton and Aldrich, 1985; Pierrot-Bults et al., 1986; Hovenkamp, 1987; Grehan, 1988; Myers and Giller, 1988; Craw, 1989a; Ladiges et al., 1991; Crisci and Morrone, 1995). However, judging by the column inches over the past two years in serials such as the *Journal of Biogeography*, *Cladistics* and *Systematic Biology*, it is clear that we have entered a new phase of methodological machination (Craw et al., 1999; Humphries and Parenti, 1999; van Veller, 2000; Bousquets and Morrone, 2001; Brooks and McLennan, 1991, 2002; Espinosa Organista et al., 2002). The time is ripe to ask, What does the proliferation of methodology really tell us? Which of the widely different theories, methods and explanations for assessing biogeographic patterns across the globe can be safely said to have moved the subject forward, and which ones should be discarded?

In trying to answer the question, it seems that historical biogeography has befallen the same fate as systematics. It has forgotten what the nature of the pursuit is. In systematics, for example, the distinction between cladograms and trees has been lost in recent years since molecular biology has come to dominate the scene, and a whole plethora of techniques from phenetics to maximum likelihood assessments for producing phylogenetic trees masquerade under the rubric of cladistics. Maybe I am old fashioned but I still consider that systematics and evolutionary studies should be kept apart (Platnick, 1979). Classifications are accessible through character analysis, but trees require information beyond that available for character analysis.

The same pattern and process divide is true in historical biogeography. One disturbing observation is that rather than creating an arena of discussion, historical biogeographers are becoming politically balkanized and talking past, or even misrepresenting, the ideas of others (see for example the papers in Nelson and Rosen, 1981; compare Craw et al., 1999 with Humphries and Parenti, 1999; and see van Veller, 2000; Van Veller et al., 2000, 2001, 2002; Brooks and McLennan, 2002). Therefore, it is possible to see some material progress, but the relative importance of pattern versus process has created a situation whereby the aims of historical biogeography are at variance with the different factions.

Thus, I agree with Crisci (2001) when he says that to understand the extraordinary proliferation of methods it would be profitable to examine the external pressures that dominate biogeographic work and the internal forces that shape divergence of opinion. The external pressures have been the changes in the understanding of geology, especially continental drift and plate tectonics, and the internal pressures are theory and methods of systematics, notably cladistics. Internal forces in this context have one major divide between panbiogeography and cladistic biogeography, and within the latter it is how the cladistic methods used in biogeography that create confusion in methodology. For example, the recent studies on three-item statements analysis, geographical paralogy and subtree analysis of Nelson and Ladiges (1996, 2001) and area cladistics of Ebach and Humphries (2002) have considered the debate to be about discovery of patterns from data, whereas current methods in phylogeography (Avise, 2000; Arbogast and Kenagy, 2001) and Brooks parsimony analysis (Brooks and McLennan, 2002; Brooks et al., 2001), for example, consider biogeography to be more about explanation of speciation and species history. Thus, it is my plan to briefly examine the external and internal forces by examining the sharpest differences of opinion and I hope to extract the milestones over the past century and especially since Croizat (1952)

---

[1] This chapter expands on an assessment of biogeography by Humphries (2000).

published his *Manual of Phytogeography* and Nelson wrote his seminal review (Nelson, 1978).

External Forces

Global Plate Tectonics

Global plate tectonics is the dominant paradigm in the geosciences, but getting there has been a long process. For almost two centuries commentators of Earth history have considered that life and Earth are separate entities. Dispersal from centers of origin and geophysical stasis, or at best gradual change, had been the enduring biogeographical paradigm from the mid-nineteenth century until well into the mid-twentieth century. Sir Charles Lyell (1830–1833), who rebelled against the prevailing biblically influenced theories of geology of the time, thought that catastrophe theories were biased, because they were based on interpretations of the writings in the *Book of Genesis*. Instead of such rapid catastrophic changes in geology in relatively short periods of time (anything between 6 days and 4000 years), he thought that it would be more practical to have gradual changes through time to vouch for the fossil remains of extinct species increasingly being unearthed. He believed that it was necessary to create a vastly long timescale for Earth's history with long periods of geological stasis.

Early in the twentieth century Taylor (1910) was the first to formulate the assumption that the Atlantic was formed by the separation of two continents that were derived from a former contiguous landmass but at a much slower rate than the early catastrophists implied.[2] In a way, this was a reincarnation of Snider-Pelligrini (1858) (Figure 10.1). Taylor made his assumption based on the similarity of the layout of the coasts on both sides of the Atlantic but also on the fact that one finds assembly lines on the continental terraces opposed to the Atlantic margins, for example in the rock types of North America and the Andes. Taylor's explanations appeared too complicated, and he did not succeed in convincing his contemporaries. Consequently, it was not until *Dei Enstehung der Kontinente und Ozeane* (1915, 4th ed. 1929) by the German geophysicist, astronomer and meteorologist, Alfred Wegener, that the idea of continental movement took hold of the imagination of biogeographers and geo-scientists alike. But Wegener's idea was still a bitter pill to swallow for the geological gradualists.

In the 1920s most geologists were not happy with continental drift because the specific mechanisms proposed by Wegener were considered unfeasible. Biogeography stuck firmly to

---

[2] Oddly enough, the catastrophe theories provided early grist to the idea of continental drift. Prior to the seventeenth century, maps of land areas on either side of the Atlantic Ocean were sufficiently clear and precise such that curious minds noted a certain parallelism in the layout of the coasts of Africa and South America (Romm, 1994). Scientists of the time tried to find an explanation for the apparent fit of the two continents. Even four centuries ago Francis Bacon, in his *Novum Organum* (1610), considered that the west coast of Africa and the east coast of South America looked as if they would fit together like pieces of a jigsaw. In 1756, Theodor Christoph Lilienthal also noted the mirrored similarities of the two continental coastlines (http://www4.geology.utoronto.ca/glg105h/lecture13/ContinentaDrift1lect.htm), and (1668) in his book, *La Corruption du Grand et du Petit Monde* (see Carozzi 1969) suggested that America was actually joined to Africa. Indeed he considered that prior to the great flood all continents on the planet were joined together. He proposed that there was one continental block and that by collapse of its center the Atlantic was created, and from it two separate terrestrial blocks emerged. Antonio Snider-Pelligrini (1858), in his book *La Création et ses Mystères Dévoilés*, wrote about continental separation and drift. According to him (and even biogeographers such as von Humboldt and Bonpland 1805), the continents of America and South America formed one block of fused rocks prior to the great flood. The flood cooled the block, and a gigantic and rapid rupture then occurred, involving the separation of the Americas and the Old World (Figure 10.1). Later, John Henry Pepper (1861) arrived at the same conclusion to explain similar coal deposit fossils in both North America and Europe. More interestingly, he came up with the notion of continental drift, and in 1881, the Reverend Osmond Fisher in his *Physics of the Earth's Crust* was the first to suggest that the continents could move around by convection currents (Fisher 1881).

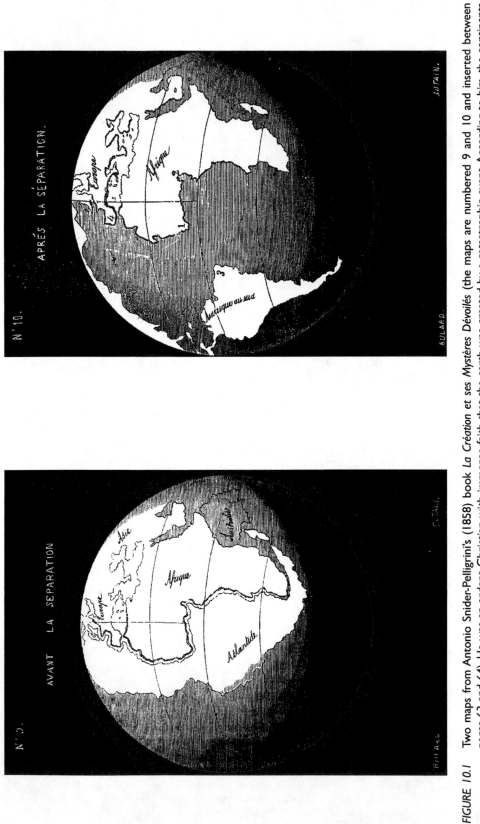

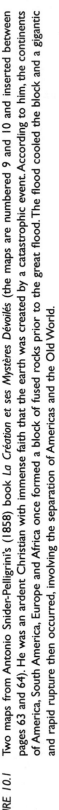

FIGURE 10.1   Two maps from Antonio Snider-Pelligrini's (1858) book *La Création et ses Mystères Dévoilés* (the maps are numbered 9 and 10 and inserted between pages 63 and 64). He was an ardent Christian with immense faith that the earth was created by a catastrophic event. According to him, the continents of America, South America, Europe and Africa once formed a block of fused rocks prior to the great flood. The flood cooled the block and a gigantic and rapid rupture then occurred, involving the separation of Americas and the Old World.

an Earth of Lyellian fixed geometry for another 50 years. Alexis du Toit (1937), Wegener's South African champion, tried a different tack. He argued for a local, radioactive melting of the oceanic floor at continental borders permitting the continents to glide through. This *ad hoc* hypothesis added no increment of plausibility to Wegener's speculation. Since drift seemed absurd in the absence of a mechanism, orthodox geologists set out to render the impressive evidence for it as a series of unconnected coincidences. Consequently, it still took a long time for continental drift and its relations to geographic disjunctions to become accepted.

However, as we all now accept, in the 1960s a whole range of geological observations coalesced to provide unequivocal evidence for continental drift. The methods included mapping of the ocean bottoms using echo-sounding devices. Dates of the north-south changes in the Earth's magnetic field were revealed by measures of magnetic memory and from radiometric dates of rocks, and finally, there was very accurate determination of the positions of earthquakes using seismometers, originally developed to detect nuclear blasts.

The concept of sea-floor spreading goes back to the work of Holmes (1931), but it was not until the classic work of Hess (1962), that geo-scientists finally came around to accept the notion of plate tectonics.[3] Interestingly there is still considerable controversy about whether or not the Earth expanding during its history would account for the anomalies of fit around the edges of continents (Carey, 1976, 1988; King, 1983; Owen, 1976). Today there is considerable support for an expanding Earth accounting especially for biogeographic patterns around the Pacific (see Nelson and Rosen, 1981; Nur and Ben Avraham, 1981; Shields, 1979, 1983, 1991, 1996), but the geophysicists argue that a mechanism to account for these patterns cannot be found. It is worth noting, however, that Croizat (1958) says that there is very little in plate tectonics that can explain the range of trans-Pacific tracks, and it is perhaps best for the biogeographer to put plate tectonics to one side (Nelson and Platnick, 1981).

The biogeographical consequences of plate movements and interactions are enormous. Whether by plate tectonics or expanding Earth, the rearrangement of continental landmasses, the formation of island arcs and the opening and closing of ocean basins have had profound effects on the history of organisms. In fact, Croizat (1952) demonstrated that Earth's evolution and the changes in spatial arrangements had a strong effect on the evolution of organisms. His most memorable maxim from *Space, Time, Form* (1964) is that "Earth and life evolve together."

Cladistics

The second external force is undoubtedly the well-documented revolution in systematics and the application of systematic thinking to historical biogeography (Brundin, 1966, 1972a,b, 1981, 1988; Rosen, 1975, 1978, 1979; Nelson, 1969, 1974, 1982, 1983, 1984, 1985; Nelson and Platnick, 1980, 1981; Platnick and Nelson, 1978). Cladistics was originally conceived in the 1930s by Willi Hennig but did not become understood until the German edition of his book appeared in 1950 and was much more widely appreciated when the English translations appeared a decade and a half later (see Hennig, 1965, 1966). Originally developed as a method for fleshing out the pattern of evolution by common descent, cladistics has, since the mid 1960s, also become the technique for showing rela-

---

[3] In 1963, the Canadian John Tuzo Wilson, postulated that volcanic oceanic islands, such as the Hawaiian archipelago, were created by the displacement of continental plates over a fixed hotspot (Wilson 1963). In 1968, the Americans Bryan Isacks, Lynn R. Sykes and Jack Oliver studied the abyssal fossils of Tonga, described the expansion and contraction of oceans and the concept of subduction (Isacks et al., 1968). Later W. Jason Morgan and Françis Xavier Le Pichon formulated the geometry of movement of lithospheric caps that covered the surface of the globe considering the magnetic anomalies compiled by the team led by Jim Heirtzler, which mapped the transformations across the globe (Heirtzler et al., 1968).

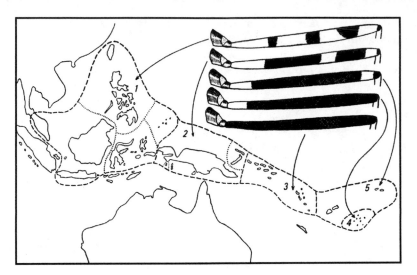

FIGURE 10.2   Map of Hennig's example of thigh color changes in different species of *Mimegralla albimana* (Diptera, Tylidae). From top to bottom, (1) western group of races (*albimana, sepsoides, galbula, palauensis*); (2) New Guinea, Bismark archipelago, Key Islands (*contraria, keiensis, striatofazsciata*); (3) Samoa (*samoana*) (= New Hebrides); (4) Tonga Islands (*tongana*); (5) New Hebrides (*extrema*) (=Samoa) The diagram was originally published in Hennig (1950:136, Figure 27). In the 1966 edition of *Phylogenetic Systematics*, Hennig reproduced the same diagram but added numbers to each area (Hennig 1966:135, Figure 39; see also Hennig 1968:185, Figure 41, 1982:135, Figure 39). The numbers misidentify Samoa as the New Hebrides and vice-versa and are hence incorrect. The confusion was first pointed out by Croizat (1975:608, 831–833; 1978:96) (Nelson, personal communication). (After Hennig, W., *Phylogenetic Systematics*, University of Illinois Press, Urbana, 1966, p. 135, Figure 39; redrawn from the 1979 reprint. Originally published in Hennig, W., *Grundzüge einer Theorie der Phylogenetischen Systematik*, Deutscher Zentralverlag, Berlin, 1950.)

tionships of organisms. Hennig himself intended that it would become the universal system of classification, which in one sense is true largely because the molecular revolution of the 1980s and 1990s has come to fruition. In many ways he was right — cladistics has evolved into a variety of methods with different underlying philosophies and a wide range of implementations,[4] especially in the different programs for computing trees (see for example Kitching et al., 1998; Schuh, 2000). Many of the developments in cladistics that have affected the thinking of cladistic biogeography are shown below.

Hennig (1966) showed only marginal interest in biogeography and was concerned mostly with the way species evolved in space. His main concern was to locate the pattern of character change in space for particular groups of flies (Figure 10.2). Because he was interested mostly in determining the centers of origin of each particular genus, he accepted the prevailing view of the time that ancestors were considered paramount for determining centers of origin. The prevailing paradigm was that phylogeny was unknowable without fossils (Patterson, 1981a,b). Indeed, as Forey (1981) noted, centers of origin were also implicitly tied to the

[4] The following URLs provide details of the different programs available:
http://evolution.genetics.washington.edu/phylip/software.html
http://tolweb.org/tree/phylogeny.html
http://tolweb.org/tree/home.pages/references.html
http://www.cladistics.org; http://www.cladistics.com
http://www.icp.ucl.ac.be/~opperd/private/phenetics.html
http://www.ossu.co.uk/usingclad/intro.shtml.

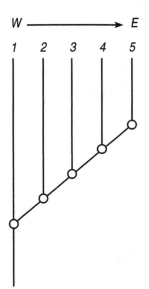

FIGURE 10.3   Diagram of the phylogenetic relationships between the five subspecies of *Mimigralla albimana* shown in Figure 10.2. The characters show increasing apomorphy in the blackening of the thigh corresponding to the direction of spreading from west to east. (After Hennig, W., *Phylogenetic Systematics*, University of Illinois Press, Urbana, 1966, p. 136, Figure 40; redrawn from the 1979 reprint).

notion of migration or dispersal from one place to another. Thus, just as characters could be optimized onto a cladogram, areas could also optimized onto the internal nodes so as to find centers of origin at the root and the migration routes along the branches (Figure 10.3).Croizat's work on panbiogeography during the 1950s and early 1960s was critical in showing that it was impossible to determine migration routes from the spatial distribution of organisms (Craw and Page, 1988). Croizat et al. (1974) showed that centers of origin were impossible to reconstruct and that only geographic links could be established between disjunct taxa (see Page, 1987). Patterson (1981a,b) showed that adding fossils into any analysis did not reveal the center of origin of any group but merely produced a greater number of areas for consideration.

Hennig's (1966) phylogenetic systematics had a dual aspect that was only resolved in the late 1970s. Platnick (1979), Wiley (1981) and Patterson (1982) all described in various ways that Hennig's method contained circular reasoning. Hypotheses about characters (synapomorphy) and hypotheses about groups (monophyly) both appealed to ancestors for their justification. Transformed cladistics (Platnick, 1979) reformulated the equation to show that hypotheses about characters (homology) gave rise to hypotheses about groups (hierarchy). Thus interpretations about ancestry could only be inferred from the cladogram using some other criteria to interpret the cladogram as a phylogenetic tree. Consequently, internal nodes were only hypothetical constructs and could not be construed as ancestors, morphotypes or archetypes.

Such a discovery, the difference between cladograms and trees, is the main reason there is a profound dichotomy between cladistic biogeography (Rosen, 1978; Nelson and Platnick, 1981; Nelson and Ladiges, 1991a,b,c) and phylogenetic biogeography (Brundin, 1966; Hennig, 1966; van Veller et al., in press). I believe that the refusal to recognize that for evolution to be studied it has to be separated from analysis has spawned so many different techniques, including ancestral areas analysis (Bremer, 1992, 1993; Ronquist, 1994, 1995; Hausdorf, 1998), Brooks parsimony analysis (BPA) (Brooks, 1981, 1985; Wiley, 1987,

1988a,b; Brooks and McLennan, 2002), parsimony analysis of endemicity (PAE) (Rosen, 1988; Craw, 1988; Morrone, 1994), event-based methods (Ronquist and Nylin, 1990; Ronquist 1997a,b; Page, 1994; Charleston, 1998) and phylogeography (Avise et al., 1987; Avise, 2000; Arbogast and Kenagy, 2001).

Although during 1970s and1980s the main competing paradigms in historical biogeography were cladistic biogeography and panbiogeography (Craw, 1983), the departures from component analysis are largely implementations that concentrate on evolutionary aspects. What is common to both panbiogeography and cladistic biogeography is the notion that life and Earth have evolved together: Biology and distributions of organisms are inextricably linked, and congruence among different generalized tracks (Croizat, 1952) or area cladograms is evidence of shared history (see for example Heads, 1989, 1990). As will be shown below, Rosen (1979) and Nelson and Platnick (1981) capitalized on this observation for the study of historical biogeography in the development of component analysis and area cladograms.

In other methods such as phylogeography and BPA the notions of dispersal and migration are used to explain departures from congruence. As with homoplasy in biological systematics, it is the signal versus noise problem that causes the most doubt in the efficacy of methods for determining congruence among the groups of taxa being compared for particular sets of areas. Methods that explain away incongruence are effectively immunizing results from criticism and interpreting homoplasy as meaningful rather than discarding incongruent patterns as being noise or constituting another pattern. The causes of incongruence include unique events such as dispersal, extinction and sympatric or parapatric speciation.

Nodes on cladograms are potentially informative about the distribution history of the organisms, although in recent years, much of the proliferation of methodology has been the development of quantitative techniques with more precise implementations of new algorithms. What has all this led to? Nelson and Ladiges (1996) noted that developments in geology and biology (plate tectonics and systematics) did not render the facts of geographic distribution more clearly but merely heightened the expectation for putting that discovery in reach of empirical investigation. Is this merely an assertion or can it be demonstrated that in the past fifty years we have maybe reached a new plateau of understanding?

Internal Forces

To examine the internal forces, I take my cue from Nelson and Ladiges' (1996) statement above. The aim is to see where efforts might be placed in future research based on assessments of the stages in historical biogeography over the past fifty years or so.

Systematics harbors rafts of different and frequently unconnected subdisciplines that have arisen because the underpinning philosophy can be seen from a variety of different standpoints; the same can be said of historical biogeography. Crisci (2001) notes that there are eight classes of methods (Table 10.1) that can be described as historical biogeography: centers of origin and dispersal, panbiogeography, phylogenetic biogeography, cladistic biogeography, phylogeography, PAE, event-based methods and ancestral areas analysis. I would dispute the rank given to these variants and say that the dichotomy is between panbiogeography and cladistic biogeography, and all of the others are derivatives of the latter. However, within these eight classes, at least 30 separate implementations have been described (Crisci, 2001). Of the 30 methods in use, 23 (76%) have been proposed in the past 15 years. This could be called a phenomenal rate of progress, but various commentators have suggested that, instead, historical biogeography is a mess, a subject looking for a method it has yet to find (Tassy and Deleporte, 1999; Humphries, 2000; Nelson and Ladiges, 2001). Crisci

TABLE 10.1   Classes of techniques for historical biogeography

Center of origin	Mathew (1915)
Panbiogeography	Croizat (1952, 1958, 1964); Craw et al. (1999)
Phylogenetic biogeography	Hennig (1966); Brundin (1966)
Ancestral areas	Bremer (1992)
Cladistic biogeography	Rosen (1978); Platnick and Nelson (1978); Nelson and Platnick (1981); Humphries and Parenti (1986, 1999); Nelson and Ladiges (1996)
Event-based methods	Ronquist and Nylin (1990); Ronquist (1997a)
Phylogeography	Avise et al. (1987); Avise (2000)
Parsimony analysis of endemicity	Rosen (1988)

Source: After Crisci, J.V., J. Biogeogr., 28, 157–168, 2001.

thinks that the internal forces shaping the arguments in historical biogeography could have given rise to this proliferation.

Conflict has always existed between pattern and process in comparative biology (Rieppel, 1988), and a similar conflict in historical biogeography exists. The distinction between ecological biogeography and historical biogeography can be attributed to de Candolle (1855) when he was deciding whether distribution patterns were the result of relatively recent ecology or geological history. de Candolle realized the difference by describing the distribution patterns as being in stations (history) or habitations (suitable ecological conditions). The distinction is about time and scale. An ecologist's view of biogeography is shaped by considerations of the factors that shape distribution on a short timescale: the effects of climate and processes such as migration and immigration in particular localities. Historical biogeographers concentrate more on the deeper history and are concerned more with the large-scale patterns over geological time and considering how the patterns of Earth history and biology coincide. Embedded within historical biogeography are arguments concerning how the patterns might be discovered, as illustrated by the squabbles between panbiogeographers and cladists (Craw et al., 1999) and even between factions within the disciplines themselves as shown by the practitioners of event-based methods and pattern methods (Brooks and McLennan, 2001; Humphries and Parenti, 1999).

Historical Biogeography

Maybe the proliferation of methodology is a problem of understanding the essential ingredients of historical biogeography. Its definition might be simple — the study of geographical distributions of organisms to determine the interrelationships of areas — but this simplicity hides the great complexity of the subject (Crisci, 2001). The title of Croizat's 1964 book, *Space, Time, Form: The Biological Synthesis,* is most apt, as it gives a clue about how to look at the situation. Throughout the history of biogeographical exposition, evolutionary narrative has been abundant, and serious attempts to cope with the modern threefold parallelism, the relationship between form, space and time, seems an endless struggle. Of the many published works since Darwin's (1859) *Origin of Species,* Croizat's (1964) *Space, Time, Form: The Biological Synthesis,* Hennig's (1966) *Phylogenetic Systematics* and Nelson and Platnick's, (1981) *Systematics and Biogeography: Cladistics and Vicariance* stand out as the major treatises in the shifting, century and a half–long, examination of the interplay between these three cornerstones. As Nelson (1978) pointed out, the two critical views that cause the dichotomy of opinion are that the Earth was either static, wherein dispersal, creation and extinction became important (see Darwin, 1859), or mobile, wherein life and Earth evolved together.

Indeed, it could be argued that in science, as in art, dialogue between modern and past contributors yields enormous dividends for determining material progress. One could say that the twentieth century contributions (Matthew, 1915, 1918; Willis, 1922; Cain, 1944; Wulff, 1950; Croizat, 1952, 1958, 1964; Simpson, 1953, 1962, 1965; Darlington, 1957, 1965; Brundin, 1966; Hennig, 1966; MacArthur and Wilson, 1967; Nelson, 1978; Nelson and Platnick, 1981; Nelson and Rosen, 1981; Hallam, 1973; Carey, 1976; Briggs, 1984, 1986, 1987a,b; LeGrand, 1988; Humphries and Parenti, 1986, 1999; Craw et al., 1999) have all considered the generalities of the space, time and form perspective in one way or another, and all could be seen as attempts to reconsider the ideas in the *Origin of Species* (Darwin, 1859).

Form, Time, Space: Centers of Origin and Dispersal Biogeography

For many, Charles Darwin is the old master whose work seems to have impinged on every field of evolutionary biology. However, it seems to me that Darwin (1859) was not terribly interested in historical biogeography, largely because he considered space in a rather static way. His theory of evolution by natural selection was embedded in ecology. He considered that species originated slowly and successively in centers of creation to then disperse from the center to other parts of the globe and continue to evolve due to changes of climate and availability of habitat. Darwin concentrated on process rather than the pattern of life. Thus, for Darwin, geography and geology are older than life. He was influenced by Sir Charles Lyell (1830–1833, 1875), who considered that although the earth changed, it did so slowly and steadily. Thus continents and oceans remained essentially static vessels in which life occurred on a globe of fixed dimensions. Consequently, new species were formed after migration across existing geographical barriers. Form was expressed as variation and selection, time as geological succession, and space as dispersal of organisms across the preformed landscape.

In many ways Darwin had not really progressed beyond Linnaeus' concept of Paradise. Organisms originate in a center of origin and migrate to another. What is all the more remarkable is that this view prevailed throughout the twentieth century (Mathew 1915; Cain, 1944; Simpson, 1953, 1962; Darlington, 1957, 1965; Brundin, 1966; Hennig, 1966; MacArthur and Wilson, 1967; Briggs, 1984, 1987a), even after plate tectonics had become accepted (e.g., Raven and Axelrod, 1974). Even more remarkable today is Bremer (1992; see also Bremer, 1995), with his ancestral areas technique, who plays on the general belief among many biologists (including him) that every group of organisms must have originated somewhere and dispersed or spread out from a center. In fact, Bremer has devised a method that actually uses cladograms and supposedly locates the origin of the group, even though it has been shown many times over that cladograms provide only details of relationships. Indeed he is not alone, as dispersal is used freely to explain complicated patterns in phylogenetic biogeography (Hennig, 1966; Brundin, 1966, 1972a,b), BPA (Wiley, 1987; Brooks and McLennan, 2002), phylogeography (Avise, 2000) and event-based methods[5] such as dispersal-vicariance analysis (Ronquist, 1997a).

Space, Time, Form: Léon Croizat

Even before plate tectonics and continental mobility became respectable, Croizat's (1964) pungent critiques of Darwin told us that rather than originating in one center by gradual speciation and then by further migration, Earth and life were forever changing through time; history forever repeats; evolution of form, or form making, took place in changing space;

---

[5] Event-based methods are becoming more prolific by the day. For discussions see Cox (1990), Cox and Moore (1993), Crisci and Morrone (1995), Enghoff (1995), Morrone and Crisci (1995), Ronquist (1997a,b, 1998) and Avise (2000).

and time represented the fusion of geography and biology expressed as panbiogeographic direction (see Craw et al., 1999). For Croizat, the only way to uncover biogeographic history was to examine wholesale distribution patterns from both a biogeographical and geological perspective (panbiogeography) and to ignore preconceived mechanisms explaining processes such as dispersal, extinction or vicariance.

In their recent reappraisal of panbiogeography, Craw et al. (1999) showed that Croizat was interested in finding geophysical and spatial homologies to classify areas of the globe in terms of their geological and biotic history. The crucial distinction between thinking in panbiogeography and other methods is that although it is recognized that organisms can and do move about by chance dispersal events, or indeed by vicariance events, the great raft of papers that have endlessly worried about the dichotomy are missing the point. Geography and evolutionary history rather than a comparison of ecological dispersal versus a biogeographical background are the focus of panbiogeography. Endlessly opposing vicariance and dispersal is considered to be futile debate (Craw et al., 1999).

Because of great difficulty in pinning down precisely the underlying theory, the method and the implementation of panbiogeography has led to considerable criticism from many biogeographers. According to Craw et al. (1999), the main criticisms have characterized panbiogeography as evolutionary metatheory (Nelson, 1989), a precursor to vicariance biogeography (Nelson and Platnick, 1981; Humphries and Parenti, 1986; Grande, 1985, 1990, 1994) or a precursor of the works of Cain (1944) and Camp (1948) (Wiley, 1981).

What is panbiogeography? Craw et al. (1999) consider it to be the reintroduction and reemphasis of life's diversity for a deeper understanding of evolutionary patterns and perhaps the processes underlying them. It emphasizes the role of place in the processes of the past as interpreted from the present. It is essentially a pattern method for uncovering the relationship of space in time. Panbiogeography has been highly successful in bringing the time and space aspects of biogeography to the forefront. However, the jury is still out because the methods required to link form with time and space seem obscure (Craw, 1978, 1979, 1982, 1988, 1989a,b; Craw and Page, 1988; Croizat, 1964; Grehan, 1988, 1990, 1991; Heads, 1989; Lovis 1989; Tangney, 1989). Ways of determining tracks, main massings, nodes and baselines, to reveal testable hypotheses of historical relationships between distributions and geological or geomorphological features are somewhat difficult to follow and inconsistent in their application.

Nevertheless, one of the most crucial outcomes of Croizat's (1952, 1958, 1964) work, and that of Craw and others (see Grehan, 1991 for a review), is that spatial homologies do not conform to present-day geology. To give a general idea, Figure 10.4 shows the generalized tracks of life, showing similar distribution patterns in many unrelated groups of organisms and showing that the relationships link parts of modern continents together with parts of other continents.

According to panbiogeographers (Craw et al., 1999), the somewhat arbitrary areas of endemism or the major biogeographical regions, as conceived by former biogeographers such as de Candolle (1820), Wallace (1876) and Sclater (1858), ignored geological and geographical conglomerates. They indicated that previous historical sutures needed to be teased apart through analysis of their biotas. Figure 10.4 shows some of the key areas that show the major biogeographic nodes (or gates in Croizat's parlance).

Modern ocean basins provided the boundaries of former history (Croizat, 1952, 1958, 1961, 1964; Figure 10.5). Dispersal across and between continents led to a wide range of concepts such as centers of origin (Adams 1902; Willis, 1922; Cain, 1944; Wulff, 1950), migrations by casual means, through corridors, filters and sweepstake routes (Simpson, 1953) and stepping stones and landbridges (Raven and Axelrod, 1972, 1974), so predominant in the books and papers of Matthew (1915), Myers (1938), Simpson (1953, 1962, 1965), Darlington (1957, 1965), Stebbins (1950) and Mayr (1963), which can all be

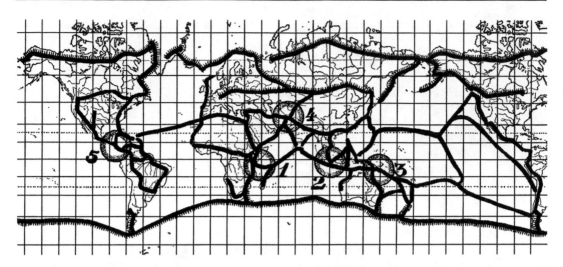

FIGURE 10.4   Croizat's tracks. These were considered to be the main channels of dispersal (plants and animals) of modern life. The hatched lines in the north and the sub-Antarctic south can be identified as boreal (Holarctic, Palaearctic, Nearctic in conventional Wallace/Sclater parlance) and austral (Antarctic, Old Oceanic, etc.). The five stippled circles, numbered 1 to 5, are the main nodes, the points of abutting tracks on an intercontinental scale. The map of the New World to the left stresses the "tracks toward the Atlantic; to the right those towards the Pacific." These particular nodes and tracks are discussed at length by Croizat (1952) and Craw et al. (1999). For example, nodes 1, 3 and 4 are considered to be of great importance to the distribution of plant life. (Ffter Croizat, L., *Panbiogeography*, published by the author, Caracas, Venezuela, 1958, p. 1018, Figure 259.)

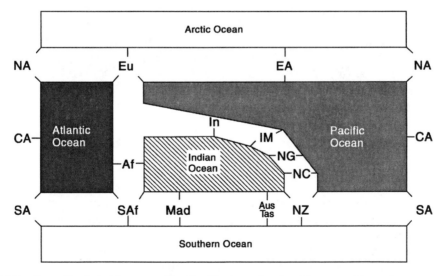

FIGURE 10.5   Formalized panbiogeographic classification of major biogeographic regions seen as ocean basins. Wallacean realms are now seen as artificial conglomerates of geology and biology located at the boundaries of tectonic basins that provide the basis for a natural concept of biogeographic homology. (After Craw, R. C. and Page, R.D.M., in *Evolutionary Processes and Metaphors*, Ho, M.-H. and Fox, S.W., Eds., John Wiley and Sons, London, 1988, Figure 12; caption based on Craw, R.C., Grehan, J.R., and Heads, M.J., *Panbiogeography: Tracking the History of Life*, Oxford University Press, New York, 1999, p. 162, Figure 6-13. Reprinted by permission of John Wiley and Sons.)

dismissed as particular explanations for observed patterns (Nelson and Ladiges, 2001). Croizat's greatest contribution was to show that time is central to an understanding of the interplay between form and space and that dispersalist and vicariant narratives are precisely that — stories devoid of empirical content (see also Craw et al., 1999 and Humphries and Parenti, 1986, 1999).

Form, Time, Space: Willi Hennig and Lars Brundin

Arguably, Hennig (1950, 1965, 1966) was perhaps the most important systematist of the twentieth century. Even though he never cited Croizat, who claimed that Hennig had rediscovered or even plagiarized the work of Daniele Rosa (1918; see Croizat, 1978, 1982), it is largely recognized among systematists that Hennig too was in some sort of dialogue with Darwin. Whereas Croizat (1964) claimed that Darwin had gotten biogeography wrong, Hennig's aim was to provide a complete approach to the problem of phylogenetic reconstruction, part of which was to flesh out an operational technique to produce phylogenetic trees (cladograms). He exposed the shortcomings of conventional typological systematic methods of the first half of the twentieth century and provides the most substantial method for reconstructing phylogenies in the framework of Darwin's evolution by common descent (Richter and Meier, 1994). Hennig provided the means of recognizing relationships of taxa and clarifying the meaning of homology (or synapomorphy) and monophyly (or hierarchy) into one rational system (see Nelson and Platnick, 1981; Humphries and Parenti, 1999). To Hennig, form was phylogenetic systematics, time was ancestor-descendant relations and space meant distribution by dispersal.

Hennig was a dispersalist and although he brought cladistics to biogeography, he did not really push biogeography in a new direction. Central to Hennig's method was to find a center of origin for a group of taxa by using the progression rule. This considered that primitive members of a group were near, or at, the center of origin and the derived taxa were furthest away from it. Quite simply, the areas occupied by terminal taxa were optimized onto the internal nodes of a cladogram to provide a particular scenario of dispersal from that center (Figure 10.3). Thus, despite a revolution in how taxa are recognized and how the relationships of organisms are revealed, Hennig's views of space were uncritical and stuck firmly in the prevailing dispersalist mode. Although Hennig invented the progression rule, it was the efforts of Lars Brundin, a Swedish dipterist, that opened the way for further developments in phylogenetic biogeography. The publication of Brundin's most famous work, *Transantarctic Relationships and their Significance as Evidenced by Midges* (1966) is one of the most important works of the twentieth century, as it influenced so many systematists and biogeographers in Europe and North America. Its publication coincided with the English translation of Hennig's *Phylogenetic Systematics* in 1966, two milestones that changed historical biogeography for good through the introduction of cladograms.

Brundin's classic studies between 1966 and 1988 showed that the southern temperate areas of South America, southern Africa, Tasmania, south-east Australia and New Zealand are inhabited by 600 to 700 species of chironomid midges. Trans-Antarctic relationships recur throughout the group as indicated by this small subclade taken from Brundin's 1981 paper (Figure 10.6). This example shows that the two small groups of Australian and New Zealand midges were considered derived from the South American taxa. In turn, the South American taxa were considered derived from an older New Zealand group that in turn originated in Laurasia. Thus up until this time dispersal was the persistent explanation for the distribution of organisms across an Earth of geological stability, even when the vicariant patterns were quite obvious.

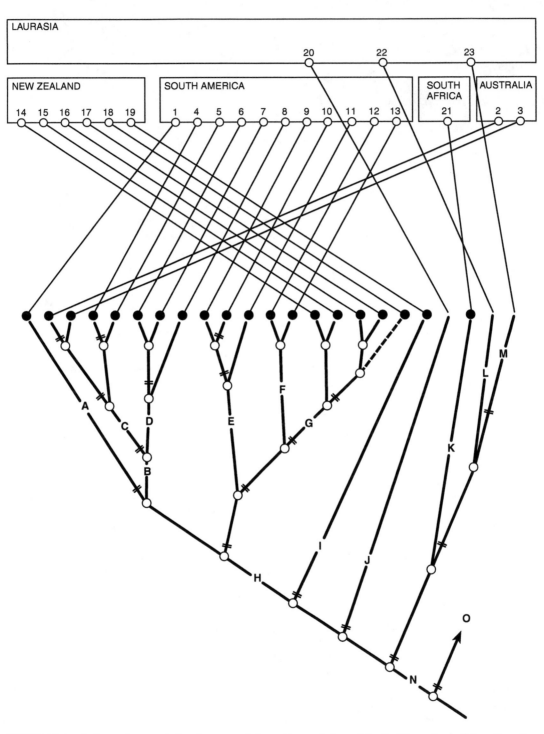

FIGURE 10.6   Relationships and distributions of the midges of the subfamily Diamesinae. (After Brundin, L., *K. Sven. Vetenskapsakad. Handl.* Ser. 4, 11, 1–472, 1966, Figure 635.)

Vicariance Biogeography

Through his memorable studies of the southern hemisphere, J. D. Hooker (1853, 1860) saw a close connection between cause and effect of repeated distribution patterns between biotas and continents. Hooker's view was that to explain the disjunctions across the modern southern hemisphere continents, at one time the continents must have been joined. His views were different from, for instance, Thistleton-Dyer (1909), who considered that all the southern hemisphere floras had originated and migrated from the northern hemisphere and that the similarities between them were merely coincidental.

Nevertheless, framing the problem of vicariance and disjunction beyond speculative scenarios came much later. In an English translation of his 1943 work, *An Introduction to Historical Plant Geography*, Russian botanist E. V. Wulff (1950) suggested that because related organisms show generalized patterns of disjunction, often between continents, then individual dispersal histories were weak assessments and could never be general explanations. He seems to have been the first to have clearly explained the idea that vicariance equated with allopatry and provide the simplest (in the sense of most parsimonious) explanation for similar continental disjunctions in ferns, bryophytes and various groups of gymnosperms and flowering plants. Slightly later, two other botanists, Stanley Cain (1844) and Léon Croizat (1952), were among the first to suggest that vicariance was equally as important to dispersal as an explanation for modern distributions and that many groups of organisms did both. However, what Croizat was really trying to say in *Space, Time, Form,* (1964) was that relationships of areas were important and causal explanations such as dispersal and vicariance were merely interpretations of the spatial pattern.

Form, Time, Space: Gareth Nelson And Norman Platnick

Nelson and Platnick (1981) capitalized on these earlier developments when they synthesized the systematics of Hennig, the panbiogeography of Croizat and the philosophy of Karl Popper into what became known as vicariance biogeography. Form was comparative anatomy and cladistics, time was ontogeny, and space was historical biogeography (relationships of areas of endemism). The book brought together ideas that they had developed during the 1970s with Donn Rosen at the American Museum. The first manifestation of vicariance biogeography came when Rosen (1979) wrote a pioneering series of papers between 1974 and 1978 on the legendary groups of fishes, *Heterandria* and *Xiphophorus*. It is these papers, together with the earlier papers of Nelson (1969, 1974, 1978, 1982, 1983, 1984, 1986), Nelson and Platnick (1980, 1981), Nelson and Rosen (1981) and Croizat et al. (1974) that generated immense interest in vicariance biogeography's application and a standard data set for trying out new implementations of vicariance biogeography.

The aim of vicariance biogeography was to provide a method that classified areas in general area cladograms based on unrelated groups of taxa occupying the same or similar areas. To Nelson and Platnick, area relationships were based on taxic relationships, and the technique clarified (for me at least) Croizat's dictum that life and Earth evolved together, especially as indicated by seeing repetitive patterns amongst disjunct (vicariant) distributions of at least two or more groups of taxa.

Area Cladograms And Components

Although Croizat denied that track analysis provided a precursor to vicariance biogeography, there is no doubt, as expressed in the papers of Croizat et al. (1974) and Rosen (1975, 1978) that tracks could be improved upon by including information about monophyletic groups and representing taxa and areas as cladograms (Figures 10.7 to 10.11). Nelson and Platnick (1980) also indicated that by distinguishing between cladograms and phylogenetic trees, it was possible to see nodes as components and provide a means of finding general area

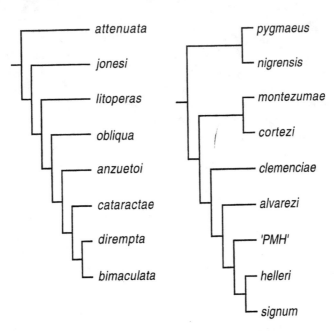

FIGURE 10.7   Rosen's cladograms for *Heterandria* and *Xiphophorus*. (From Rosen, D.E., *Bull. Am. Mus. Nat. Hist.*, 162, 267–376, 1979. With permission.)

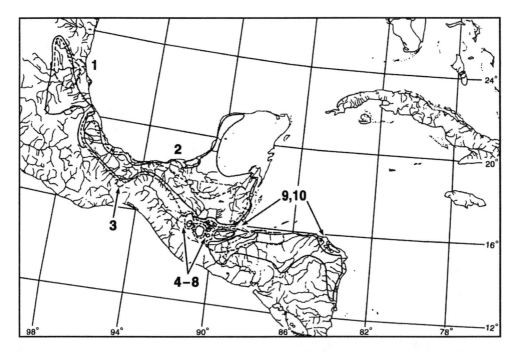

FIGURE 10.8   Combined distribution maps for *Heterandria* and *Xiphophorus* in Central America depicting 10 numbered areas of endemism occupied by the two groups.(From Rosen, D.E., *Bull. Am. Mus. Nat. Hist.*, 162, 267–376, 1979. With permission.)

cladograms from individual area cladograms through component analysis. Rosen's (1979) application to the two groups of fishes, *Heterandria* and *Xiphophorus*, exploited component analysis only to find that because of slight differences between the distribution and species

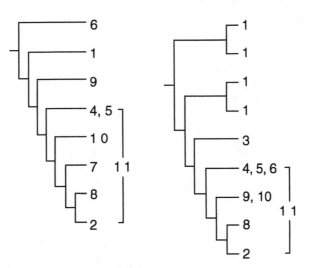

FIGURE 10.9   Area cladograms for *Heterandria* and *Xiphophorus* obtained by substituting species names for areas of endemism occupied by both groups.

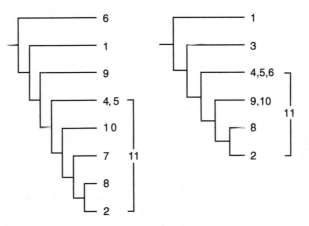

FIGURE 10.10   Reduced area cladograms for *Heterandria* and *Xiphophorus* obtained by removing redundant (paralogous areas).

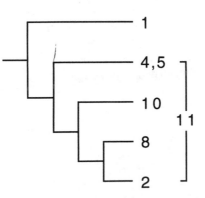

FIGURE 10.11   Pruned consensus tree for *Heterandria* and *Xiphophorus*, showing four nodes of agreement obtained by removing unique and conflicting area positions from the two reduced area cladograms shown in Figure 10.9.

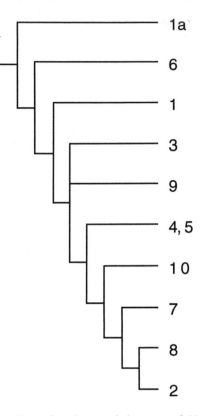

FIGURE 10.12 Consensus tree for the reduced area cladograms of *Heterandria* and *Xiphophorus* using component and analysis and assumption 2. (After Platnick, N.I., in *Advances in Cladistics: Proceedings of the First Meeting of the Willi Hennig Society*, Funk, V.A. and Brooks, D.R., Eds., New York Botanical Garden, New York, 1981, pp. 223–227.)

relationships of the two groups, it was not possible to obtain an unequivocal congruent general area cladogram (see Figures 10.7 to 10.11 for details). Central to this thinking was the idea that taxa and areas should be treated as similar units of comparison, and information regarding interrelationships of taxa should say something about the interrelationships of areas. The problems arose because on comparison, the species of each group of fishes were represented in some areas and not others (i.e., some areas were missing and several species were present in more than one area of endemism [i.e., the species were widespread]). Rosen's solution was to simply extract the pattern common to both groups of taxa (Figure 10.11). Naturally, this was incomplete, so the stages of converting the taxon cladograms into a general area cladogram lost information about the interrelationships of areas.

Assumptions 1 and 2

Platnick and Nelson (1978) made a great intuitive leap when they realized that a cladogram showing the interrelationships of areas could be derived from the same cladogram showing the interrelationships of taxa, but at the same time they were not necessarily saying the same thing with respect to relationships of areas. Taxa and areas could be treated separately. Interestingly, both Wiley (1980, 1981) and Platnick (1981) responded to Rosen's fish studies by noting that unique areas (i.e., those found for one species only) and unresolved nodes either could not be incongruent with cladograms of other taxa without the unique areas or could be resolved for particular relationships.

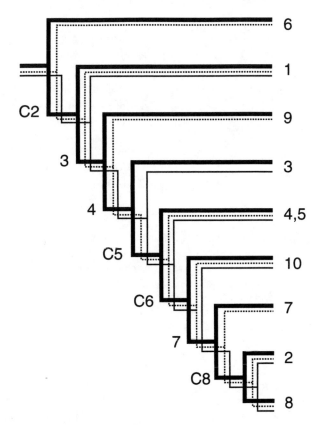

FIGURE 10.13 Reconciliation between the reduced area cladograms and general area consensus tree obtained using component v. 2.0. The solid lines represent the area cladograms; the reduced area cladograms are depicted as a dotted line for *Heterandria* and a thin solid line for *Xiphophorus*. The internal nodes are labeled, and those with a capital C are the areas in common discovered by Rosen in his pruned consensus tree (see Figure 10.10). (After Page, R.D.M., *Cladistics*, 5, 167–182, 1989b.)

Rosen and Nelson and Platnick identified ways of producing general area cladograms for all of the areas concerned, but Platnick's method based on component analysis, has proven to be of greater generality (Figures 10.12 and 10.13; Page, 1989a,b, 1993a). The work of Rosen and Nelson and Platnick was critical to understanding how to deal with spatial relationships. If anything was lacking, it was the ability to deal with spatial proximity. Biogeographers consider that the space-time processes that can most modify the distribution of organisms are speciation, extinction, dispersal and vicariance. Because these processes can affect unrelated groups of organisms quite differently, there is inevitable conflict between the two or more groups of organisms occurring in similar areas. To overcome this problem, the importance of component analysis was enhanced by recognizing that the interrelationships of areas, even though derived from the taxon cladograms, might be quite different from those initially expected.

Nelson and Platnick, by developing the rather cryptic "assumptions 1 and 2" meant that component analysis could resolve the conflicts caused by different processes. This meant that conflict was resolved using cladogram logic rather than by invoking evolutionary interpretations or *a priori* principles that so bedevil BPA and other methods. The importance of this milestone was that it represented a purely pattern cladistic approach, which was a brand new protocol for systematic biogeography in the late 1970s. A simple account of the technique is given in Figures 10.7 to 10.11.

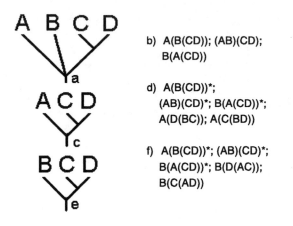

b) A(B(CD)); (AB)(CD);
    B(A(CD))

d) A(B(CD))*;
    (AB)(CD)*; B(A(CD))*;
    A(D(BC)); A(C(BD))

f) A(B(CD))*; (AB)(CD)*;
    B(A(CD))*; B(D(AC));
    B(C(AD))

FIGURE 10.14   To resolve relationships in the light of ambiguity and partial resolution Platnick and Nelson (1978) and Nelson and Platnick (1981) devised a technique known as assumptions 1 and 2 that showed what allowable or disallowable solutions are possible in the light of new evidence: (a) a simple example of partial resolution where areas A and B can only be further resolved by additional information from other groups of taxa. (b) Assumption 1 states that whatever is true of the relationship between area A in terms of areas C and D is also true of area B. The allowable resolutions under assumption 1 are shown. (c) Assumption 2 states that whatever is true for the relationship between area A in terms of areas C and D is not necessarily the same for area B. (d) The five allowable resolutions for area B under assumption 2. (e) It follows that assumption 2 also states that whatever is true for the relationship between area B in terms of areas C and D is not necessarily the same for area A. (f) The five allowable resolutions for area A. The asterisks show the common solutions solving for areas A and B under assumption 2. Comparing the two rows of allowable solutions under assumption 2 shows that for one unresolved node for areas A and B in the original cladogram (a), there are seven unique solutions. By comparison, under assumption 1 there are only three solutions (b).

Permutations of Component Analysis

The papers of Nelson and Platnick spawned two decades of appraisal and attempts to quantify component analysis and various debates as to the good and evil of the new paradigm. The first critiques appeared in 1981 (Nelson and Rosen, 1981), although they were mostly reactionary points of view by panbiogeographers (Craw and Weston, 1984; Grehan, 1991), phylogenetic biogeographers (Brundin, 1966, 1988) and dispersalists (e.g., Koopman, 1981). Notable exceptions, however, showing that congruence is critical for showing patterns of interrelationships between areas, came from Patterson (1981a), Edmunds (1981) and Parenti (1981). Among the first attempts at quantifying vicariance biogeography, Mickevich (1981) is notable for devising an implementation that utilized Farris optimization and unrooted trees to connect areas of endemism in an effort to show that historical biogeography was more about maps and graphical representation than mere cladistics.

The year 1981 also saw the first efforts of Brooks (see also Brooks, 1985) to provide an analytical procedure for a matrix representation of cladograms and a parsimony method (BPA) to construct a single tree from several matrix-coded trees (Brooks, 1990). Initially this technique showed immense promise on two counts: (1) It suggested that the protocols for hosts and associates, whether they were hosts and parasites, taxa and areas or gene trees and species trees (Page, 1993b), could be analyzed from a matrix using a parsimony method; and (2) standard parsimony techniques such as Hennig86 (Farris, 1988) could be used to calculate the trees by coding two or more groups of areas x taxa in the same matrix. Later, after many experiments in the use of BPA and other matrix methods (see Myers and Giller, 1988; Kluge, 1988; Humphries et al., 1988, Humphries, 1992; Cracraft, 1989), Page (1994)

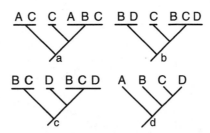

FIGURE 10.15 Nelson's (1984) example of component analysis. (a–c) The three cladograms represent three different groups of four taxa occurring for each example in three of four areas (A, B, C and D). In each of the examples, one area is missing, one species is endemic, and two species of each group area widespread in two and three areas respectively. (d) The example, considered by Nelson as intuitively opaque as can be to visual analysis, yields one allowable solution, A(B(CD)) for the four areas when calculated using a combination of assumption 2 and Nelson consensus (From Humphries and Parenti 1986, 1999).

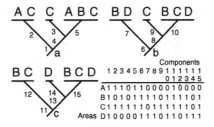

FIGURE 10.16 Brooks parsimony analysis. (a–c) Three cladograms indicating the terminal and internal components (see Figure 10.14). The matrix is coded for analysis using assumption 0 and Brooks parsimony analysis.

and Humphries and Parenti (1999) showed that BPA could produce spurious trees (because of conflicting components) as a result of the differences in the fundamental trees. The errors were presumably the result of dispersal, extinction and failures to respond to tectonics. Nelson (1984) argued the case for component analysis most eloquently by his "intuitively opaque" example wherein everything but component analysis using assumption 2 (Figure 10.14) and Nelson consensus produced a satisfying result (Figure 10.15).

With the benefit of hindsight, it is now possible that the parting of the ways of phylogenetic biogeography and component analysis began to emerge. Zandee and Roos (1987) developed a method called "component compatibility." Being somewhat similar to BPA, it is a matrix method that states clearly that distributions of organisms should be taken as read, and assumptions 1 and 2 should not be applied (Figure 10.16). In the spirit of Nelson and Platnick's renditions of assumptions 1 and 2, Zandee and Roos called the application of the given distributions assumption 0. Figures 10.16 and 10.17 show how the method fails using Nelson's (1984) intuitively opaque example (Figure 10.15).

Wiley (1987) emphasized even further the dichotomy between the pattern and process approaches of Nelson and Platnick by elaborating BPA to optimize taxa back onto area trees, a procedure that is analogous to numerical cladistics. Treating the cladograms as representing species history meant that homoplasy was meaningful; all kinds of dispersals, extinctions and separate origins were employed to explain the distributions deviating from congruence. Geographical homoplasy led to range of contortions and explanations for species to provide explanations for dispersal, extinction and multiple origins (Page, 1989a,b, 1990). The recent work of van Veller, Brooks and colleagues has revived and elaborated on Wiley's technique, calling geographical homoplasy secondary analysis or secondary BPA

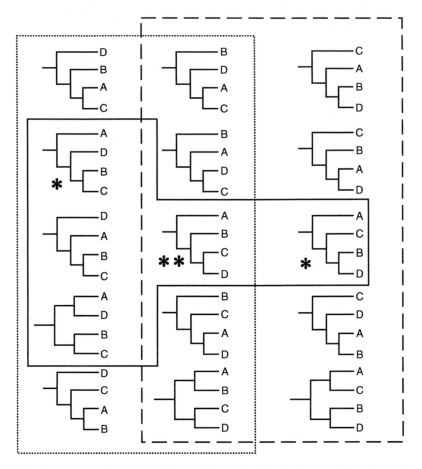

FIGURE 10.17 For four areas there are 15 possible resolvable cladograms. The three clusters of five
cladograms show the possible outcomes for each of the three cladograms in Figure 10.14.
The degree of overlap shows that some solutions are unique to each cladogram, some occur
in two groups of taxa, and only one, A(B(CD)), indicated by two asterisks, fits all three of
the fundamental cladograms (Figure 10.14a–c). Brooks parsimony analysis (based on an all-0
outgroup for the matrix in Figure 10.15) is flawed in the sense that it recovers two solutions,
as shown by the two cladograms marked with one asterisk. However, in terms of parsimony,
the two solutions fit only two of the fundamental cladograms each, neither agree and neither
form part of the general solution obtained by component analysis.

(Brooks et al., 2001; Brooks and McLennan, 2002; van Veller, 2000, 2003; van Veller and
Brooks, 2001; van Veller et al., 1999, 2000, 2001, 2002). Despite the amount written on
the topic, this is a fine example of the emperor's new clothes. The justification for the method
is that the trees derived from secondary BPA produce species histories rather than relationships of areas.

Form, Space, Time: Chris Humphries and Lynne Parenti)

By the end of the 1970s the great realization that became obvious to all involved in
systematics was that ancestors, even hypothetical ones, were unimportant. Ancestry becomes
a matter of interpretation rather than being inherent to the method as in Hennig's approach.
Pattern cladistics modified the practicalities of empirical endeavor (e.g., Humphries and
Parenti, 1999) such that form and space represent the measurable aspects, and ancestry and

time are merely inferences. The logical sequence for the threefold parallelism is form, space and time. Form is comparative morphology and anatomy, space is historical biogeography, and time is an inference to be tested with external knowledge. It follows that historical biogeography is about classification of areas among biological and spatial coordinates and not about dispersal, individual historical scenarios or even vicariance to explain the shared area patterns of organisms. To put that into a broader context, "the geographic distributions of organisms are coherent patterns related to, and explained by, historical processes of geographic change" (Nelson and Ladiges, 1996) but without having to involve historical processes as part of the method. The space, time, form parallelism can still be recharacterized in terms of theory, method, implementation and results.

A Widening Rift

Looking back into the immediate past, the only interesting theoretical developments to have occurred in biogeography during the 1990s are three-item statements analysis, the removal of geographical paralogy and subtree analysis. These are improvements in area pattern cladistics and by their very nature have departed greatly from phylogenetic biogeography of the kinds described by Brooks, van Veller and colleagues.

Three-Item Analysis, Geographical Paralogy and Subtree Analysis

De Pinna (1996) wrote, "The most interesting idea in mainstream theoretical systematics in recent years is the so-called three-item analysis." As Williams (1998) noted, "This may strike cladists as a curious statement, given that most commentary relating to three-item statements analysis has been decidedly negative" (Harvey, 1992; Kluge, 1993, 1994; Farris et al., 1995). De Pinna went on to point out that "a major success of the cladistic approach was the recognition that ancestor-descendant relationships among taxa cannot be objectively proposed or tested, only sister-group relationships." Nelson and Platnick (1981) built upon the idea that one taxon cannot give rise to another. Rather than character coding with the recognition of transformation series and the use of character optimization, Nelson and Platnick pushed component analysis to its limit by adopting Hennig's (1966) idea that the minimum required to express a relationship is three taxa. Thus, given the example AB(CD), there are two possible expressions of relationship, given that there is only one informative component: A(CD), B(CD). Their reasoning was that given the idea that ancestral taxa have no place in analysis, then the same must be true for ancestral characters.

This idea is clearly embedded, but not particularly elaborated, by Nelson and Platnick (1981) in the sense that ancestors, whether taxa or characters, appear to be based on the absence of evidence. All cladists agree that cladistics is about grouping by synapomorphy and that synapomorphy is evidence of relationship. Three-item statements analysis codes data in the sense that *taxon* and *homology* represent the same relationship (Nelson and Ladiges, 1994), and unlike all other methods, it focuses on the smallest possible unit of relationship, the three-item statement, and then finds the tree to accommodate the most statements (Nelson and Ladiges, 1991d, 1994, 1995). Viewed this way, three-item statements analysis shows an entirely different appreciation of data and accordingly, it improves the precision of parsimony (Nelson and Platnick, 1991; Morrone and Carpenter, 1994).

Three-item analysis in biogeography has its roots in component analysis and assumption 2. Component analysis and assumption 2 are about allowable and nonallowable relationships discovered by pure cladogram logic. Coupled with subtree analysis and the understanding that problems of paralogy in molecular biology have their equivalent in geography (Page 1993a,b), Nelson and Ladiges (1996) realized that the problems of overlapping areas within and between groups and seemingly incongruent patterns, caused presumably by dispersal, extinction and speciation, can be resolved. By using subtree analysis, it is possible to extract the cladistic signal from a sea of paralogous noise (Figure 10.18). Figure 10.19

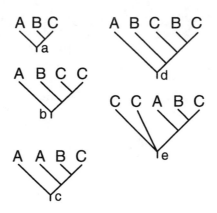

FIGURE 10.18 Three-item statements analysis, geographical paralogy and subtrees. (a) Cladogram for three areas, showing the relationship among three areas of endemism, A(BC). (b–e) Four cladograms showing varying degrees of paralogy (repetitive occurrences) for the three areas A, B and C. Coding for components gives the same result for all five cladograms — only the component BC is informative. Coding for three-item statements gives informative statements for A(BC) for cladograms b to d and two informative statements, A(BC) and C(AB) for cladogram e and is thus more sensitive to information content than components alone. Removal of geographical paralogy results in one topology for all five cladograms, A(BC).

shows one example of this. Application of the method to the biogeography of *Nothofagus* (see, for example, Romero, 1986) shows that rather than the group being the key to southern hemisphere biogeography, there are only two informative nodes worthy of further discussion.

Discussion

Narrative Scenarios

Sadly, among the proliferating methods (see Crisci, 2001) the pattern/process dichotomy persists, especially because dispersal biogeography continues to reinvent itself. As we have seen, phylogenetic biogeography, as described by Hennig (1966) and Brundin (1966), attempted to determine centers of origin by invoking gradual progression and morphological deviation from the proposed center as depicted at the roots of branching diagrams. More recently, Bremer (1992) and Swenson and Bremer (1998) stated that we need to locate centers of origin because organisms must have originated somewhere. They used an optimization procedure based on gains and losses of areas on area cladograms to determine centers of origin and provide dispersal scenarios to explain present-day biogeographic patterns for particular groups of organisms. Despite its quantitative appeal, this method is yet another reworking of the idea that the "tracks of life" (Craw et al., 1999) are independent from geography and that each group of organisms has its own story to tell.

The various critiques of Craw et al.'s paper should have laid the matter to rest, but dispersal scenarios are being published at a rate even faster than before (see, for example, Swenson et al., 2001; Arbogast and Kenagy, 2001). Most of the papers published over the past 12 years or so under the banner of phylogeography (Avise, 2000) (a subject defined as the study of biogeography as revealed by a comparison of estimated phylogenies of populations or species with their geographic distributions) are cases in point. Few phylogeographers are interested in general patterns of area relationships. By studying organisms at the population level, most papers are revitalized descendants (by the addition of cladistics and molecular biology) of the new systematics that now, as then, blur the boundaries of ecology, population biology, genetics and systematics. There is no distinction between pattern and

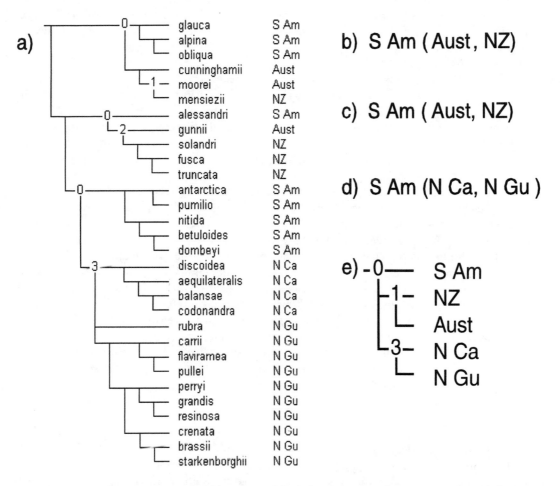

FIGURE 10.19 (a) Interrelationships of 30 species of *Nothofagus*, showing the species names and the native areas in which they occur: S Am, South America; Aust, Australia; NZ, New Zealand; N Ca, New Caledonia; N Gu, New Guinea. (b–d) The three (only) informative nodes on the *Nothofagus* cladogram. Nodes marked 0 are basal for paralogy-free subtrees. Geographically informative nodes are labeled 1 to 3; all other nodes are geographically paralogous. (e) The combined (consensus) tree for the informative nodes on the *Nothofagus* cladogram. (Figure 10.19a after Linder, H.P. and Crisp, M.D., *Cladistics*, 11, 5–32, 1995(1996); Nelson, G.J. and Ladiges, P.Y., *Austr. J. Bot.*, 49, 389–409, 2001.)

process (Rieppel, 1988), and hence phylogeographers consider space to be older than form rather than both part of the same system. BPA is somewhat similar in that it attempts to reveal the processes of speciation, using paralogy to interpret sympatry, extinction and dispersal to map taxa onto trees.

Analytical Methods

Both cladistic biogeography and panbiogeography address the present and historical debates over pattern and process (Morrone, 2002). Problems of dispersal and vicariance that have dogged historical biogeography since the early nineteenth century should now have been firmly put in their place. Panbiogeography sees itself as a subject that utilizes a form-making model that recognizes both "dispersal and vicariance as important processes by which organisms achieve their geographic distributions" (Craw et al., 1999). Its supposed strength is to recognize that it is the distributions of the organisms themselves that diagnose the areas to be classified. By using track analysis and the coincidence of individual tracks to become generalized tracks, it is considered possible to reconstruct ancestral biotas that have become modified through time by geological and geographical change.

Initially, cladistic analyses were considered unnecessary to undertake biogeographic analysis. The latter has an independent criterion for relationship, which in panbiogeography is spatial homology. In fact, should one have undertaken a phylogenetic analysis according to Grehan (pers. comm.), the biological relationships could be reassessed according to their spatial relationships through criteria such as common baseline, spatial geometry (nodes and gates) and main massings. Any of these spatial characters may lead to the reassessment of the biological hypothesis, just as the biological hypothesis itself introduces constraints on the spatial analysis in the first place. Grehan considers this as reciprocal illumination, but it looks like circular reasoning. My problem with panbiogeography is that generalized tracks are just that — generalized — and the tracks do not make clear statements as to what constitutes relationship between different areas on the earth.

Cladistic biogeography has overcome the problem of the pattern/process debate by separating the empirical work from inference. It is about classification of areas of endemism — an analogy to systematics on how areas and the relationships of areas are recognized. Thus, by assuming clear correspondence between systematic relationships in different taxa in similar areas, cladistics provides biogeographically informative relationships of the biotas in those areas. Relationships of areas of endemism are identified from the internal nodes of cladograms, with the geographic distributions of taxa placed at the terminal nodes. The historical relationships of the areas of endemism are inferred through congruence of pattern among two or more groups of taxa, and common history is represented in branching diagrams and area cladograms — hierarchical relationships of areas derived from cladograms of taxa. Biogeographic homology is thus based on biological homology and area congruence of unrelated groups of taxa.

So what can be said about the different classes of methods that Crisci (2001; Table 10.1) described? Some are about resolving the problems of conflicting data, some are explanations of pattern, and others variations on cladistic biogeography implementing optimizing principles, implementing quantitative methods, considering spatial models and identifying areas of endemism and how to recognize relationships between them. Most of them are lost causes. Techniques such as parsimony analysis of endemicity (PAE) (Rosen, 1988), which classifies areas by their shared taxa, and event-based methods (Hovenkamp, 2001), which rely on known geological events to determine the meaning of nodes on cladograms, are misguided. Event-based methods involve circular reasoning. PAE is a parody of systematic biogeography because it has corrupted the meaning of area homology. It is an attempt to understand the relationships between areas without knowing the relationships between taxa. Nevertheless, it does raise the question of how to deal with biogeographical relationships rendered equivocal by widespread taxa across areas of endemism and by different patterns among different taxa brought about by a range of processes from extinction to different responses to earth history events. The subject is complicated by the apparent lack of congruence among different taxa (see Humphries and Parenti, 1999 for details), for which there have been a number of attempted solutions. Although others would disagree, I still consider that the greatest success in solving this problem has come from component analysis (Nelson and Platnick, 1981; Humphries, 1989), subtree analysis and the removal of geographical paralogy.

Recently, these techniques have been put together as area cladistics (Ebach and Humphries, 2002), which embodies the rationale of cladistic biogeography. Allopatry is the underlying theory of cladistic biogeography, congruence is the consequence, the relationships of areas equate to geographical proximity if one accepts that nested sets equate to areas, and sampling of different data sets gives consistent results. Anomalies due to a variety of historical reasons can be catered for without assuming any dispersal, vicariant events or extinction events. Removal of area paralogy has greatly simplified the discovery of area relationships from area cladograms rife with geographic duplication (Nelson and Ladiges, 1996).

Despite these developments, there is a growing realization that a number of problems have yet to be solved in historical biogeography. For example, Hovenkamp (2001) suggests that it is not necessary to predefine areas of endemism, with all its associated problems (Platnick, 1991; Axelius, 1991; Harold and Mooi, 1994; Morrone, 1994; Platnick and Nelson, 1984). Also, widespread species should be considered on equal footing with less widespread or endemic ones when evaluating their contribution to geographical and geological boundaries. Thus of the eight classes of historical biogeography recognized by Crisci (2001), cladistic biogeography clearly has considered the essential ingredients for uncovering congruent patterns. It is the only one that has attempted to develop a purely pattern approach that has tried to overcome the problems of dispersal, vicariance and extinction by systematic means rather considering them as part of the method or implementation (as in BPA, for example). There is still more to do about what areas of endemism really mean, but judging from the plethora of methods, there is renewed interest in the subject.

Conclusion: The Future

I believe that although there is still a very long way to go in uncovering biogeographic patterns, there is just one underlying sequence of geographical and geological evolution. When this is adequately investigated it will explain most, if not all, the large-scale distribution patterns. Critical changes in biogeographic methodology suggest that the underlying theory of change will be based on an exhaustive analysis of biological and geological cladograms (Humphries and Parenti, 1999; Ebach and Humphries, 2002). This would be more satisfying than a continuous outpouring of individual stories for every group of organisms and techniques for overcoming the need for taxic homology. I do recognize that there are problems with combining taxa into general area patterns, with the delineation and recognition of areas and with finding a sequence of events found by aligning the multiple simple sequences that can be derived from cladograms (Hovenkamp, 1997, 2001). Nevertheless, I consider that cladistic biogeography takes the subject several steps beyond its main rivals such as BPA and the panbiogeography program that ultimately sees "biodiversity ... made up of ... tracks and nodes of life (Craw et al., 1999)." This represents only half of the equation — the recognition of geographic homology. If ultimately the aim is to provide a general system of classification for all of the places on earth, then area homologies, the relationships of areas, have to be considered in terms of the essential biological component — the homologies of taxa.

Acknowledgments

I thank Peter Forey and David Williams for inviting me to participate in this symposium and allowing me to procrastinate longer than usual in contributing to a Systematics Association volume.

References

Adams, C.C., Southeastern United States as a center of geographical distribution of fauna and flora, *Biol. Bull. Mar. Biol. Lab. Woods Hole*, 3, 115–131, 1902.

Arbogast, B.S. and Kenagy, G.J., Comparative phylogeography as an integrative approach to historical biogeography, *J. Biogeogr.*, 28, 819–825, 2001.

Avise, J.C., *Phylogeography: The History and Formation of Species*, Harvard University Press, Cambridge, MA, 2000.

Avise J.C., Arnold, J., Ball, R.M., Bermingham, E., Lamb, T., Neigl, J.E., Reeb, C.A., and Saunders, N.C., The mitochondrial DNA bridge between populations, genetics and systematics, *Annu. Rev. Ecol. Syst.*, 3, 457–498, 1987.

Axelius, B., Areas of distribution and areas of endemism, *Cladistics*, 7: 197–199, 1991.

Bacon, F., *Novum Organum Scientiarum*, in *The Works*, B. Montague, Ed., Parry & MacMillan, Philadelphia, 1610 (1854).

Ball, I.R., Nature and formulation of biogeographic hypotheses, *Syst. Zool.*, 24, 407–430, 1976.

Bousquets, J.L. and Morrone, J.J., Eds., *Introduccion a la Biogeographia en Latinoamerica: Teorias, Conceptos, Metodos y Aplicaciones*, Universidad Nacional Autonoma de Mexico, D.F., 2001.

Bremer, K., Ancestral areas: a cladistic reinterpretation of the center of origin concept, *Syst. Biol.*, 41, 436–445, 1992.

Bremer, K., Intercontinental relationships of African and South American Asteraceae: a cladistic biogeographic analysis, in *Biological Interrelationships between Africa and South America*, Goldbatt, P., Ed., Yale University Press, New Haven, CT, 1993, pp. 105–135.

Bremer, K., Ancestral areas: optimisation and probability, *Syst. Biol.*, 44, 255–259, 1995.

Briggs, J.C., *Centres of Origin in Biogeography*, University of Leeds, Leeds, UK, 1984.

Briggs, J.C., Introduction to the zoogeography of North American fishes, in *The Zoogeography of North American Freshwater Fishes*, Hocutt, C. and Wiley, E.O., Eds., John Wiley and Sons, New York, 1986, pp. 1–16.

Briggs, J.C., *Biogeography and Plate Tectonics*, Elsevier, Amsterdam, 1987a.

Briggs, J.C., Antitropical distribution and evolution in the Indo-West Pacific Ocean, *Syst. Zool.*, 36, 237–247, 1987b.

Brooks, D.R., Hennig's parasitological method: a proposed solution, *Syst. Zool.*, 30, 229–249, 1981.

Brooks, D.R., Historical ecology: a new approach to studying the evolution of ecological associations, *Ann. Mo. Bot. Gard.*, 72, 660–680, 1985.

Brooks, D.R., Parsimony analysis in historical biogeography and coevolution: methodological and theoretical update, *Syst. Zool.*, 39, 14–30, 1990.

Brooks, D.R. and McLennan, D.A., *Phylogeny, Ecology, and Behavior*, University of Chicago Press, Chicago, 1991.

Brooks, D.R. and McLennan, D.A., A comparison of discovery-based and an event-based method of historical biogeography, *J. Biogeogr.*, 28, 757–767, 2001.

Brooks, D.R. and McLennan, D.A., *The Nature of Diversity*, University of Chicago Press, Chicago, 2002.

Brooks, D.R., van Veller, M., and McLennan, D.A., How to do BPA, really, *J. Biogeogr.*, 28, 343–358, 2001.

Brundin, L., Transantarctic relationships and their significance as evidenced by midges, *K. Sven. Vetenskapsakad. Handl.* Ser. 4, 11, 1–472, 1966.

Brundin, L., Evolution, causal biology, and classification, *Zool. Scr.*, 1, 107–120, 1972a.

Brundin, L., Phylogenetics and biogeography, *Syst. Zool.*, 21, 69–79, 1972b.

Brundin, L., Croizat's panbiogeography versus phylogenetic biogeography, in *Vicariance Biogeography: A Critique*, Nelson, G. and Rosen, D.E., Eds., Columbia University Press, New York, 1981, pp. 94–158.

Brundin, L., Phylogenetic biogeography in *Analytical Biogeography*, Myers, A.A. and Giller, P.S., Chapman and Hall, London, 1988, pp. 343–369.

Cain, S.A., *Foundations of Plant Geography*, Harper and Row, New York, 1944.

Carey, S.W., *The Expanding Earth*, Elsevier, Amsterdam, 1976.

Carey, S.W., *Theories of the Earth and Universe*, Stanford University Press, Palo Alto, CA, 1988.

Carozzi, A.V., A propos de l'origine de la théorie des derives continentales: Francis Bacon (1620), François Placet (1668), A. von Humboldt (1801) et A. Snider (1858), *Comptes Rendus Soc. Phys. Hist. Nat. Genève*, 4, 171–179, 1969.

Charleston, M.A., Jungles: a new solution to the host/parasite phylogeny reconciliation problem, *Math. Biosci.*, 149, 191–223, 1998.

Cox, C.B., New geological theories and old biogeographical problems, *J. Biogeogr.*, 17, 117–130, 1990.

Cox, C.B., The biogeographical regions reconsidered, *J. Biogeogr.*, 28, 511–523, 2001.

Cox, C.B. and Moore, P.D., *Biogeography: An Ecological and Evolutionary Approach*, Blackwell Scientific, Oxford, 1993.

Cracraft, J., Geographic differentiation, cladistics, and vicariance biogeography: reconstructing the tempo and mode of evolution, *Am. Zool.*, 22, 411–424, 1982.

Cracraft, J., Biological diversification and its causes, *Ann. Mo. Bot. Gard.*, 72, 794–822, 1985.

Cracraft, J., Deep-history biogeography: retrieving the historical pattern of evolving continental biotas, *Syst. Zool.*, 37, 221–236, 1989.

Craw, R.C., Two biogeographic frameworks: implications for the biogeography of New Zealand, *Tuatara*, 23, 81–114, 1978.

Craw, R.C., Generalized tracks and dispersal in biogeography: a response to R. M. McDowall, *Syst. Zool.*, 28, 99–107, 1979.

Craw, R.C., Phylogenetics, areas, geology and the biogeography of Croizat: a radical view *Syst. Zool.*, 31, 304–316, 1982.

Craw, R.C., Panbiogeography and vicariance cladistics: are they truly different?, *Syst. Zool.*, 32, 431–438, 1983.

Craw, R.C., Continuing the synthesis between panbiogeography, phylogenetic systematics and geology as illustrated by empirical studies on the biogeography of New Zealand and the Chatham Islands, *Syst. Zool.*, 37, 291–310, 1988.

Craw, R.C., New Zealand biogeography: a panbiogeographic approach, *N.Z. J. Zool.*, 16, 527–547, 1989a.

Craw, R.C., Quantitative panbiogeography: introduction to methods, *N.Z. J. Zool.*, 16, 485–494, 1989b.

Craw, R.C., Grehan, J.R., and Heads, M.J., *Panbiogeography: Tracking the History of Life*, Oxford biogcography series 11. Oxford University Press, New York, 1999.

Craw, R. C. and Page, R.D.M., Panbiogeography: method and metaphor in the new biogeography, in *Evolutionary Processes and Metaphors*, Ho, M.-H. and Fox, S.W., Eds., John Wiley & Sons, London, 1988, pp. 163–189.

Craw, R.C. and Weston, P.H., Panbiogeography: a progressive research program? *Syst.-Zool.*, 33, 1–13, 1984.

Crisci, J.V., The voice of historical biogeography, *J. Biogeogr.*, 28, 157–168, 2001.

Crisci, J.V. and Morrone, J.J., Historical biogeography: introduction to methods, *Annu. Rev. Ecol. Syst.*, 26, 373–401, 1995.

Croizat, L., *Manual of Phytogeography*, W. Junk, The Hague, The Netherlands, 1952.

Croizat, L., *Panbiogeography*, published by the author, Caracas, Venezuela, 1958.

Croizat, L., *Principia Botanica*, published by the author, Caracas, Venezuela, 1961.

Croizat, L., *Space, Time, Form: The Biological Synthesis*, published by the author, Caracas, Venezuela, 1964.

Croizat, L., Biografía analítica y sintectica ("panbiogeografía") de las Américas, *Boll. Acad. Cienc. Fís. Mat. Nat. Caracas*, 35(103–106), 10–890, 1975.

Croizat, L., Hennig (1966) entre Rosa (1918) y Løvtrup (1977): medio siglo de sistemática filogenética, *Boll. Acad. Cienc. Fís. Mat. Nat. Caracas*, 38, 59–147, 1976.

Croizat, L., Vicariance, vicariism, panbiogeography, 'vicariance biogeography', etc. a clarification, *Syst. Zool.*, 31, 291–304, 1982.

Croizat, L., Nelson, G., and Rosen, D.E., Centers of origin and related concepts, *Syst. Zool.*, 23, 265–287, 1974.

Darlington P.J., Jr., *Zoogeography: The Geographical Distribution of Animals*, John Wiley and Sons, New York, 1957.

Darlington P.J., Jr., *Biogeography of the Southern End of the World*, Harvard University Press, Cambridge, MA, 1965.

Darwin, C., *On the Origin of Species by Means of Natural Selection, or the Preservation of Favoured Races in the Struggle for Life*. John Murray, London, 1859.

de Candolle, A.P., Essai elementaire de geographie botanique, *Dict. Sci. Nat.*, 18, 359–422, 1820.

de Candolle A., *Géographie Botanique Raisonée*, Paris, 1855.

de Pinna, M., Comparative biology and systematics: some controversies in retrospective, *J. Comp. Biol.*, 1, 3–16, 1996.

du Toit, A.L., *Our Wandering Continents: An Hypothesis of Continental Drifting*, Oliver and Boyd, Edinburgh, 1937.

Ebach, M.C. and Humphries, C.J., Cladistic biogeography and the art of discovery, *J. Biogeogr.*, 20, 427–444, 2002.

Edmunds, G.F., Discussion [of vicarious plant distributions and paleogeography of the Pacific region], in *Vicariance Biogeography: A Critique*, Nelson, G. and Rosen, D.E., Eds., Columbia University Press, New York, 1981, pp. 287–97.

Enghoff, H., Historical biogeography of the holarctic: area relationships, and dispersal of marine mammals, *Cladistics*, 11, 223–263, 1995(1996).

Espinosa Organista, D., Morrone, J.J., Bousquets, J.J., and Flores Villela, O., *Introduccion al Analisis de Patrones en Biogeografia Historica*, Las Prensas de Ciencias, Facultad de Ciencias, Universidad Nacional Autonoma de Mexico, D.F., 2002.

Farris, J.S., *Hennig86: An Interactive DOS Program for Cladistic Analysis*, published by the author, Port Jefferson Station, New York, 1988.

Farris, J.S., Källersjö, M., Albert, V.A., Allard, M., Anderberg, A., Bowditch, B., Bult, C., Carpenter, J.M., Crowe, T.M., De Laet, J., Fitzhugh, K., Frost, D., Goloboff, P., Humphries, C.J., Jondelius, U., Judd, D, Karis, P.O., Lipscomb, D., Luckow, M., Mindell, D., Muona, J., Nixon, K., Presch, W., Seberg, O., Siddall, M.E., Struwe, L., Tehler, A., Wenzel, J., Wheeler, Q., and Wheeler, W., Explanation, *Cladistics*, 11, 211–218, 1995.

Fisher, O., *Physics of the Earth's Crust*, Macmillan, London, 1881.

Forey, P.L., Biogeography, in *Chance, Change and Challenge*, Vol. 2. *The Evolving Biosphere*, Forey, P.L., Ed., British Museum (Natural History), London and Cambridge University Press, Cambridge, 1981, pp. 241–245.

Grande, L., The use of paleontology in systematics and biogeography, and a time control refinement for historical biogeography, *Paleobiology*, 11, 234–243, 1985.

Grande, L., Vicariance biogeography, in *Palaeobiology: A Synthesis*, Briggs, D.E. and Crowther, P.R., Eds., Blackwell Scientific, Oxford, 1990, pp. 448–451.

Grande, L., Repeating patterns in nature, predictability, and 'impact' in science, in *Interpreting the Hierarchy of Nature, from Systematic Patterns to Coevolutionary Process Theories*, Grande, L. and Rieppel, O., Eds., Academic Press, London, 1994, pp. 61–84.

Grehan, J.R., Panbiogeography: evolution in space and time, *Riv. Biol. Biol. Forum*, 81, 469–485, 1988.

Grehan, J.R., Panbiogeography: past, present and future, *N.Z. J. Zool.*, 16, 513–525, 1990.

Grehan, J.R., Panbiogeography 1981-91: development of an earth/life synthesis, *Progr. Phys. Geogr.*, 15, 331–363, 1991.

Hallam, A., *A Revolution in the Earth Sciences: From Continental Drift to Plate Tectonics*, Clarendon Press, Oxford, 1973.

Harold, A.S. and Mooi, R.D., Areas of endemism: definition and recognition criteria, *Syst. Biol.*, 43, 261–266, 1994.

Harvey, A.W., Three-taxon statements: more precisely, an abuse of parsimony?, *Cladistics*, 8, 345–354, 1992.

Hausdorf, B., Weighted ancestral area analysis and a solution of the redundant distribution problem, *Syst. Biol.*, 47, 445–456, 1988.

Heads, M.J., Integrating earth and life sciences in New Zealand natural history: the parallel arcs model, *N.Z. J. Zool.*, 16, 549–585, 1989.

Heads, M.J., Mesozoic tectonics and the deconstruction of biogeography: a new model of Australasian biology, *J. Biogeogr.*, 17, 223–225, 1990.

Heirtzler, J.R., Dickson, G.O., Herron, E.M., Pitman, W.C., and Le Pichon, X., Marine magnetic anomalies, geomagnetic field reversals, and motions of the ocean floor and continents, *J. Geophys. Res.*, 73, 2119–2139, 1968.

Hennig, W., *Grundzüge einer Theorie der Phylogenetischen Systematik*, Deutscher Zentralverlag, Berlin, 1950.

Hennig, W., Phylogenetic systematics, *Annu. Rev. Entomol.*, 10, 97–116, 1965.

Hennig, W., *Phylogenetic Systematics*, University of Illinois Press, Urbana, 1966.

Hennig, W., *Elementos de una Sistemática Filogenética*, Editorial Universitaria, Buenos Aires, 1968.

Hennig, W., *Phylogenetische Systematik*, Pareys Studientexte 34, Paul Parey, Berlin, 1982.

Hess, H.H., History of ocean basins, in *Petrologic Studies: A Volume in Honour of A. F. Buddington*, Engel, A.E., James, J.H.L., and Leonard, B.F., Eds., Geological Society of America, Boulder, Colorado, 1962, pp. 599–620.

Holmes, A., Radioactivity and earth movements, *Trans. Geol. Soc. Glasgow*, 18, 559–606, 1931.

Hooker, J.D., *The Botany of the Antarctic Voyage of H.M. Discovery Ships Erebus and Terror in the Years 1839-1843. Vol. II. Flora Novae Zelandiae*, Part 1. *Flowering Plants*, Lovell Reeve, London, 1853.

Hooker, J.D., *Botany of the Antarctic Voyage of H.M. Discovery Ships Erebus and Terror in the Years 1839-1843*, Vol. III. *Flora Tasmaniae*, Lovell Reeve, London, 1860.

Hovenkamp, P., Ed., *Systematics and Evolution: A Matter of Diversity*, University of Utrecht, Utrecht, The Netherlands, 1987.

Hovenkamp, P., Vicariance events, not areas, should be used in biogeographical analysis, *Cladistics*, 13, 67–79, 1997.

Hovenkamp, P., A direct method for the analysis of vicariance patterns, *Cladistics*, 17, 260–265, 2001.

Humphries, C.J., Any advance on Assumption 2?, *J. Biogeogr.*, 16, 101–102.

Humphries, C.J., Cladistic biogeography, in *Cladistics: A Practical Course in Systematics*, Forey, P.L., Humphries, C.J., Kitching, I.J., Scotland, R.W., Siebert, D.J., and Williams, D.M., Systematics Association. Publication 10, Oxford University Press, Oxford, 1992, pp. 137–159.

Humphries, C.J., Form, space and time: which comes first?, *J. Biogeogr.*, 27, 11–15, 2000.

Humphries, C.J., Ladiges, P.Y., Roos, M., and Zandee, M., Cladistic biogeography, in *Analytical Biogeography*, Myers, A.A. and Giller, P.S., Eds., Chapman and Hall, London, 1988, pp. 371–404.

Humphries, C.J. and Parenti, L.R., *Cladistic Biogeography*, Clarendon Press, Oxford, 1986.

Humphries, C.J. and Parenti, L.R., *Cladistic Biogeography: Interpreting Patterns of Plant and Animal Distributions*, 2nd ed., Oxford University Press, Oxford, 1999.

Isacks, B., Oliver, J., and Sykes, L.R., Seismology and the new global tectonics, *J. Geophys. Res.*, 73, 5855–5899, 1968.

King, L.C., *Wandering Continents and Spreading Sea Floors on an Expanding Earth*, John Wiley and Sons, Chichester, UK, 1983.

Kitching, I.J., Forey, P.L., Humphries, C.J., and Williams, D.M., *Cladistics: The Theory and Practice of Parsimony Analysis*, 2nd ed., Systematics Association Publication 10, Oxford University Press, Oxford, 1998.

Kluge, A.G., Parsimony in vicariance biogeography: a quantitative method and a Greater Antillean example, *Syst. Zool.*, 37, 315–328, 1988(1989).

Kluge, A.G., Three-taxon transformation in phylogenetic inference: ambiguity and distortion as regards explanatory power, *Cladistics*, 9, 246–259, 1993.

Kluge, A.G., Moving targets and shell games, *Cladistics*, 10, 403–413, 1994.

Koopman, K.F., Discussion [of Taxon pulses, vicariance and dispersal], in *Vicariance Biogeography: A Critique*, Nelson, G. and Rosen, D.E., Eds., Columbia University Press, New York, 1981, pp. 184–187.

Ladiges, P.Y., Humphries, C.J., and Martinelli, L.W., Eds., *Austral Biogeography*, CSIRO, Melbourne, 1991.
LeGrand, H.E., *Drifting Continents and Shifting Theories*, Cambridge University Press, Cambridge, 1988.
Leviton, A.E. and Aldrich, M.L., Plate tectonics and biogeography, *Earth Sci. Hist.*, 4, 91–196, 1985(1986).
Linder, H.P. and Crisp, M.D., *Nothofagus* and Pacific biogeography, *Cladistics*, 11, 5–32, 1995(1996).
Linnaeus, C., *Selected Dissertations from the Amoenitates Academicae*, Robeson and Robeson, London, 1781.
Lovis, J.D., Timing, exotic terranes, angiosperm diversification, and panbiogeography, *N.Z. J. Zool.*, 16, 713-729, 1989.
Lyell, C., *Principles of Geology*, 1st ed., John Murray, London, 1810–1833.
Lyell, C., *Principles of Geology*, 12th ed., John Murray, London, 1875.
MacArthur, R.H. and Wilson, E.O., *The Theory of Island Biogeography*, Princeton University Press, Princeton, NJ, 1967.
Matthew, W.D., Climate and evolution, *Ann. NY Acad. Sci.*, 24, 171–318, 1915.
Matthew, W.D., Affinities and origin of the Antillean mammals, *Geol. Soc. Am. Bull.*, 29, 657–666, 1918.
Mayr, E., *Animal Species and Evolution*, Harvard University Press, Cambridge, MA, 1963.
Mickevich, M.F., Quantitative phylogenetic biogeography, in *Advances in Cladistics: Proceedings of the First Meeting of the Willi Hennig Society*, Funk, V.A. and Brooks, D.R., Eds., New York: New York Botanical Garden, 1981, pp. 209–222.
Morrone, J.J., On the identification of areas of endemism, *Syst. Biol.*, 43, 438–441, 1994.
Morrone, J.J., Biogeographical regions under track and cladistic scrutiny, *J. Biogeogr.*, 29, 149–152, 2002.
Morrone, J.J. and Carpenter, J.M., In search of a method for cladistic biogeography: an empirical comparison of component analysis, Brooks parsimony analysis, and three-area statements, *Cladistics*, 10, 99–153, 1994(1995).
Morrone, J.J. and Crisci, J.V., Historical biogeography: introduction to methods, *Annu. Rev. Ecol. Syst.*, 26, 373–401, 1995.
Myers, A.A. and Giller, P.S., Eds., *Analytical Biogeography: An Integrated Approach to the Study of Animal and Plant Distributions*, Chapman and Hall, London, 1988.
Myers, G.S., Fresh-water fishes and West Indian zoögeography, in *Annual Report of the Board of Regents of the Smithsonian Institution for* 1937, Smithsonian Institution, Washington, DC, 1938, pp. 339–364.
Nelson, G.J., The problem of historical biogeography, *Syst. Zool.*, 18, 243–246, 1969.
Nelson, G.J., Historical biogeography: an alternative formalization, *Syst. Zool.*, 23, 555–558, 1974
Nelson, G.J., From Candolle to Croizat: comments on the history of biogeography, *J. Hist. Biol.*, 11, 269–305, 1978.
Nelson, G.J., Cladistique et biogeographie, *Comptes Rendus Soc. Biogeogr.*, 58, 75–94, 1982.
Nelson, G.J., Vicariance and cladistics: historical perspectives with implications for the future, in *Evolution, Time and Space: The Emergence of the Biosphere*, Sims, R.W., Price, J.H., and Whalley, P.E.S., Eds., Academic Press, London, 1983, pp. 469–492.
Nelson, G.J., Cladistics and biogeography, in *Cladistics: Perspectives on the Reconstruction of Evolutionary History*, Duncan, T. and Stuessy, T.F., Eds., Columbia University Press, New York, 1984, pp. 273–293.
Nelson, G.J., A decade of challenge: the future of biogeography, *Earth Sci. Hist.* 4, 187–196, 1985(1986).
Nelson, G.J., Models and prospects of biogeography, in *Pelagic Biogeography: Proceedings of an International Conference, The Netherlands 29 May–5 June 1985*, Pierrot-Bults, A.C., van der Spoel, S., Zahuranec, B.J., and Johnson, R.K., Eds., 1986, pp. 214–218, ICoPBII, Proceedings of the Second International Conference IOC/UNESCO, Paris.

Nelson, G.J., Cladistics and evolutionary models, *Cladistics, 5*, 275–289, 1989.

Nelson, G.J. and Ladiges, P.Y., Standard assumptions for biogeographic analysis, *Austr. Syst. Bot.*, 4, 41–58, 1991a.

Nelson, G.J. and Ladiges, P.Y., Three-area statements: standard assumptions for biogeographic analysis, *Syst. Zool.*, 40, 470–485, 1991b.

Nelson, G.J. and Ladiges, P.Y., *TAS: MSDOS Program for Cladistic Biogeography*, published by the authors, New York, 1991c.

Nelson, G.J. and Ladiges, P.Y., *TAX: MSDOS Program for Cladistic Systematics*, published by the authors, New York, 1991d.

Nelson, G.J. and Ladiges, P.Y., Missing data and three-item analysis, *Cladistics* 9, 111–113, 1993.

Nelson, G.J. and Ladiges, P.Y., *TASS Versions 1.4, 1.5: Three Area Subtrees*, published by the authors, New York, 1994.

Nelson, G.J. and Ladiges, P.Y., *TASS Version 2.0: Three Area Subtrees*, published by the authors, New York, 1995.

Nelson, G.J. and Ladiges, P.Y., Paralogy in cladistic biogeography and analysis of paralogy-free subtrees, *Am. Mus. Novit.*, 3167, 1–58, 1996.

Nelson, G.J. and Ladiges, P.Y., Gondwana, vicariance biogeography and the New York school revisited, *Austr. J. Bot.*, 49, 389–409, 2001.

Nelson, G.J. and Platnick, N.I., A vicariance approach to historical biogeography, *Biosci.*, 30, 339–343, 1980.

Nelson, G.J. and Platnick, N.I., *Systematics and Biogeography: Cladistics and Vicariance*, Columbia University Press, New York, 1981.

Nelson, G.J. and Platnick, N.I., Three-taxon statements: a more precise use of parsimony?, *Cladistics, 7*, 351–366, 1991.

Nelson, G.J. and Rosen, D.E., Eds., *Vicariance Biogeography: A Critique*, Columbia University Press, New York, 1981.

Nur, A. and Ben Avraham, Z., Lost Pacifica continent: a mobilistic speculation, in *Vicariance Biogeography: A Critique*, Nelson, G. and Rosen, D.E., Eds., Columbia University Press, New York, 1981, pp. 341–358.

Owen, H.G., Continental displacement and expansion of the Earth during the Mesozoic and Cenozoic, *Philos. Trans. R. Soc.*, 281, 223–290, 1976.

Page, R.D.M., Graphs and generalized tracks: quantifying Croizat's panbiogeography, *Syst. Zool.*, 36, 1–17, 1987.

Page, R.D.M., Quantitative cladistic biogeography: constructing and comparing area cladograms, *Syst. Zool.*, 37, 254–270, 1988(1989).

Page, R.D.M., *COMPONENT Version 1.5*, University of Auckland, Auckland, NZ, 1989a.

Page, R.D.M., Comments on component-compatibility in historical biogeography, *Cladistics, 5*, 167–182, 1989b.

Page, R.D.M., Component analysis: a valiant failure?, *Cladistics, 6*, 119–136, 1990.

Page, R.D.M., *COMPONENT, Version 2.0*, The Natural History Museum, London, 1993a.

Page, R.D.M., Genes, organisms, and areas: the problem of multiple lineages, *Syst. Biol.*, 42, 77–84, 1993b.

Page, R.D.M., Maps between trees and cladistic analysis of historical associations among genes, organisms and areas, *Syst. Biol.*, 43: 58–77, 1994.

Parenti, L.R., Discussion [of methods of paleobiogeography], in *Vicariance Biogeography: A Critique*, Nelson, G. and Rosen, D.E., Eds., Columbia University Press, New York, 1981, pp. 490–497.

Patterson, C., The development of the North American fish fauna: a problem of historical biogeography, in *Chance, Change and Challenge*, Vol. 2. *The Evolving Biosphere*, Forey, P.L., Ed., British Museum (Natural History), London and Cambridge University Press, Cambridge, 1981a, pp. 265–281.

Patterson, C., Methods of paleobiogeography, in *Vicariance Biogeography: A Critique*, Nelson, G. and Rosen, D.E., Eds., Columbia University Press, New York, 1981b, pp. 446–489.

Patterson, C., Morphological characters and homology, in *Problems of Phylogenetic Reconstruction*, Joysey, K.A., and Friday, A.E., Eds., Academic Press, London, 1982, pp. 21–74.

Pepper, J.H., *The Playbook of Metals Including Personal Narratives of Visits to Coal, Lead, Copper, and Tin Mines*, George Routledge & Sons, London, 1861.

Pierrot-Bults, A.C., van der Spoel, S., Zahuranec, B.J., and Johnson, R.K., Eds., *Pelagic Biogeography: Proceedings of an International Conference, the Netherlands 29 May–5 June 1985*, UNESCO technical papers in marine science 49, ICoPBII, Proceedings of the Second International Conference IOC/UNESCO, Paris, 1986.

Platnick, N.I., Philosophy and the transformation of cladistics, *Syst. Zool.*, 28, 537–546, 1979.

Platnick, N.I., Widespread taxa and biogeographic congruence, in *Advances in Cladistics: Proceedings of the First Meeting of the Willi Hennig Society*, Funk, V.A. and Brooks, D.R., Eds., New York Botanical Garden, New York, 1981, pp. 223–227.

Platnick, N.I., Commentary: areas of endemism, *Austr. Syst. Bot.*, 4, xi–xii, 1991.

Platnick, N.I. and Nelson, G., A method of analysis for historical biogeography, *Syst. Zool.*, 27, 1–16, 1978.

Platnick, N.I. and Nelson, G., Composite areas in vicariance biogeography, *Syst. Zool.*, 33, 328–335, 1984.

Raven, P.H. and Axelrod, D.I., Plate tectonics and Australasian paleobiogeography: the complex biogeographic relations of the region reflect its geologic history, *Science*, 176, 1379–1386, 1972.

Raven, P.H. and Axelrod, D.I., Angiosperm biogeography and past continental movements, *Ann. Mo. Bot. Gard.*, 61, 539–673, 1974.

Richter, S. and Meier, R., The development of phylogenetic concepts in Hennig's early theoretical publications (1947-1966), *Syst. Biol.*, 43, 212–221, 1994.

Rieppel, O., *Fundamentals of Comparative Biology*, Birkhäuser, Basel, Switzerland, 1988.

Romero, E.J., Fossil evidence regarding the evolution of *Nothofagus* Blume, *Ann. Mo. Bot. Gard.*, 73, 276–283, 1986.

Romm, J., A new forerunner to continental drift, *Nature*, 367, 407–408, 1994.

Ronquist, F., Ancestral areas and parsimony, *Syst. Biol.*, 43, 267–274, 1994.

Ronquist, F., Ancestral areas revisited, *Syst. Biol.*, 44, 572–575, 1995.

Ronquist, F., Dispersal-vicariance analysis: a new approach to quantification historical biogeography, *Syst. Biol.*, 46, 195–203, 1997a.

Ronquist, F., Phylogenetic approaches in coevolution and biogeography, *Zool. Scr.*, 26, 313–322, 1997b.

Ronquist, F., Three-dimensional cost-matrix optimization and maximum co-speciation, *Cladistics*, 14, 167–172, 1998.

Ronquist, F. and Nylin, S., Process and pattern in the evolution of species associations, *Syst. Zool.*, 39, 323–344, 1990.

Rosa, D., *Ologenesi: Nuova Teoria dell'Evoluzione e della Distribuzione Geografica dei Viventi*, Bemporad e Figlio, Florence, 1918.

Rosen, B.R., From fossils to earth history: applied historical biogeography, in *Analytical Biogeography: An Integrated Approach to the Study of Animal and Plant Distributions*, Myers, A.A. and Giller, P.S., Eds., Chapman and Hall, London, 1988, pp. 437–481.

Rosen, D.E., A vicariance model of Caribbean biogeography, *Syst. Zool.*, 24, 431–464, 1975(1976).

Rosen, D.E., Vicariant patterns and historical explanation in biogeography, *Syst. Zool.*, 27, 159–188.

Rosen, D.E., Fishes from the uplands and intermontane basins of Guatemala: revisionary studies and comparative geography, *Bull. Am. Mus. Nat. Hist.*, 162, 267–376, 1979.

Schuh, R.T., *Biological Systematics Principles and Applications*, Comstock, London and Cornell University Press, Ithaca, NY, 2000.

Sclater, P.L., On the general geographical distribution of the members of the class Aves, *J. Linn. Lond. Zool.*, 2, 130–145, 1858.

Shields, O., Evidence for the initial opening of the Pacific Ocean in the Jurassic, *Palaeogeogr. Palaeoclimatogr. Palaeoecol.*, 26, 181–220, 1979.

Shields, O., Trans-Pacific biotic links that suggest earth expansion, in *Expanding Earth Symposium*, Carey, S.W., Ed., University of Tasmania, Hobart, 1983, pp. 199–205.

Shields, O., Pacific biogeography and rapid earth expansion, *J. Biogeogr.*, 18, 583–585, 1991.

Shields, O., Plate tectonics or an expanding earth?, *J. Geol. Soc. Ind.*, 47, 399–408, 1996.

Simpson, G.G., *The Major Features of Evolution*, Columbia University Press, New York, 1953.

Simpson, G.G., *Evolution and Geography*, Oregon State System of Higher Education, Eugene, 1962.

Simpson, G.G., *The Geography of Evolution*, Chilton, Philadelphia, 1965.

Snider-Pelligrini, A., *La Création et ses Mystères Dévoilés*. Paris, 1858.

Stebbins, G.L., Jr., *Variation and Evolution in Plants*, Columbia University Press, New York, 1950.

Swenson, U., Backlund, A., McLoughlin, S. and Hill, R.S., *Nothofagus* biogeography revisited with special emphasis on the enigmatic distribution of subgenus *Brassospora* in New Caledonia, *Cladistics*, 17, 28–47, 2001.

Swenson, U. and Bremer, K., Pacific biogeography of the Asteraceae genus *Abrotanella* (Senecioneae, Blemnospermatinae), *Syst. Bot.*, 22, 493–508, 1998.

Tangney, R.S., Moss biogeography in the Tasman Sea region, *N.Z. J. Zool.*, 16, 665–678.

Tassy, P. and Deleporte, P., Hennig XVII, a time for integration, 21-25 Septembre 1998, Sao Paulo (Brésil), *Bull. Soc. Fr. Syst.*, 21, 13–14, 1999.

Taylor, F.B., Bearing of the Tertiary mountain belt on the origin of the Earth's plan, *Bull. Geol. Soc. Am.*, 21, 179–226, 1910.

Thistleton-Dyer, W., Geographical distribution of plants, in *Darwin and Modern Science*, Seward, A.C., Ed., Cambridge: University Press, 1909, pp. 298–316.

van Veller, M.G.P., *Unveiling Vicariant Methodologies in Vicariance Biogeography*, Universiteit Leiden, Leiden, The Netherlands, 2000.

van Veller, M.G.P., Methodology and implementation of component analysis in coevolutionary and biogeographical studies, in *All Things Considered: A Phylogenetic Study of Historical Associations*, van Veller, M.G.P., McLennan, D.A., Mattern, M.Y., Klassen, G., Hoberg, E.P., and Brooks, D.R., Eds., University of Chicago Press, Chicago, in press-a.

van Veller, M.G.P., Methodology and implementation of three area statement analysis in coevolutionary and biogeographical studies, in *All Things Considered: A Phylogenetic Study of Historical Associations*, van Veller, M.G.P., McLennan, D.A., Mattern, M.Y., Klassen, G., Hoberg, E.P., and Brooks, D.R., Eds., University of Chicago Press, Chicago, in press-b.

van Veller, M.G.P. and Brooks, D.R., When simplicity is not parsimonious: *a priori* and *a posteriori* approaches to historical biogeography, *J. Biogeogr.*, 28, 1–11, 2001.

van Veller, M.G.P., Brooks, D.R., and Zandee, M., Cladistic and phylogenetic biogeography: the art and the science of discovery, *J. Biogeogr.*, 30, 319–329, 2003.

van Veller, M.G.P., Kornet, D.J., and Zandee, M., Methods in vicariance biogeography: assessment of the implementations of assumptions zero, 1 and 2, *Cladistics*, 16, 319–345, 2000.

van Veller, M.G.P., Kornet, D.J., and Zandee, M., A posteriori and a priori methodologies for testing hypotheses of causal processes in vicariance biogeography, *Cladistics*, 18, 207–217, 2002.

van Veller, M.G.P., Zandee, M., and Kornet, D.J., Two requirements for obtaining valid common patterns under different assumptions in vicariance biogeography, *Cladistics*, 15, 393–406, 1999.

van Veller, M.G.P., Zandee, M., and Kornet, D.J., Measures for obtaining inclusive sets of area cladograms under assumption zero, 1 and 2 with different methods for vicariance biogeography, *Cladistics*, 17, 248–259.

von Humboldt, F.A. and Bonpland, A.J.A., *Essai sur la Geographie des Plantes*, Levrault, Schoell, Paris, 1805.

Wallace, A.R., *The Geographical Distribution of Animals*, Macmillan, London, 1876.

Wegener, A., *Die Enstehung der Kontinente und Ozeane*, F. Vieweg und Sohn, Braunschweig, Germany, 1915.

Wegener, A., *Die Enstehung der Kontinente und Ozeane*, 4th ed., F. Vieweg und Sohn, Braunschweig, Germany, 1929.

Wiley, E.O., Phylogenetic systematics and vicariance biogeography, *Syst. Bot.*, 5, 194–220, 1980.

Wiley, E.O., *Phylogenetics: The Theory and Practice of Phylogenetic Systematics,* John Wiley and Sons, New York, 1981.

Wiley, E.O., Methods in vicariance biogeography in *Systematics and Evolution: A Matter of Diversity*, Hovenkamp, P. Ed., University of Utrecht, Utrecht, The Netherlands, 1987, pp. 283–306.

Wiley, E.O., Vicariance biogeography, *Annu. Rev. Ecol. Syst.*, 19, 513–542, 1988a.

Wiley, E.O., Parsimony analysis and vicariance biogeography, *Syst. Zool.*, 37, 271–290, 1988(1989).

Williams, D.M., Three-item statements analysis, in *Cladistics: The Theory and Practice of Parsimony Analysis*, 2nd ed., Kitching, I.J., Forey, P.L., Humphries, C.J. and Williams, D.M., Systematics Association Publication 11, Oxford University Press, Oxford, 1998, pp. 168–186.

Willis, J.C., *Age and Area: A Study in Geographical Distribution and Origin of Species*, Cambridge University Press, Cambridge, 1922.

Wilson, J.T., A possible origin of the Hawaiian Islands, *Can. J. Phys.*, 41, 863–870, 1963.

Wulff, E.V., *An Introduction to Historical Plant Geography* (trans. from Russian by E. Brissenden), Chronica Botanica, Waltham, MA, 1950.

Zandee, M. and Roos, M., Component compatibility in historical biogeography, *Cladistics,* 3, 305–332, 1987.

Chapter 11

The Fall and Rise of Evolutionary Developmental Biology

Peter W.H. Holland

Abstract

In recent years, there has been considerable research activity at the interface between evolutionary biology and developmental biology. This research area has its roots in the nineteenth century, when evolution and embryology were heavily interdependent subjects. Through most of the twentieth century, the two disciplines drifted apart; developmental biology and evolutionary biology now addressed different questions, using different methodologies. In the late 1970s and early 1980s, there was a resurgence of interest in comparative developmental biology, driven primarily by the discovery of developmental genes conserved between *Drosophila* and vertebrates, a series of technical advances and the rise of molecular phylogenetics. In the 1990s, the field advanced still further, such that it now embraces a wider range of taxa, interpreted within phylogenetic frameworks. Here I review the history of evolutionary developmental biology and identify a series of milestones that stimulated the recent growth of the discipline.

Introduction

Embryonic development has long been compared between species. Such comparisons have contributed substantially to our knowledge about development and the evolutionary relationships between taxa. Comparisons have also thrown up unexpected similarities or differences that have taken decades or more to understand. This article, however, is not primarily about the scientific insights that have come from comparing embryos and embryonic development. Instead, I intend to chart the history of comparative developmental biology from its beginnings in the early nineteenth century to its emergence as a fully fledged scientific discipline in the late twentieth century. This account will pass swiftly through the era of Darwin and the early evolutionary biologists (when embryology and evolution became closely intertwined subjects) and the subsequent separation of these subjects between the 1930s and 1970s (when evolutionary biology and developmental biology parted company to become separate sciences). For these historical aspects, I am indebted to the accounts and insights of Gilbert et al. (1996) and Arthur (1997).

I will examine in most detail the modern era, which saw evolutionary biology and developmental biology coming close together again, giving birth to a new scientific discipline at their interface. This rise of evolutionary developmental biology began around 1978, took off dramatically in 1984 and underwent a second phase of growth from 1990 onward. My principal goal is to illuminate the key discoveries and technological advances that stimulated this emergence and growth; I see these as milestones in evolutionary developmental biology. Needless to say, the events I have selected as milestones represent a personal perspective on this topic. Other people would doubtless select other milestones, although I hope there would be agreement concerning certain key events (such as the discovery of the homeobox). My selection was introduced briefly in Holland 1999; this article expands on this selection and explains my reasoning.

What is Evolutionary Developmental Biology?

It can be unwise to compartmentalize science into distinct disciplines. This is partly because boundaries between disciplines change with time, but more importantly because undue emphasis on recognized disciplines can inadvertently hold back research at their interfaces. With that caveat in mind, I would define evolutionary developmental biology as a new scientific discipline that has emerged at the interface between two (or possibly three) established disciplines. The two long-established topics are developmental biology and evolutionary biology, with the third being molecular biology, which already had a healthy interface with each of the two others. At the time of this writing, there are three specialist international journals devoted to publishing research articles in the new subject area. These are *Development, Genes and Evolution,* launched in 1996 as a refocused *Roux's Archives of Developmental Biology; Evolution and Development,* founded in 1999; and *Molecular and Developmental Evolution,* launched in 1999 as a specialist section of the *Journal of Experimental Zoology.*

The central goal of evolutionary developmental biology is to understand how genetic change is manifest in morphological diversity, through the intermediary of development. Within this broad goal, one can identify more focused objectives, some developmental in nature, some more evolutionary. In the first category, for example, developmental processes and control mechanisms can be compared between species to identify similarities and differences and thereby appreciate the extent to which developmental insights from one species can be extrapolated to other taxa. The role of Hox genes in anteroposterior patterning is a clear example. Current comparative data suggest it is safe to extrapolate this role across all or most species of bilateral animals, and perhaps further (Holland, 1999; Ferrier and Holland, 2001). We can then ask why some developmental processes are more highly conserved than others. For example, the processes that establish Hox gene expression domains vary greatly between taxa, in contrast to the Hox genes themselves. As for differences in developmental process between species, it is important to determine which of these relate to differences in adult (or larval) body form and which are alternative means to achieve the same end.

From the evolutionary side, we can use gene expression patterns in embryos and larvae to assist our detection of anatomical homology (that is, to distinguish common descent of structure from convergent evolution) and then use these homologies as indicators of phylogenetic relationships or as clues to help reconstruct putative ancestral states. This application has proved rather successful; for example, in the detection of homologies between the brains of all chordates (Wada et al., 1998), between the branchial slits of hemichordates and chordates (Ogasawara et al., 1999) and between the head segments of chelicerates and insects (Damen et al., 1998; Telford and Thomas, 1998; but see Abzhanov et al., 1999). It is much harder, however, to use this approach to disprove homology. For example, gene expression gives no support for Bateson's (1884) proposed homology between the enteropneust stomachord and the chordate notochord (Peterson et al., 1999) but has not yet ruled it out. Homology detection is useful in discerning (or at least postulating) scenarios of what happened in morphological evolution. It is a far more difficult task to address the question of "how did it happen?" This involves identifying the loci responsible for morphological change in evolution, tracking down mutations that might have caused such change and deducing whether particular mutations were necessary or sufficient for that change (for example, to what extent is gene duplication necessary for the evolution of new characters?). To my knowledge, no direct, causative associations between mutation and major morphological evolution in animals have been demonstrated to date. Until they are, a final challenge of evolutionary developmental biology — to deduce whether macroevolution and microevolution involve different types of genetic change — remains out of reach.

Merging Together and Drifting Apart

Karl Ernst von Baer was one of the first scientists to compare explicitly the embryonic development of different animal species. In his treatise *Entwicklungsgeschichte der Thiere: Beobachtung und Reflexion* (1828), he noted that vertebrate embryos are all rather similar in shape and form at their early stages but become progressively more different as development proceeds. This law still has some validity, although several authors have pointed out that it applies only to postgastrulation or postneurulation stages (Slack et al., 1993; Duboule, 1994). Like Richard Owen's concepts of the archetype and homology (Owen, 1848), von Baer's law was not expressed in evolutionary terms but was an observed pattern of similarity and variation.

After publication of *On the Origin of Species* (Darwin, 1859), comparative biology became closely intertwined with evolutionary thinking. Indeed, Darwin himself used embryonic similarities as one of many lines of evidence to argue in favor of descent with modification. The use of developmental similarities as evidence for evolution was a principle accepted by Ernst Haeckel, but one that he largely turned on its head. For example, his influential (though overly simplistic) argument that ontogeny recapitulates phylogeny (Haeckel, 1866) sees evolutionary history as a sufficient explanation for much of embryonic form. In essence, Darwin used embryology as support for evolution, while for Haeckel evolution explained embryology.

The marriage of evolutionary biology and developmental biology, almost complete by the end of the nineteenth century (Table 11.1), was heading for a messy divorce. It is not my intention to chart the series of events that resulted in the two disciplines drifting apart, but the end result was clear. From the 1930s, right through to the 1970s, evolutionary biology and developmental biology each made enormous strides forward, but in quite different directions. Developmental biology was now a science in search of mechanisms, with experimentation and manipulation having taken the place of description (Gilbert et al., 1996). Furthermore, the approaches used to investigate development had become specialized for each species. Analysis of mutants became paramount in *Drosophila*, cell culture and the study of cell differentiation became a cornerstone of mammalian developmental biology, while experimental transplantation and the search for cell interactions and elusive inducing factors was central to amphibian embryo research.

Evolutionary biology, growing in parallel, concentrated on the detection and quantification of parameters such as selection coefficients, allele frequencies and genetic drift. It was now a quantitative science, with no place for qualitative morphological descriptions or near-mythical inducing factors. The two disciplines were not only addressing different goals, but they were using divergent approaches built on distinct philosophies and rules of proof (Gilbert et al., 1996). This is not to say that there was no exchange of ideas between the two fields; indeed several important works at the interface were published in this period (Table 11.1). It is fair to say, however, that mainstream research in evolutionary biology, with its modern synthesis, kin selection and selfish genes, treated development as an unopened black box — if it considered development at all.

TABLE 11.1   Milestones in evolutionary developmental biology

Phase 1: Comparative Embryology

1828	von Baer	Progressive deviation of embryonic forms
1848	Owen	Archetype and homology
1859	Darwin	Origin of species
1866	Haeckel	Ontogeny recapitulates phylogeny

Phase 2: Drifting Apart

1917	D'Arcy Thompson	On growth and form
1940	Goldschmidt	Hopeful monsters
1940	de Beer	Embryos and ancestors

Phase 3: Mice and Flies: Comparative Developmental Biology

1978	Lewis	Genetics of BX-C
1980	Nusslein-Volhard and Weischaus	Segmentation cascade
1983	Bender et al.	BX-C genomic walk
1983	Scott et al.	ANT-C genomic walk
1983	Akam; Hafen et al; Levine et al.	Drosophila embryo *in situ* hybridization
1984	McGinnis et al.; Scott and Weiner	Discovery of homeobox
1984	McGinnis et al.	Homeobox "zoo-blots"
1984	Carrasco et al.	Vertebrate homeobox cloned
1985	Awgulewitsch et al., 1986; Gaunt et al., 1986	Mouse homeobox *in situ* hybridization
1989	Duboule and Dolle; Graham et al.	Hox/HOM cluster homology
1994	Quiring et al.	*eyeless/Pax-6* orthology

Phase 4: Embracing Diversity

1988	Field et al.	18S rRNA metazoan phylogeny
1988	Saiki et al.	PCR with thermostable polymerase
1989	Kocher et al.	PCR with universal primers
1989	Tautz and Pfeifle	Whole mount *in situ* hybridization
1989	Patel et al.	Engrailed antibody 4D9
1992	Holland et al.	Amphioxus Hox gene expression
1994	Garcia-Fernàndez & Holland	Amphioxus Hox gene cluster map
1997	Aguinaldo et al.	18S rDNA metazoan phylogeny
1998	Brooke et al.	Amphioxus *ParaHox* gene cluster

The Rise of Evolutionary Developmental Biology

The 1980s saw a dramatic resurgence of interest and research activity at the interface between evolutionary biology and developmental biology. Indeed by 1990, almost every developmental biology conference included a session on evolution, while evolution meetings incorporated development talks. The total scale of research activity is hard to gauge, but the trend can be discerned by looking at the number of scientific papers published containing the word evolution (or phylogeny or their derivatives) and the word development (or embryo or their derivatives) in the title (Figure 11.1). Although this is probably a gross underestimate of the numbers of papers published in the field, it suggests a steady output through the 1980s and a further increase in the 1990s.

What stimulated the rise of evolutionary developmental biology in the 1980s? One factor may have been the publication of a seminal book on the subject in 1983 by Rudy Raff and Thom Kaufman (Raff and Kaufman, 1983). Certainly, I recall reading the book cover to

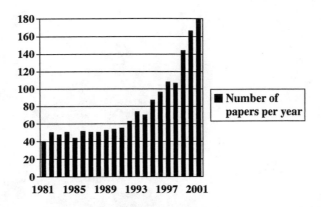

FIGURE 11.1   Annual number of scientific papers in the Science Citation Index with (evolution* OR phylogen*) and (development* OR embryo*) in the title. The general trend is for a steady increase in research output post-1990. The figures are crude estimates of research activity because many studies omit these terms from the title, while a minority of papers with these terms actually deal with unrelated topics (e.g., computing, chemistry, sedimentology). Data from ISI Web of Science.

cover upon its publication when I was an undergraduate. While this book may have introduced the scientific questions and concepts, such as homology, heterochrony and conservation, it could not offer many routes to their further investigation. Too many technical and methodological obstacles stood in the way. The invention and refinement of new techniques applicable to a diversity of species (not just traditional laboratory model systems) was going to be crucial. These technical advances are outlined below. Two other crucial factors emerged as a result of these new techniques: the rise of metazoan molecular phylogenetics and the discovery of conserved developmental genes (notably homeobox genes). These are outlined in subsequent sections.

Technical Advances

With the benefit of hindsight, three key technical advances provided the principal tools with which evolutionary developmental biology was founded. The first was low-stringency Southern hybridization, famously used by McGinnis et al. (1984a,b) to detect homeobox sequences in animals other than *Drosophila*. This may sound like a rather trivial advance; after all, Southern hybridization was introduced in 1975, and by 1984 it was one of the most widely used techniques in molecular biology (Southern, 1975). Low-stringency hybridization involves carefully altering conditions so that imperfect DNA:DNA double helices are stable, thereby allowing related (but nonidentical) genes to be detected without excessive spurious probe binding. This is trivial in theory, but very difficult in practice. I have used low stringency hybridization extensively since 1984, and I have yet to find any protocol that comes close to McGinnis' in terms of consistency, lack of background and sensitivity (Figure 11.2a). Low-stringency hybridization was used in many of the key experiments that marked the rise of evolutionary developmental biology, most notably the initial cloning of vertebrate homologues of *Drosophila* homeotic and segmentation genes. With later modifications allowing oligonucleotides to be used as probes, it also allowed cloning of nematode and *Hydra* homeobox genes, which had proved difficult with cloned DNA probes (Bürglin et al., 1989; Schummer et al. 1992).

The technique was soon complemented by the polymerase chain reaction (PCR), using partially degenerate primers. The PCR strategy has two principal advantages. First, it requires a lower overall degree of DNA sequence conservation. Second, it can use much

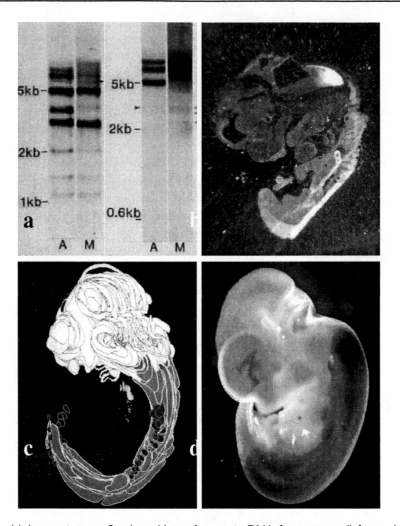

FIGURE 11.2   (a) Low-stringency Southern blots of genomic DNA from mouse (left two lanes) and sea urchin (right two lanes), hybridized with two homeobox probes, A and M. Dark bands indicate putative homeobox sequences. (From Holland, P.W.H. and Hogan, B.L.M., *Nature*, 321, 251–253, 1986. with permission.) (b–d) Improving methods for visualizing homeobox gene expression in mouse embryos. Each figure shows expression of the same gene, *Hoxb-5*, in a 12.5 day mouse embryo. (b) The first Hox gene *in situ* hybridization I performed, June 1986, using a radioactive probe to tissue sections. The distribution of *Hoxb-5* RNA is indicated by the white silver grains produced after autoradiography and dark-field microscopy. Note the triangular patch of silver grains in the hindbrain, plus additional signal in the lung (center right, figure of eight) and spinal cord (bottom right). Note the poor cellular resolution achieved by this method and the difficulty of tissue identification. (c) Three-dimensional reconstruction using serial sections hybridized as in (b). Expression artificially colored as gray is hindbrain and spinal cord, black is spinal ganglia, and light gray is the nodose ganglion. Digitization has simplified display, because individual organs or tissue layers can be shown independently. In this example, only the nervous system is shown. Cellular resolution is severely compromised, however. (d) Whole mount *in situ* hybridization using a nonradioactive probe, performed by Kathryn Brennan. Black staining in hindbrain, spinal cord and visceral organs indicates distribution of *Hoxb-5* RNA. This newer method is simpler, allows three-dimensional visualization and has cellular resolution. Whole mounts can also be sectioned histologically for internal examination.

smaller amounts of starting tissue and does not require construction of DNA libraries (to be fair, library screening is still widely used to isolate the rest of the gene, using the PCR fragment as a probe, although this can be overcome by using more complex methods such as rapid amplification of cDNA ends-PCR [RACE-PCR], litigation-anchored-PCR [LA-PCR] or inverse-PCR). Degenerate PCR is thereby applicable to any living species. The idea of using one set of PCR primers in a range of taxa was first used in population biology (Kocher et al., 1989). By 1990, we, and others, were using PCR routinely to clone developmental gene homologues from a wide range of obscure animals, ranging from brachiopods to Brazilian crab-eating foxes (e.g., Holland and Williams, 1990; Holland et al., 1991; Lanfear and Holland, 1991).

Cloning related genes from a range of species can give some evolutionary insights, such as extent of sequence conservation or patterns of gene duplication, but very often this is not enough. Many evolutionary questions require comparison of sites of gene expression between embryos from different species. Northern blotting and RNase protection are the traditional tools used by molecular biologists to examine gene expression, but these lack fine-scale spatial resolution because they rely on hybridization to purified RNA extracted from cells. In the early 1980s, *in situ* hybridization was introduced to overcome this limitation; this involves hybridization of probes to messenger RNA still within the (fixed) cells of an organ or tissue. By 1983, techniques had been developed for *in situ* hybridization to embryos of sea urchins and *Drosophila*, using radioactively labeled probes hybridized to tissue sections (Angerer and Angerer, 1981; Lynn et al., 1983; Akam, 1983; Hafen et al., 1983; Levine et al., 1983). Some of these early *in situ* experiments generated wonderfully informative patterns, such as broad domains of homeotic gene activity for *Ubx* and *Antp* (Akam, 1983; Hafen et al., 1983; Levine et al., 1983) or beautiful reiterated stripes for the pair-rule gene *fushi tarazu* (Hafen et al., 1984). The latter image, adorning the front cover of *Cell* in September 1984, was particularly striking and must have captured the attention of all developmental biologists. I recall seeing these images as a new PhD student and, influenced by the elegance of these pictures, I decided to abandon my planned project and instead try to adapt *in situ* hybridization to look at *Hox* gene expression in vertebrate embryos.

I was not alone in this decision. From 1985 onward many laboratories published descriptions of spatial gene expression in mouse embryos. At first these focused on abundant transcripts such as expressed repetitive elements (Brûlet et al., 1985) or alpha-fetoprotein (Holland and Hogan, unpublished 1985). This was quickly followed in March 1986 by Awgulewitsch et al. (1986), who reported spatially restricted expression of a *Hox* gene in newborn mice, visualized by *in situ* hybridization. By the end of the same year, Gaunt et al. (1986) had reported similar data for very early mouse embryos. These were extremely influential papers, since they gave the first evidence that *Hox* gene roles might be similar in flies and mice. My own mouse embryo *in situ* results did not appear in print until early 1987 (Krumlauf et al., 1987), although in fact my first successful mouse embryo *in situs* were performed in 1985, and by June 1986 we had clear patterns of *Hox* gene expression at 12.5 days of embryonic development (e.g., Figure 11.2b shows our very first result).

The impact of *in situ* hybridization on the emerging field of comparative developmental biology was enormous. Using this visualization method in combination with low stringency screening (or, much later, PCR) made it possible to clone similar genes from widely different taxa and compare their spatial expression patterns in embryos. The usefulness and wide applicability of *in situ* hybridization was to rise still further, however, with the advent of methods that were directly applicable to whole embryos, rather than to tissue sections (Tautz and Pfeifle, 1989). Before introduction of this method, there had been several attempts to reconstruct three-dimensional gene expression patterns by digitizing tissue sections (after *in situ* hybridization) prior to reconstructing virtual embryos (e.g., Wilkinson et al., 1987; Holland and Hogan, 1988; Figure 11.2c). These methods were time-consuming, demanded computer power that was not readily available and lacked cellular-level resolution. In contrast, the whole-mount *in situ* hybridization method introduced by Tautz and Pfeifle (1989) was simple and reproducible and applicable even to minute embryos or species for which embryonic anatomy was largely unknown. Consequently, description and interpretation of complex spatial expression patterns suddenly became much simpler and faster. Whole-mount *in situ* hybridization remains a central tool of evolutionary developmental biology today (Figure 11.2d).

I have argued that low stringency screening, PCR and *in situ* hybridization were the three principal technological advances behind the rise of this field, but they are not the only methods in use. Cross-reacting monoclonal antibodies, detecting a particular protein in one species and related proteins in divergent taxa, also gave important insights. The most influential of these studies was that of Patel et al. (1989), in which a monoclonal antibody to the engrailed homeodomain protein was isolated and used to compare spatial expression of En class products between embryos from several phyla. More recently, Panganiban et al. (1995) used a cross-reactive antibody to Dll class homeodomain proteins to similar effect. While not disputing the importance of the scientific insights obtained from these studies, this method was not central to the growth of the field, due to the limited availability of cross-reacting antibodies. In contrast, *in situ* hybridization is applicable to every gene family. As for the future, additional techniques are set to revolutionize further comparative developmental studies. One of the most important is likely to be RNA interference (RNAi), a technique for disrupting the activity of a chosen gene in almost any species through use of specific double-stranded RNA molecules (for review, see Gura, 2000). The influence of new technologies on the growth of the field looks set to continue.

Molecular Phylogenetics

The comparison of protein, RNA and DNA sequences between species as a method to investigate evolutionary relationships dates to well before the rise of evolutionary developmental biology. Application of these methods to invertebrate phylogeny had been very limited, however, until the publication of a seminal paper by Field at al. (1988). These authors determined nucleotide sequences from the 18S rRNA molecule, chosen for its ubiquitous distribution and the fact that its high concentration in eukaryotic cells meant that it could be sequenced directly (using reverse transcriptase) without prior cloning. The Field paper was exceptionally influential, not just because it started to challenge some long-standing views about the relationships between major invertebrate groups. It was just as important that the authors had applied molecular biological techniques to a huge diversity of invertebrates, thereby helping to build a sound bridge between classical zoology and molecular biology. The paper stimulated an avalanche of further studies describing new or improved methods for sequence analysis or addition of extra taxa to the original data set produced by Field and colleagues.

The advent of PCR (described above) greatly simplified the determination of these additional sequences, since by PCR amplification of the rDNA gene (rather than analysis of the rRNA product), sensitivity was increased and smaller or rarer samples could be analyzed. I will not recount the stepwise additions to our knowledge afforded by these numerous papers, but simply draw attention to one of the more recent analyses of invertebrate rDNA data. In 1997, Aguinaldo and colleagues used 18S rDNA data to propose a radical reorganization of animal phylogeny, proposing three "great clades" of bilaterians: the Ecdysozoa, Lophotrochozoa and Deuterostomia (Aguinaldo et al., 1997). The implications of this phylogenetic framework for comparative developmental studies are explored by Adoutte et al. (1999).

Discovery of Conserved Developmental Genes

If any year is to be heralded as the most significant in the rise of evolutionary developmental biology, it must be 1984. In short, 1984 was the year of the homeobox. To understand why this discovery (or rather series of discoveries) was such a dramatic catalyst for the field, one must first step back a few years. Throughout the 1950s, 1960s and 1970s, Ed Lewis and colleagues had been undertaking genetic analysis of the so-called homeotic genes: a series of loci that seemed to play particularly important roles in formation of the body plan of *Drosophila*. When mutated, homeotic genes cause body structures to develop inappropriately, such that they develop into the likeness of structures found elsewhere along the head to tail (anteroposterior) axis. In a seminal paper that emerged from this work, Lewis (1978) described how several homeotic loci map to the same chromosomal location yet can be distinguished as genetic neighbors: they constitute a gene complex (the bithorax complex or BX-C). Lewis noted that the gene order of BX-C loci matched the order of their activity along the anteroposterior axis of the fly, such that perhaps these genes could act combinatorially to specify spatial location of developing cells and hence direct correct development. From an evolutionary perspective, Lewis suggested that since the loci have similar roles and map to adjacent positions they may share common ancestry, perhaps having evolved by gene duplication.

The BX-C work cannot be considered in isolation, however. At around the same time, Nusslein-Volhard and Wieschaus (1980) completed an exhaustive genetic screen for *Drosophila* mutations that disrupted normal segmental patterning and postulated the existence of a cascade of segmentation genes. Kaufman and colleagues were undertaking characterization of another set of homeotic loci with roles rather similar to BX-C genes, but primarily in the head: These comprise the Antenapedia complex or ANT-C (Kaufman et al., 1980). Molecular cloning of these genes was not far behind; for example, Bender et al. (1983) published cloning of the BX-C DNA, while Scott et al. (1983) cloned much of the ANT-C.

In 1984, two research groups independently published a remarkable discovery, made using the cloned DNA from these loci. In Walter Gehring's lab in Switzerland, and also in Thomas Kaufman's lab in the United States, it was found that a region of DNA, 180 nucleotides in length, was remarkably similar between a gene from the BX-C, a gene from the ANT-C and one of the segmentation cascade genes (McGinnis et al., 1984a; Scott and Weiner, 1984). The existence of this conserved region, later dubbed the homeobox, had many implications. From an evolutionary point of view, its existence confirmed Lewis' suggestion that the BX-C genes had a common ancestry, particularly as the homeobox was subsequently found in all three protein-coding transcripts of the BX-C. The implications were far wider, however. In particular, the existence of a homeobox in ANT-C genes (in all ANT-C homeotic genes, it later transpired) and some segmentation genes suggested these genes also had a common ancestry.

The discovery of homeobox was certainly intriguing and provided new ways to think about *Drosophila* development and its evolution. Even greater excitement, however, was generated by the next wave of experiments published by Gehring, McGinnis and their colleagues. The first of these papers used homeobox sequences, under reduced stringency conditions, to probe Southern blots prepared from DNA of several animal species (McGinnis et al., 1984b). The detection of multiple bands strongly suggested that homeobox genes were present in a range of insects, plus vertebrates and an annelid. This raised the possibility of wide conservation across the Metazoa, although a possible restriction to metamerically segmented animals eventually turned out to be unfounded (McGinnis, 1985; Holland and Hogan, 1986). Bands on Southern blots are clues to a gene's existence, rather than proof. The confirmation that vertebrates do indeed have true homeobox sequences came very quickly — indeed in the same issue of *Cell*. This report (Carrasco et al., 1984) described an amphibian homeobox gene; reports followed rapidly of human and mouse homeobox genes.

These reports opened up a new comparative era for developmental biology. The immediate thought was that if homeobox genes are master regulators of developmental decisions in *Drosophila*, then perhaps they have similar roles in vertebrates. Would the homeobox sequence provide a technical shortcut enabling rapid cloning of regulatory genes controlling vertebrate development? How similar might the genes, and their roles, be between insects and vertebrates? If there were differences, how and when did these arise in evolution?

This initial excitement was not shared by all. The developmental biology community became somewhat polarized into fervent adherents of homeobox fever and sceptics. Nonetheless, homeobox fever did sweep across developmental biology, generating an avalanche of data for analysis, consideration and general pontification. These data cannot be reviewed here; suffice to say that the homeobox genes detected in vertebrates did turn out to be very close homologues of the insect homeotic genes, with similar developmental regulatory roles and with similar properties of colinearity and gene clustering (reviewed by McGinnis and Krumlauf, 1992).

Equally important was the change in mindset caused by these discoveries. It was no longer adequate to think of "*Drosophila* genes" or "mouse genes": The concept of ancient genes with conserved roles had been born. From now on, every time an important gene was cloned from *Drosophila* (whether encoding a transcription factor, a signalling molecule, a receptor or anything else), a search was made for its vertebrate homologues. The search very occasionally moved in the opposite direction, from vertebrates to flies (such as the *Brachyury* gene first cloned in mouse and then searched for in other taxa), and in a few cases independent research in flies and mice converged on identification of the same gene. The most famous example is undoubtedly the pair of genes known as *Pax-6* (in vertebrates) or *eyeless* (in flies). These were independently cloned in the two taxa, in each case as a key gene involved in eye development; they turned out to be an orthologous pair of genes with a conserved developmental role (for review, Callaerts et al., 1997).

By the start of the 1990s, there were long lists of genes or gene families shared between distant members of the animal kingdom. In addition to *Pax-6/eyeless* and the homeotic genes (now called Hox genes), there were other genes with homeobox sequences (such as *engrailed* in flies and *En* in vertebrates), other transcription factors, conserved signalling molecules (such as *hedgehogs*, *Wnts* and *BMPs*) and conserved receptors. Many, but certainly not all, had similar roles in flies and mice, suggesting conservation of developmental role across hundreds of millions of years of divergent evolution. The broad-brush picture emerging was of a genetic toolkit that was shared between insects and vertebrates and thereby dated at least to their most recent common ancestor.

From Pairwise Comparison to Evolutionary Biology: 1990 Onward

From the discovery of the homeobox in 1984 to around 1990 can be viewed as the first phase of the new evolutionary developmental biology. This period might be more accurately called comparative developmental biology, however, because evolutionary thinking had not pervaded far into the field. The main problem was that almost all studies focused on just two terminal taxa — *Drosophila* and the house mouse — leaving little scope for going beyond the cataloging of similarity versus difference. There were a few striking exceptions, of course, the most impressive being the paper by Patel et al. (1989), in which a cross-reactive antibody to En class proteins was applied to embryos of *Drosophila*, locust, crayfish, leech, zebrafish and chick.

By 1990, however, it was possible to bring together all three of the advances discussed above. First, technical advances (notably PCR and whole-mount *in situ* hybridization) meant that it was possible to clone homologous genes from several taxa and compare developmental expression, or indeed to target species chosen for phylogenetic significance rather than laboratory convenience. Second, molecular phylogenetics had provided a useful framework within which data could be interpreted, while simultaneously reawakening an interest in the diversity of the animal kingdom. Third, and most important, the discovery of a conserved genetic toolkit meant that comparing gene activities between taxa was no longer an absurd idea, since developmental evolution was presumably acting through modification of this toolkit.

Consequently, evolutionary developmental biology, therefore, has been experiencing dramatic growth over the past 10 years. It is now commonplace to read papers in which particular genes, or even interacting sets of genes, are compared between species within an explicit phylogenetic framework, whether that is for chordates, arthropods, cnidarians, echinoderms or indeed any other phylum. Some researchers have argued that large-scale comparisons between phyla or classes may not have the power to detect the actual variation exploited by evolution; hence, complementary research programs have started that focus on finer-grained developmental differences. A third strategy has been to choose one species on the basis of phylogenetic significance and then to extract maximal information regarding genes, gene families, gene expression, gene pathways and genomic organization. Here, accurate phylogenetics and systematics play an important role in directing developmental biologists to the most appropriate species for investigation. This is the approach we have taken in my laboratory with the cephalochordate amphioxus, chosen for its pivotal phylogenetic position as the sister group to the vertebrates. The quantity of data that can be generated by concentrating on one species is substantial, enabling useful comparison to the established model systems. For example, we reported the cloning and expression of the first amphioxus *Hox* gene in 1992 (Holland et al., 1992), reported a physical map of the Hox cluster two years later (Garcia-Fernàndez and Holland 1994), described colinearity of Hox genes (Wada et al., 1998) and analyzed Hox *cis*-regulatory elements (Manzanares et al., 2000).

In parallel to this work, we have looked at brain and nervous system development using amphioxus Otx, Emx, Msx, Mnx and Krox genes (Williams and Holland, 1996, 2000; Sharman et al., 1999; Knight et al., 2000; Ferrier et al., 2001), used amphioxus genes to examine chordate genome evolution (Patton et al., 1998; Furlong and Holland, 2001) and discovered a new cluster of homeobox genes implicated in gut development (Brooke et al., 1998). The pace of discovery is fast, principally because there are major technical advantages afforded by concentrating on one species. For example, the same DNA libraries and protocols can be used to tackle many different biological questions.

Conclusions

In this article, I have divided the history of evolutionary developmental biology into four phases. The first encompassed most of the nineteenth century and saw comparative biology being applied to embryos, culminating in embryology and evolution becoming intertwined subjects. The second phase encompassed the twentieth century up to 1970s, during which time the study of embryos and the study of evolution were forging ahead in different directions. The third phase began around 1978 or 1980 and covered a particularly exciting decade, during which conserved developmental genes were discovered and their expression patterns compared between vertebrates and insects. After 1990, technical developments had advanced sufficiently that gene expression patterns could be examined in almost any species, while molecular phylogenetics was providing a framework within which to interpret characters such as expression patterns, gene numbers or DNA sequences. This fourth phase, therefore, witnessed a gradual shift to a more wide-ranging evolutionary developmental biology, encompassing an A to Z of the animal kingdom: amphioxus, butterflies, cavefish, dogfish and more.

The future of the subject is hard to predict, although we can be sure that new disciplines such as genomics and new techniques such as RNAi and expressed sequenced tag (EST) sequencing will have a big impact (Holland, 1999). Indeed, genomics is already throwing up some dramatic exceptions to the rule that genetic toolkits are conserved between divergent animal taxa. For example, the nuclear hormone receptor genes duplicated extensively in the nematode lineage (Ruvkun and Hobert, 1998), while the zinc-finger genes show complex patterns of gene duplication in different phyla (Knight and Shimeld, 2001). The expected avalanche of genetic and developmental data is sure to have repercussions for systematics and phylogenetics. In particular, knowing which taxa possess or lack particular genes or genetic pathways could prove to be a useful source of phylogenetic information. I suggest, therefore, that the arrival of genomics heralds the fifth phase in the history of evolutionary developmental biology: a period in which we have very high hopes for a deeper understanding of the links between genomes, embryos and evolution.

Acknowledgments

This paper derives from a contribution to the Systematics Association Third Biennial meeting; I thank Peter Forey and David Williams for the invitation to participate. I also thank Brigid Hogan for being a sufficiently open-minded PhD supervisor to allow me to switch my research topic to homeobox genes 18 years ago. The author's work in this field is funded predominantly by the Biotechnology and Biological Sciences Research Council.

References

Abzhanov, A., Popadic, A., and Kaufman, T.C., Chelicerate *Hox* genes and homology of arthropod segments, *Evol. Dev.*, 1, 77–89, 1999.

Adoutte, A., Balavoine, G., Lartillot, N., and de Rosa, R., Animal evolution: the end of the intermediate taxa?, *Trends Genet.*, 15, 104–108, 1999.

Aguinaldo, A.M., Turbeville, J.M., Linford, L.S., Rivera, M.C., Garey, J.R., Raff, R.A., and Lake, J.A., Evidence for a clade of nematodes, arthropods and other moulting animals, *Nature*, 387, 489–493, 1997.

Akam, M., The location of *Ultrabithorax* transcripts in *Drosophila* tissue sections, *EMBO J.*, 2, 2075–2084, 1983.

Angerer, L. and Angerer, R.C., Detection of poly A+RNA in sea urchin eggs and embryos by quantitive *in situ* hybridisation, *Nucleic Acids Res.*, 9, 2819–2840, 1981.

Arthur, W., *The Origin of Animal Body Plans: A Study in Evolutionary Developmental Biology*, Cambridge University Press, Cambridge, 1997.

Awgulewitsch, A., Utset, M.F., Hart, C.P., McGinnis, W., and Ruddle, F.H., Spatial restriction of a mouse homeobox locus within the central nervous system, *Nature*, 320, 328–335, 1986.

Bateson, W., The early stages in the development of *Balanoglossus* (sp. incert.), *Q. J. Microsc. Sci.*, 24, 208–236, 1884.

Bender, W., Akam, M., Karch, F., Beachy, P.A., Pfeifer, M., Spierer, P., Lewis, E.B., and Hogness, D.S., Molecular genetics of the Bithorax complex in *Drosophila*, *Science*, 221, 23–29, 1983.

Brooke, N.M., Garcia-Fernàndez, J., and Holland, P.W.H., The *ParaHox* gene cluster is an evolutionary sister of the *Hox* gene cluster, *Nature*, 392, 920–922, 1998.

Brûlet, P., Condamine, H., and Jacob, F., Spatial distribution of transcripts of the long repeated E.Tn sequence during early mouse embryogenesis, *Proc. Natl. Acad. Sci. USA*, 82, 2054–2058, 1985.

Bürglin, T.R., Finney, M., Coulson, A., and Ruvkun, G., *Caenorhabditis elegans* has scores of homoeobox-containing genes, *Nature*, 341, 239–243, 1989.

Callaerts, P., Halder, G., and Gehring, W.J., Pax-6 in development and evolution, *Annu. Rev. Neurosci.*, 20, 483–532, 1997.

Carrasco, A.E., McGinnis, W., Gehring, W.J., and De Robertis, E.M., Cloning of a *X. laevis* gene expressed during early embryogenesis coding for a peptide region homologous to *Drosophila* homeotic genes, *Cell*, 37, 409–414, 1984.

Damen, W.G., Hausdorf, M., Seyfarth, E.-A., and Tautz, D., A conserved mode of head segmentation in arthropods revealed by the expression pattern of *Hox* genes in a spider, *Proc. Natl. Acad. Sci. USA*, 95, 10665–10670, 1998.

Darwin, C., *On the Origin of Species by Means of Natural Selection, or, the Preservation of Favoured Races in the Struggle for Life,* John Murray, London, 1859.

de Beer, G.R., *Embryos and Ancestors*, Oxford University Press, Oxford, 1940.

Duboule, D., Temporal colinearity and the phylotypic progression: a basis for stability of the vertebrate Bauplan and the evolution of morphologies through heterochrony, *Development* (Suppl.):135–142, 1994.

Duboule, D. and Dolle, P., The structural and functional organization of the murine Hox gene family resembles that of Drosophila, *EMBO J.* 8, 1497-1505, 1989.

Ferrier, D.E.K., Brooke N.M., Panopoulou G., and Holland P.W.H., The Mnx homeobox gene class defined by *HB9, MNR2* and amphioxus *AmphiMnx*, *Dev. Genes Evol.*, 211, 103–107, 2001.

Ferrier, D.E.K. and Holland, P.W.H., Ancient origins of the *Hox* gene cluster, *Nat. Rev. Genet.*, 2, 33–38, 2001.

Field, K.G., Olsen, G.J., Lane, D.J., Giovannoni, S.J., Ghiselin, M.T., Raff, E.C., Pace, N.R., and Raff, R.A., Molecular phylogeny of the animal kingdom, *Science,* 239, 748–753, 1988.

Furlong, R.F. and Holland, P.W.H., Were vertebrates octoploid?, *Philos. Trans. R. Soc. B* 357, 531–544, 2001.

Garcia-Fernàndez, J. and Holland, P.W.H., Archetypal organization of the amphioxus *Hox* gene cluster, *Nature*, 370, 563–566, 1994.

Gaunt, S.J., Miller, J.R., Powell, D.J., and Duboule, D., Homeobox gene expression in mouse embryos varies with position by the primitive streak stage, *Nature*, 324, 662–664, 1986.

Gilbert, S.F., Opitz, J.M., and Raff, R.A., Resynthesizing evolutionary and developmental biology, *Dev. Biol.,* 173, 357–372, 1996.

Goldschmidt, R.B., *The Material Basis of Evolution*, Yale University Press, New Haven, CT, 1940.

Graham, A., Papalopulu, N., and Krumlauf, R., The murine and Drosophila homeobox gene complexes have common features of organization and expression, *Cell* 57, 367-378, 1989.

Gura, T., A silence that speaks volumes, *Nature*, 404, 804–808, 2000.

Haeckel, E., *Generelle Morphologie der Organismen*, 2 volumes. G. Reimer, Berlin, 1866.

Hafen, E., Kuroiwa, A., and Gehring, W.J., Spatial distributions of transcripts form the segmentation gene fushi-tarazu during *Drosophila* embryonic development, *Cell*, 37, 833–841, 1984.

Hafen, E., Levine, M., Gerber, R.L., and Gehring, W.J., An improved *in situ* hybridisation method for the detection of cellular RNAs in Drosophila tissue sections and its application for localising transcripts of the homeotic Antenapedia gene complex, *EMBO J.,* 2, 617–623, 1983.

Holland, P.W.H. and Hogan, B.L.M., Spatial expression of the alpha-fetoprotein gene in mouse extra-embryonic membranes, Unpublished, 1985.

Holland, P.W.H., The future of evolutionary developmental biology, *Nature,* 402, C41–C44, 1999.

Holland, P.W.H. and Hogan, B.L.M., Phylogenetic distribution of Antennapedia-like homeo boxes, *Nature*, 321, 251–253, 1986.

Holland, P.W.H. and Hogan, B.L.M., Spatially restricted patterns of expression of the homeobox-containing gene *Hox 2.1.* during mouse embryogenesis, *Development*, 102, 159–174, 1988.

Holland, P.W.H., Holland, L.Z., Williams, N.A., and Holland, N.D., An amphioxus homeobox gene: sequence conservation, spatial expression during development and insights into vertebrate evolution, *Development*, 116, 653–661, 1992.

Holland, P.W.H. and Williams, N.A., Conservation of engrailed-like homeobox sequences during vertebrate evolution, *FEBS Lett.*, 277, 250–252, 1990.

Holland, P.W.H., Williams, N.A., and Lanfear, J., Cloning of segment polarity gene homologues from the unsegmented brachiopod *Terebratulina retusa* (Linnaeus), *FEBS Lett.*, 291, 211–213, 1991.

Kaufman, T.C., Lewis, R., and Wakimoto, B., Cytogenetic analysis of chromosome 3 in *Drosophila melanogaster:* the homeotic gene complex in polytene chromosome interval 84A-B, *Genetics*, 94, 115–133, 1980.

Kocher, T.D., Thomas, W.K., Meyer, A., Edwards, S.V., Paado, S., Villablanca, F.X., and Wilson, A.C., Dynamics of mitochondrial DNA evolution in animals: amplification and sequencing with conserved primers, *Proc. Natl. Acad. Sci. USA*, 86, 6196–6200, 1989.

Knight, R.D., Panapoulou, G., Holland, P.W.H., and Shimeld, S.M., An amphioxus Krox gene: insights into vertebrate hindbrain evolution, *Dev. Genes Evol.*, 210, 518–521, 2000.

Knight, R.D. and Shimeld, S.M., Identification of conserved C2H2 zinc-finger gene families in the Bilateria, *Genome Biol.*, 2, 0016.1–0016.8, 2001.

Krumlauf, R., Holland, P.W.H., McVey, J.H., and Hogan, B.L.M., Developmental and spatial patterns of expression of the mouse homeobox gene, *Hox 2.1*, *Development*, 99, 603–617, 1987.

Lanfear, J. and Holland, P.W.H., The molecular evolution of ZFY-related genes in birds and mammals, *J. Mol. Evol.*, 32, 310–315, 1991.

Levine, M., Hafen, E., Gerber, R.L., and Gehring, W.J., Spatial distribution of *Antennapedia* transcripts during *Drosphila* development, *EMBO J.*, 2, 2037–2046, 1983.

Lewis, E.B., A gene complex controlling segmentation in *Drosophila*, *Nature*, 276, 565–570, 1978.

Lynn, D.L., Angerer, L.M., Bruskin, A.M., Klein, W.H., and Angerer, R.C., Localization of a family of mRNAs in a single cell type and its precursors in sea urchin embryos, *Proc. Natl. Acad. Sci. USA*, 80, 2656–2660, 1983.

Manzanares, M., Wada, H., Itasaki, N., Trainor, P.A., Krumlauf, R., and Holland, P.W.H., Conservation and elaboration of *Hox* gene regulation during evolution of the vertebrate head, *Nature*, 408, 854–857, 2000.

McGinnis, W., Homeo box sequences of the Antennapedia class are conserved only in higher animal genomes, *Cold Spring Harbor Symp. Quant. Biol.*, 50, 263–270, 1985.

McGinnis, W., Garber, R.L., Wirz, J., Kuroiwa, A., and Gehring, W.J., A homologous protein-coding sequence in *Drosophila* homeotic genes and its conservation in other metazoa, *Cell*, 37, 403–408, 1984b.

McGinnis, W. and Krumlauf, R., Homeobox genes and axial patterning, *Cell*, 68, 283–302, 1992.

McGinnis, W., Levine, M.S., Hafen, E., Kuroiwa, A., and Gehring, W.J., A conserved DNA sequence in homeotic genes of *Drosophila* Antennapedia and Bithorax complexes, *Nature*, 308, 428–433, 1984a.

Nusslein-Volhard, C. and Wieschaus, E., Mutations affecting segment number and polarity in *Drosophila*, *Nature*, 287, 795–801, 1980.

Ogasawara, M., Wada, H., Peters, H., and Satoh, N., Developmental expression of *Pax1/9* genes in urochordate and hemichordate gills: insight into function and evolution of the pharyngeal epithelium, *Development*, 126, 2539–2550, 1999.

Owen, R., *On the Archetype and Homologies of the Vertebrate Skeleton*, John van Voorst, London, 1848.

Panganiban, G., Sebring, A., Nagy, L., and Carroll, S.B., The development of crustacean limbs and the evolution of arthropods, *Science*, 270, 1363–1366, 1995.

Patel, N.H., Martin-Blanco, E., Coleman, K.G., Poole, S.J., Ellis, M.C., Kornberg, T.B., and Goodman, C.S., Expression of engrailed proteins in arthropods, annelids, and chordates, *Cell*, 58, 955–968, 1989.

Patton, S.J., Luke, G.N., and Holland, P.W.H., Complex history of a chromosomal paralogy regions: insights from amphioxus aromatic amino acid hydroxylase genes and insulin-related genes, *Mol. Biol. Evol.*, 15, 1373–1380, 1998.

Peterson, K.J., Cameron, R.A., Tagawa, K., Satoh, N., and Davidson, E.H., A comparative molecular approach to mesodermal patterning in basal deuterostomes: the expression pattern of Brachyury in the enteropneust hemichordate *Ptychodera flava*, *Development*, 126, 85–95, 1999.

Quiring R., Walldorf, U., Kloter, U. and Gehring, W.J., Homology of the eyeless gene of Drosophila to the Small eye gene in mice and Aniridia in humans, *Science*, 265, 785–789, 1994.

Raff, R.A. and Kaufman,T.C., *Embryos, Genes and Evolution*, MacMillan, New York, 1983.

Ruvkun, R. and Hobert, O., The taxonomy of developmental control in *Caenorhabditis elegans*, *Science*, 282, 2033–2041, 1998.

Saiki, R.K., Grelfland, D.H., Stoffel, S., Scharf, S., Higuchi, R.H., Horn, G.T., Mullis, K.B. and Erlich, H.A., Primer-directed enzymatic amplification of DNA with a thermostable DNA polymerase, *Science* 239, 487-491, 1988.

Schummer, M., Scheurlen, I., Schaller, C., and Galliot, B., HOX/HOM homeogenes are present in Hydra (*Chlorohydra viridissima*) and are differentially expressed during regeneration, *EMBO J.*, 11, 1815–23, 1992.

Scott, M.P. and Weiner, A., Structural relationships among genes that control development: sequence homology between the *Antennapedia*, *Ultrabithorax* and *fushi tarazu* loci of Drosophila, *Proc. Natl. Acad. Sci. USA*, 81, 4115–4119, 1984.

Scott, M.P., Weiner, A.J., Hazelrigg, T.I., Polisky, B.A., Pirrotta, V., Scalenghe, F., and Kaufman, T.C., The molecular organisation of the *Antenapedia* locus of Drosophila, *Cell*, 35, 763–776, 1983.

Sharman, A.C., Shimeld, S.M., and Holland, P.W.H., An amphioxus Msx gene expressed predominantly in the dorsal neural tube, *Dev. Genes Evol.*, 209, 260–263, 1999.

Slack, J.M.W., Holland, P.W.H., and Graham, C.F., The zootype and the phylotypic stage, *Nature*, 361, 490–492, 1993.

Southern, E., Detection of specific sequences among DNA fragments separated by gel electrophoresis, *J. Mol. Biol.*, 98, 503–517, 1975.

Tautz, D. and Pfeifle, C., Non-radioactive *in situ* hybridisation methods for the localization of specific RNAs in *Drosophila* embryos reveals translational control of the segmentation gene hunchback, *Chromosoma*, 98, 81–85, 1989.

Telford, M.J. and Thomas, R.H., Expression of homeobox genes shows chelicerate arthropods retain their duetocerebral segment, *Proc. Natl. Acad. Sci. USA*, 95, 10671–10675, 1998.

Thompson, D'A. *On Growth and Form*, Cambridge University Press, Cambridge, 1917.

von Baer, K.E., *Entwicklungsgeschichte der Thiere: Beobachtung und Reflexion*, Bornträger, Königsberg, 1828.

Wada, H., Saiga, H., Satoh, N., and Holland. P.W.H., Tripartite organization of the ancestral chordate brain and the antiquity of placodes: insights from ascidian *Pax-2/5/8*, Hox and Otx genes, *Development*, 125, 1113–1122, 1998.

Wilkinson, D.G., Bailes, J.A., and McMahon, A.P., Expression of the protooncogene *int-1* is restricted to specific neural cells in the developing mouse embryo, *Cell*, 50, 79–88, 1987.

Williams, N.A. and Holland, P.W.H., Old head on young shoulders, *Nature*, 383, 490, 1996.

Williams, N.A. and Holland, P.W.H., An amphioxus *Emx* homeobox gene reveals duplication in the vertebrate *Emx* gene family, *Mol. Biol. Evol.*, 17, 520–528, 2000.

Index

Systematics Association Publications

1. Bibliography of Key Works for the Identification of the British Fauna and Flora, 3rd edition (1967)†
Edited by G.J. Kerrich, R.D. Meikie and N. Tebble
2. Function and Taxonomic Importance (1959)†
Edited by A.J. Cain
3. The Species Concept in Palaeontology (1956)†
Edited by P.C. Sylvester-Bradley
4. Taxonomy and Geography (1962)†
Edited by D. Nichols
5. Speciation in the Sea (1963)†
Edited by J.P. Harding and N. Tebble
6. Phenetic and Phylogenetic Classification (1964)†
Edited by V.H. Heywood and J. McNeill
7. Aspects of Tethyan biogeography (1967)†
Edited by C.G. Adams and D.V. Ager
8. The Soil Ecosystem (1969)†
Edited by H. Sheals
9. Organisms and Continents through Time (1973)†
Edited by N.F. Hughes
10. Cladistics: A Practical Course in Systematics (1992)*
P.L. Forey, C.J. Humphries, I.J. Kitching, R.W. Scotland, D.J. Siebert and D.M. Williams
11. Cladistics: The Theory and Practice of Parsimony Analysis (2nd edition)(1998)*

I. J. Kitching, P.L. Forey, C.J. Humphries and D.M. Williams
*Published by Oxford University Press for the Systematics Association
†Published by the Association (out of print)

Systematics Association Special Volumes

1. The New Systematics (1940)
Edited by J.S. Huxley (reprinted 1971)
2. *Chemotaxonomy and serotaxonomy (1968)**
Edited by J.C. Hawkes
3. Data Processing in Biology and Geology (1971)*
Edited by J.L. Cutbill
4. Scanning Electron Microscopy (1971)*
Edited by V.H. Heywood
5. Taxonomy and Ecology (1973)*
Edited by V.H. Heywood
6. The Changing Flora and Fauna of Britain (1974)*
Edited by D.L. Hawksworth
7. Biological Identification with Computers (1975)*
Edited by R.J. Pankhurst
8. Lichenology: Progress and Problems (1976)*
Edited by D.H. Brown, D.L. Hawksworth and R.H. Bailey
9. Key Works to the Fauna and Flora of the British Isles and Northwestern Europe, 4th edition (1978)*
Edited by G.J. Kerrich, D.L. Hawksworth and R.W. Sims
10. Modern Approaches to the Taxonomy of Red and Brown Algae (1978)
Edited by D.E.G. Irvine and J.H. Price

11. Biology and Systematics of Colonial Organisms (1979)*
Edited by C. Larwood and B.R. Rosen
12. The Origin of Major Invertebrate Groups (1979)*
Edited by M.R. House
13. Advances in Bryozoology (1979)*
Edited by G.P. Larwood and M.B. Abbott
14. Bryophyte Systematics (1979)*
Edited by G.C.S. Clarke and J.G. Duckett
15. The Terrestrial Environment and the Origin of Land Vertebrates (1980)
Edited by A.L. Pachen
16. Chemosystematics: Principles and Practice (1980)*
Edited by F.A. Bisby, J.G. Vaughan and C.A. Wright
17. The Shore Environment: Methods and Ecosystems (2 volumes)(1980)*
Edited by J.H. Price, D.E.C. Irvine and W.F. Farnham
18. The Ammonoidea (1981)*
Edited by M.R. House and J.R. Senior
19. Biosystematics of Social Insects (1981)*
Edited by P.E. House and J.-L. Clement
20. Genome Evolution (1982)*
Edited by G.A. Dover and R.B. Flavell
21. Problems of Phylogenetic Reconstruction (1982)
Edited by K.A. Joysey and A.E. Friday
22. Concepts in Nematode Systematics (1983)*
Edited by A.R. Stone, H.M. Platt and L.F. Khalil
23. Evolution, Time And Space: The Emergence of the Biosphere (1983)*
Edited by R.W. Sims, J.H. Price and P.E.S. Whalley
24. Protein Polymorphism: Adaptive and Taxonomic Significance (1983)*
Edited by G.S. Oxford and D. Rollinson
25. Current Concepts in Plant Taxonomy (1983)*
Edited by V.H. Heywood and D.M. Moore
26. Databases in Systematics (1984)*
Edited by R. Allkin and F.A. Bisby
27. Systematics of the Green Algae (1984)*
Edited by D.E.G. Irvine and D.M. John
28. The Origins and Relationships of Lower Invertebrates (1985)‡
Edited by S. Conway Morris, J.D. George, R. Gibson and H.M. Platt
29. Infraspecific Classification of Wild and Cultivated Plants (1986)‡
Edited by B.T. Styles
30. Biomineralization in Lower Plants and Animals (1986)‡
Edited by B.S.C. Leadbeater and R. Riding
31. Systematic and Taxonomic Approaches in Palaeobotany (1986)‡
Edited by R.A. Spicer and B.A. Thomas
32. Coevolution and Systematics (1986)‡
Edited by A.R. Stone and D.L. Hawksworth
33. Key Works to the Fauna and Flora of the British Isles and Northwestern Europe, 5th edition (1988)‡
Edited by R.W. Sims, P. Freeman and D.L. Hawksworth
34. Extinction and Survival in the Fossil Record (1988)‡
Edited by G.P. Larwood
35. The Phylogeny and Classification of the Tetrapods (2 volumes)(1988)‡
Edited by M.J. Benton

36. Prospects in Systematics (1988)‡
Edited by J.L. Hawksworth
37. Biosystematics of Haematophagous Insects (1988)‡
Edited by M.W. Service
38. The Chromophyte Algae: Problems and Perspective (1989)‡
Edited by J.C. Green, B.S.C. Leadbeater and W.L. Diver
39. Electrophoretic Studies on Agricultural Pests (1989)‡
Edited by H.D. Loxdale and J. den Hollander
40. Evolution, Systematics, and Fossil History of the Hamamelidae (2 volumes)(1989)‡
Edited by P.R. Crane and S. Blackmore
41. Scanning Electron Microscopy in Taxonomy and Functional Morphology (1990)‡
Edited by D. Claugher
42. Major Evolutionary Radiations (1990)‡
Edited by P.D. Taylor and G.P. Larwood
43. Tropical Lichens: Their Systematics, Conservation and Ecology (1991)‡
Edited by G.J. Galloway
44. Pollen and Spores: Patterns and Diversification (1991)‡
Edited by S. Blackmore and S.H. Barnes
45. The Biology of Free-Living Heterotrophic Flagellates (1991)‡
Edited by D.J. Patterson and J. Larsen
46. Plant–Animal Interactions in the Marine Benthos (1992)‡
Edited by D.M. John, S.J. Hawkins and J.H. Price
47. The Ammonoidea: Environment, Ecology and Evolutionary Change (1993)‡
Edited by M.R. House
48. Designs for a Global Plant Species Information System (1993)‡
Edited by F.A. Bisby, G.F. Russell and R.J. Pankhurst
49. Plant Galls: Organisms, Interactions, Populations (1994)‡
Edited by M.A.J. Williams
50. Systematics and Conservation Evaluation (1994)‡
Edited by P.L. Forey, C.J. Humphries and R.I. Vane-Wright
51. The Haptophyte Algae (1994)‡
Edited by J.C. Green and B.S.C. Leadbeater
52. Models in Phylogeny Reconstruction (1994)‡
Edited by R. Scotland, D.I. Siebert and D.M. Williams
53. The Ecology of Agricultural Pests: Biochemical Approaches (1996)**
Edited by W.O.C. Symondson and J.E. Liddell
54. Species: the Units of Diversity (1997)**
Edited by M.F. Claridge, H.A. Dawah and M.R. Wilson
55. Arthropod Relationships (1998)**
Edited by R.A. Fortey and R.H. Thomas
56. Evolutionary Relationships among Protozoa (1998)**
Edited by G.H. Coombs, K. Vickerman, M.A. Sleigh and A. Warren
57. Molecular Systematics and Plant Evolution (1999)
Edited by P.M. Hollingsworth, R.M. Bateman and R.J. Gornall
58. Homology and Systematics (2000)
Edited by R. Scotland and R.T. Pennington
59. The flagellates: Unity, Diversity and Evolution (2000)
Edited by B.S.C. Leadbeater and J.C. Green
60. Interrelationships of the Platyhelminthes (2001)
Edited by D.T.J. Littlewood and R.A. Bray
61. Major Events in Early Vertebrate Evolution (2001)

*Published by Academic Press for the Systematics Association
†Published by the Palaeontological Association in conjunction with Systematics Association
‡Published by the Oxford University Press for the Systematics Association
**Published by Chapman & Hall for the Systematics Association
§Published by CRC Press for the Systematics Association